Chemometrics and Numerical Methods in LIBS

Chemometrics and Numerical Methods in LIBS

Edited by

Vincenzo Palleschi
Applied and Laser Spectroscopy Laboratory, Institute of Chemistry of Organometallic Compounds, Research Area of National Research Council Pisa, Italy

This edition first published 2023
© 2023 John Wiley & Sons Ltd

All rights reserved. No part of this publication may be reproduced, stored in a retrieval system, or transmitted, in any form or by any means, electronic, mechanical, photocopying, recording or otherwise, except as permitted by law. Advice on how to obtain permission to reuse material from this title is available at http://www.wiley.com/go/permissions.

The right of Vincenzo Palleschi to be identified as the author of the editorial material in this work has been asserted in accordance with law.

Registered Offices
John Wiley & Sons, Inc., 111 River Street, Hoboken, NJ 07030, USA
John Wiley & Sons Ltd, The Atrium, Southern Gate, Chichester, West Sussex, PO19 8SQ, UK

Editorial Office
The Atrium, Southern Gate, Chichester, West Sussex, PO19 8SQ, UK

For details of our global editorial offices, customer services, and more information about Wiley products visit us at www.wiley.com.

Wiley also publishes its books in a variety of electronic formats and by print-on-demand. Some content that appears in standard print versions of this book may not be available in other formats.

Limit of Liability/Disclaimer of Warranty
The contents of this work are intended to further general scientific research, understanding, and discussion only and are not intended and should not be relied upon as recommending or promoting scientific method, diagnosis, or treatment by physicians for any particular patient. In view of ongoing research, equipment modifications, changes in governmental regulations, and the constant flow of information relating to the use of medicines, equipment, and devices, the reader is urged to review and evaluate the information provided in the package insert or instructions for each medicine, equipment, or device for, among other things, any changes in the instructions or indication of usage and for added warnings and precautions. While the publisher and authors have used their best efforts in preparing this work, they make no representations or warranties with respect to the accuracy or completeness of the contents of this work and specifically disclaim all warranties, including without limitation any implied warranties of merchantability or fitness for a particular purpose. No warranty may be created or extended by sales representatives, written sales materials or promotional statements for this work. The fact that an organization, website, or product is referred to in this work as a citation and/or potential source of further information does not mean that the publisher and authors endorse the information or services the organization, website, or product may provide or recommendations it may make. This work is sold with the understanding that the publisher is not engaged in rendering professional services. The advice and strategies contained herein may not be suitable for your situation. You should consult with a specialist where appropriate. Further, readers should be aware that websites listed in this work may have changed or disappeared between when this work was written and when it is read. Neither the publisher nor authors shall be liable for any loss of profit or any other commercial damages, including but not limited to special, incidental, consequential, or other damages.

Library of Congress Cataloging-in-Publication Data
Names: Palleschi, V., editor.
Title: Chemometrics and numerical methods in LIBS / edited by Vincenzo
   Palleschi.
Description: Hoboken, NJ : Wiley, 2023. | Includes bibliographical
   references and index.
Identifiers: LCCN 2022030870 (print) | LCCN 2022030871 (ebook) | ISBN
   9781119759584 (cloth) | ISBN 9781119759560 (adobe pdf) | ISBN
   9781119759577 (epub)
Subjects: LCSH: Laser-induced breakdown spectroscopy. | Chemometrics.
Classification: LCC QD96.A8 C44 2022 (print) | LCC QD96.A8 (ebook) | DDC
   543/.52–dc23/eng20220919
LC record available at https://lccn.loc.gov/2022030870
LC ebook record available at https://lccn.loc.gov/2022030871

Cover Design: Wiley
Cover Image by © kagankiris/Gettyimages; © ntmw/Gettyimages; © Fernando Cortes/Shutterstock

Set in 9.5/12.5pt STIXTwoText by Straive, Pondicherry, India
Printed and bound by CPI Group (UK) Ltd, Croydon, CR0 4YY

Contents

List of Contributors *xiii*
Preface *xvii*

Introduction and Brief Summary of the LIBS Development *1*

Part I Introduction to LIBS *5*

1 LIBS Fundamentals *7*
Mohamad Sabsabi
1.1 Interaction of Laser Beam with Matter *8*
1.2 Basics of Laser–Matter Interaction *9*
1.3 Processes in Laser-Produced Plasma *10*
1.4 Factors Affecting Laser Ablation and Laser-Induced Plasma Formation *11*
1.4.1 Influence of Laser Parameters on the Laser-Induced Plasmas *11*
1.4.2 Laser Wavelength (λ) *12*
1.4.3 Laser Pulse Duration (τ) *12*
1.4.4 Laser Energy (E) *13*
1.4.5 Influence of Ambient Gas *13*
1.5 Plasma Properties and Plasma Emission Spectra *14*
References *15*

2 LIBS Instrumentations *19*
Mohamad Sabsabi and Vincenzo Palleschi
2.1 Basics of LIBS instrumentations *19*
2.2 Lasers in LIBS Systems *20*
2.3 Desirable Requirements for Atomic Emission Spectrometers/Detectors *22*
2.4 Spectrometers *23*
2.4.1 Czerny–Turner Optical Configuration *23*
2.4.2 Paschen–Runge Design *24*
2.4.3 Echelle Spectrometer Configuration *25*
2.5 Detectors *26*
2.5.1 Photomultiplier Detectors *26*

2.5.2	Solid-State Detectors	27
2.5.3	The Interline CCD Detectors	27
2.5.3.1	The Image Intensifier	28
	References	29

3 Applications of LIBS *31*
Vincenzo Palleschi and Mohamad Sabsabi

3.1	Industrial Applications	31
3.1.1	Metal Industry	31
3.1.2	Energy Production	34
3.2	Biomedical Applications	34
3.3	Geological and Environmental Applications	36
3.4	Cultural Heritage and Archaeology Applications	37
3.5	Other Applications	37
	References	38

Part II Simplications of LIBS Information *45*

4 LIBS Spectral Treatment *47*
Sabrina Messaoud Aberkane, Noureddine Melikechi and Kenza Yahiaoui

4.1	Introduction	47
4.2	Baseline Correction	47
4.2.1	Polynomial Algorithm	48
4.2.2	Model-free Algorithm	49
4.2.3	Wavelet Transform Model	52
4.3	Noise Filtering	55
4.3.1	Wavelet Threshold De-noising (WTD)	55
4.3.2	Baseline Correction and Noise Filtering	59
4.4	Overlapping Peak Resolution	60
4.4.1	Curve Fitting Method	61
4.4.2	The Wavelet Transform	64
4.5	Features Selection	66
4.5.1	Principal Component Analysis	68
4.5.2	Genetic Algorithm (GA)	68
4.5.3	Wavelet Transformation (WT)	68
	References	71

5 Principal Component Analysis *81*
Mohamed Abdel-Harith and Zienab Abdel-Salam

5.1	Introduction	81
5.1.1	Laser-Induced Breakdown Spectroscopy (LIBS)	81
5.2	The Principal Component Analysis (PCA)	82
5.3	PCA in Some LIBS Applications	83
5.3.1	Geochemical Applications	83

5.3.2	Food and Feed Applications	*85*
5.3.3	Microbiological Applications	*88*
5.3.4	Forensic Applications	*91*
5.4	Conclusion	*94*
	References	*94*

6 Time-Dependent Spectral Analysis *97*
Fausto Bredice, Ivan Urbina, and Vincenzo Palleschi

6.1	Introduction	*97*
6.2	Time-Dependent LIBS Spectral Analysis	*98*
6.2.1	Independent Component Analysis	*98*
6.2.2	3D Boltzmann Plot	*102*
6.2.2.1	Principles of the Method	*103*
6.3	Applications	*109*
6.3.1	3D Boltzmann Plot Coupled with Independent Component Analysis	*109*
6.3.2	Analysis of a Carbon Plasma by 3D Boltzmann Plot Method	*109*
6.3.3	Assessment of the LTE Condition Through the 3D Boltzmann Plot Method	*114*
6.3.4	Evaluation of Self-Absorption	*114*
6.3.5	Determination of Transition Probabilities	*118*
6.3.6	3D Boltzmann Plot and Calibration-free Laser-induced Breakdown Spectroscopy	*121*
6.4	Conclusion	*123*
	References	*123*

Part III Classification by LIBS *127*

7 Distance-based Method *129*
Hua Li and Tianlong Zhang

7.1	Cluster Analysis	*132*
7.1.1	Introduction	*132*
7.1.2	Theory	*133*
7.1.2.1	K-means Clustering	*133*
7.1.2.2	Hierarchical Clustering	*134*
7.1.3	Application	*135*
7.2	Independent Components Analysis	*138*
7.2.1	Introduction	*138*
7.2.2	Theory	*138*
7.2.3	Application	*140*
7.3	K-Nearest Neighbor	*143*
7.3.1	Introduction	*143*
7.3.2	Theory	*143*
7.3.3	Application	*145*
7.4	Linear Discriminant Analysis	*145*

7.4.1	Introduction	145
7.4.2	Theory	148
7.4.2.1	The Calculation Process of LDA (Two Categories)	148
7.4.3	Application	151
7.5	Partial Least Squares Discriminant Analysis	153
7.5.1	Introduction	153
7.5.2	Theory	155
7.5.3	Application	157
7.6	Principal Component Analysis	161
7.6.1	Introduction	161
7.6.2	Theory	164
7.6.3	Application	166
7.7	Soft Independent Modeling of Class Analogy	174
7.7.1	Introduction	174
7.7.2	Theory	175
7.7.3	Application	177
7.8	Conclusion and Expectation	180
	References	181

8 Blind Source Separation in LIBS 189
Anna Tonazzini, Emanuele Salerno, and Stefano Pagnotta

8.1	Introduction	189
8.2	Data Model	193
8.3	Analyzing LIBS Data via Blind Source Separation	193
8.3.1	Second-order BSS	193
8.3.2	Maximum Noise Fraction	194
8.3.3	Independent Component Analysis	196
8.3.4	ICA for Noisy Data	197
8.4	Numerical Examples	197
8.5	Final Remarks	206
	References	207

9 Artificial Neural Networks for Classification 213
Jakub Vrábel, Erik Képeš, Pavel Pořízka, and Jozef Kaiser

9.1	Introduction and Scope	213
9.2	Artificial Neural Networks (ANNs)	214
9.3	Cost Functions and Training	216
9.4	Backpropagation	219
9.5	Convolutional Neural Networks	221
9.6	Evaluation and Tuning of ANNs	224
9.7	Regularization	227
9.8	State-of-the-art LIBS Classification Using ANNs	229
9.9	Summary	233
	Acknowledgments	234
	References	234

10	**Data Fusion: LIBS + Raman** *241*	
	Beatrice Campanella and Stefano Legnaioli	
10.1	Introduction *241*	
10.2	Data Fusion Background *242*	
10.3	Data Treatment *244*	
10.4	Working with Images *245*	
10.4.1	Vectors Concatenation *246*	
10.4.2	Vectors Co-addition *246*	
10.4.3	Vectors Outer Sum *246*	
10.4.4	Vectors Outer Product *247*	
10.4.5	Data Analysis *247*	
10.5	Applications *248*	
10.6	Conclusion *253*	
	References *253*	

Part IV Quantitative Analysis *257*

11	**Univariate Linear Methods** *259*	
	Stefano Legnaioli, Asia Botto, Beatrice Campanella, Francesco Poggialini, Simona Raneri, and Vincenzo Palleschi	
11.1	Standards *259*	
11.2	Matrix Effect *260*	
11.3	Normalization *261*	
11.4	Linear vs. Nonlinear Calibration Curves *264*	
11.5	Figures of Merit of a Calibration Curve *267*	
11.5.1	Coefficient of Determination *270*	
11.5.2	Root Mean Squared Error of Calibration *270*	
11.5.3	Limit of Detection *270*	
11.6	Inverse Calibration *273*	
11.7	Conclusion *274*	
	References *274*	

12	**Partial Least Squares** *277*	
	Zongyu Hou, Weiran Song, and Zhe Wang	
12.1	Overview *277*	
12.2	Partial Least Squares Regression Algorithms *278*	
12.2.1	Nonlinear Iterative PLS *278*	
12.2.2	SIMPLS Algorithm *279*	
12.2.3	Kernel Partial Least Squares *279*	
12.2.4	Locally Weighted Partial Least Squares *280*	
12.2.5	Dominant Factor-based Partial Least Squares *281*	
12.3	Partial Least Squares Discriminant Analysis *282*	
12.4	Results of Partial Least Squares in LIBS *283*	
12.4.1	Coal Analysis *283*	

12.4.2	Metal Analysis *285*
12.4.3	Rocks, Soils, and Minerals Analysis *285*
12.4.4	Organics Analysis *291*
12.5	Conclusion *291*
	References *295*

13 Nonlinear Methods *303*
Francesco Poggialini, Asia Botto, Beatrice Campanella, Stefano Legnaioli, Simona Raneri, and Vincenzo Palleschi

13.1	Introduction *303*
13.2	Multivariate Nonlinear Algorithms *304*
13.2.1	Artificial Neural Networks *304*
13.2.1.1	Conventional Artificial Neural Networks *304*
13.2.1.2	Convolutional Neural Networks *310*
13.2.2	Other Nonlinear Multivariate Approaches *312*
13.2.2.1	The Franzini–Leoni Method *312*
13.2.2.2	The Kalman Filter Approach *313*
13.2.2.3	Calibration-Free Methods *314*
13.3	Conclusion *315*
	References *316*

14 Laser Ablation-based Techniques – Data Fusion *321*
Jhanis Gonzalez

14.1	Introduction *321*
14.2	Data Fusion of Multiple Analytical Techniques *322*
14.2.1	Low-level Fusion *322*
14.2.2	Mid-level Fusion *323*
14.2.3	High-level Fusion *324*
14.3	Data Fusion of Laser Ablation-Based Techniques *324*
14.3.1	Introduction *324*
14.3.2	Classification of Edible Salts *326*
14.3.2.1	LIBS and LA-ICP-MS Measurements of the Salt Samples *327*
14.3.2.2	Mid-Level Data Fusion of LIBS and LA-ICP-MS of Salt Samples *327*
14.3.2.3	PLS-DA Classification Model for Salt Samples *333*
14.3.3	Coal Discrimination Analysis *334*
14.3.3.1	LIBS and LA-ICP-TOF-MS Measurements of the Coal Samples *335*
14.3.3.2	Mid-Level Data Fusion of LIBS and LA-ICP-TOF-MS of Coal Samples *335*
14.3.3.3	PCA Combined with K-means Cluster Analysis for Coal Samples *338*
14.3.3.4	PLS-DA and SVM for Coal Samples Analysis *340*
14.4	Comments and Future Developments *341*
	Acknowledgments *343*
	References *343*

Part V Conclusions *347*

15 **Conclusion** *349*
Vincenzo Palleschi

Index *351*

List of Contributors

Mohamed Abdel-Harith
National Institute of Laser-Enhanced Science
Cairo University
Cairo, Egypt

Zienab Abdel-Salam
National Institute of Laser-Enhanced Science
Cairo University
Cairo, Egypt

Sabrina Messaoud Aberkane
Ionized Media and Lasers Division
Center for Development of Advanced Technologies
Algiers, Algeria

Asia Botto
Applied and Laser Spectroscopy Laboratory
Institute of Chemistry of Organometallic Compounds
Research Area of the National Research Council
Pisa, Italy

Fausto Bredice
Atomic Spectroscopy Laboratory
Centro de Investigaciones Ópticas (CIOp)
La Plata, Argentina

Beatrice Campanella
Applied and Laser Spectroscopy Laboratory
Institute of Chemistry of Organometallic Compounds
Research Area of the National Research Council
Pisa, Italy

Jhanis Gonzalez
Applied Spectra, Inc.
West Sacramento, CA, USA
and
Lawrence Berkeley National Laboratory
Berkeley, CA, USA

Zongyu Hou
State Key Lab of Power Systems
Department of Energy and Power Engineering
Tsinghua University
Beijing, China
and
Shanxi Research Institute for Clean Energy
Tsinghua University
Taiyuan, China

List of Contributors

Jozef Kaiser
CEITEC BUT
Brno University of Technology
Brno, Czech Republic
and
Institute of Physical Engineering
Brno University of Technology
Brno, Czech Republic

Erik Képeš
Brno University of Technology
Brno, Czech Republic
and
Institute of Physical Engineering
Brno University of Technology
Brno, Czech Republic

Stefano Legnaioli
Applied and Laser Spectroscopy Laboratory
Institute of Chemistry of Organometallic Compounds
Research Area of the National Research Council
Pisa, Italy

Hua Li
Key Laboratory of Synthetic and Natural Functional Molecule of the Ministry of Education
College of Chemistry and Materials Science
Northwest University
Xi'an, China

Noureddine Melikechi
Department of Physics and Applied Physics
Kennedy College of Sciences
University of Massachusetts Lowell,
MA, USA

Stefano Pagnotta
Department of Earth Sciences
University of Pisa
Pisa, Italy

Vincenzo Palleschi
Applied and Laser Spectroscopy Laboratory
Institute of Chemistry of Organometallic Compounds
Research Area of National Research Council
Pisa, Italy

Francesco Poggialini
Applied and Laser Spectroscopy Laboratory
Institute of Chemistry of Organometallic Compounds
Research Area of the National Research Council
Pisa, Italy

Pavel Pořízka
CEITEC BUT
Brno University of Technology
Brno, Czech Republic
and
Institute of Physical Engineering
Brno University of Technology
Brno, Czech Republic

Simona Raneri
Applied and Laser Spectroscopy Laboratory
Institute of Chemistry of Organometallic Compounds
Research Area of the National Research Council
Pisa, Italy

Mohamad Sabsabi
Conseil national de recherches Canada
Boucherville, Quebec, Canada

Emanuele Salerno
Institute of Information Science and Technologies
National Research Council of Italy
Pisa, Italy

Weiran Song
State Key Lab of Power Systems
Department of Energy and Power Engineering
Tsinghua University
Beijing, China
and
Shanxi Research Institute for Clean Energy
Tsinghua University
Taiyuan, China

Anna Tonazzini
Institute of Information Science and Technologies
National Research Council of Italy
Pisa, Italy

Ivan Urbina
Atomic Spectroscopy Laboratory, Centro de Investigaciones Ópticas (CIOp)
La Plata, Argentina

Jakub Vrábel
CEITEC BUT
Brno University of Technology
Brno, Czech Republic
and
Institute of Physical Engineering
Brno University of Technology
Brno, Czech Republic

Zhe Wang
State Key Lab of Power Systems
Department of Energy and Power Engineering
Tsinghua University
Beijing, China
and
Shanxi Research Institute for Clean Energy
Tsinghua University
Taiyuan, China

Kenza Yahiaoui
Ionized Media and Lasers Division
Center for Development of Advanced Technologies
Algiers, Algeria

Tianlong Zhang
Key Laboratory of Synthetic and Natural Functional Molecule of the Ministry of Education
College of Chemistry and Materials Science
Northwest University
Xi'an, China

Preface

The laser-induced breakdown spectroscopy (LIBS) technique was born with this name in 1981, when Loree and Radziemski for the first time published two companion papers proposing a new laser spectroscopy technique for material analysis, capable of operating at a distance, without physical contact, on gaseous, liquid, and solid samples.

Although the use of a laser for spectrochemical analysis of materials was already proposed twenty years earlier, immediately after the invention of the laser itself, and the spectral analysis of the optical emission of high temperature samples, by arks, sparks, or flames, predates the laser of a couple of centuries, at least, the 1981 papers of Loree and Radziemski presented all the characteristics of "modern" LIBS, where the laser is used for ablating a tiny amount of matter from the solid samples and, at the same time, for bringing it at a temperature high enough to observe a strong atomic emission.

The first papers on LIBS also evidenced the highly dynamic nature of the laser-induced plasmas, which is probably the most limiting features of LIBS with respect to other similar spectro-analytical techniques, such as inductively coupled plasma-optical emission spectroscopy (ICP-OES), where the optical signal is stable over a long time. In the following years, it become evident in the analytical applications of the technique the need for acquiring many LIBS spectra, for sampling a significant amount of matter from the sample and reducing the effects produced by random or systematic variations of measurement parameters such as the laser energy, for example.

At the same time, the LIBS community realized that the high spatial resolution of LIBS (in principle only limited by optical diffraction and, hence, of the order of the laser wavelength), coupled to the intrinsic speed of the technique, would have made LIBS the techniques of election for the elemental analysis and mapping of complex inhomogeneous materials, as geological or biological samples.

In both cases, either for doing significant quantitative analysis on homogeneous samples or for mapping inhomogeneous materials, the LIBS applications often lead to the accumulation of a huge quantity of spectral information, which needs to be quickly processed, with a speed adequate to the short times needed for its accumulation.

One of the many revolutions in LIBS research that have characterized the explosion of new applications of the technique at the end of the last century has been the introduction in the analysis of the LIBS data of the same statistical methods that were developed in that times in Information Science for the treatment of what would have been later defined as "Big Data."

After many years of wide applications in LIBS of these methods, which in analytical chemistry go under the general denomination of chemometrics, it is emblematic the absence of a reference text describing the principles of chemometrics for LIBS, in relation to the problems of simplifying the spectral information, using it for classifying large numbers of samples or obtaining quantitative information about their elemental compositions.

There are many books on chemometrics, discussing methods that are also currently applied on LIBS, but none of them is specifically devoted to LIBS. There are, on the other hand, several review papers or book chapters on this topic, but they cannot be considered as exhaustive.

This book aims to fill the present gap between the diffuse use of chemometric methods in LIBS and the knowledge of the concepts at the basis of these methods. This knowledge should not imply a detailed understanding of the mathematical algorithms underlying the many different statistical methods used for the analysis of the LIBS spectra, but rather the awareness of their potential (and limits) in the typical applications. There is still some confusion on the proper application of chemometric methods in LIBS and, most of all, on the procedures that must be applied for the validation of the results obtained. The popular representation of these methods as "black boxes", based on mechanisms as inscrutable as the ones regulating human intelligence, must not make us forget that we are just dealing with computer algorithms, whose predictions can (and must) be verified against our chemical and physical knowledge of the system under study.

The LIBS community is constantly growing, and the generation of old school spectroscopists is being quickly replaced by a new generation of young researchers that are, by nature, extremely attracted by the potential great advantages given by chemometrics in LIBS analysis. This book is thus mainly aimed to young researchers working in public or private institutions and Ph.D. students who need a clear and practical information on the application of advanced statistical methods for the analysis of LIBS spectra. This practical approach would be useful also to more experienced researchers, wanting to stay updated on the most recent development in LIBS spectral analysis, as well as to the specialists, since the number of chemometric techniques is so large that no one can honestly say of mastering all the aspects and all the methods at the same level. Lecturers of chemistry, physics, and engineering courses at university could propose the book among the teaching material in courses on spectroanalytical techniques involving LIBS.

Introduction and Brief Summary of the LIBS Development

The reader can find the concept of laser-induced breakdown spectroscopy (LIBS) described in almost any LIBS paper. Everyone knows that when we focus a laser beam on a sample, the irradiation in the focal volume leads to local heating of the material. When the irradiance of the laser pulse exceeds the threshold of material ablation ($>$MW/cm^2), there is vaporization, and a hot ionized gas (called a plasma) is formed. In this plasma, atoms and ions are in excited states that emit light by radiative decay. Quantitative and qualitative analyses can be carried out by collecting and spectrally analyzing the plasma light and monitoring the spectral line emission positions and intensities. The technique based on that approach is called LIBS.

The LIBS technique is a form of atomic emission spectroscopy of plasma generated by a laser focused on the material to be analyzed. It is similar to other optical emission spectroscopy techniques based on plasmas, such as spark ablation, glow discharge, inductively coupled plasma, or arc plasma techniques. However, these techniques use an adjacent physical device (electrodes or a coil) to produce the plasma, whereas LIBS uses the laser-generated plasma as the hot vaporization, atomization, and excitation source. This gives LIBS the advantage that it can interrogate samples at a distance and analyze the material without contact, independent of the nature of the sample, thus making it suitable for in-the-field and real-time analysis of any type of material, whether in the solid, liquid, slurry, or gas phase. The capabilities of LIBS to effectively carry out fast, in situ, real-time, and remote spectrochemical analysis with minimal sample preparation, and its potential applications to detect traces of a wide variety of materials, make it an extremely versatile analytical technique. These attributes of LIBS attracted the interest of spectroscopists, analytical chemists, and physicists since the invention of the laser in the 1960s. Indeed, the first work on "LIBS" appeared in 1962. Since then, according to the Scopus database, more than 14 000 papers have been published in the field of LIBS, covering fundamentals, instrumentation, and applications. Figure 1 reveals the significant increase in the annual number of LIBS papers in recent decades, from a few in the 1960s to an annual rate of more than 900 today. Moreover, the field is still growing.

When we look at the development of the technique, we need to consider that the LIBS plasma is quite simple and yet complicated at the same time. You need a laser as a source of energy to generate the plasma. The plasma formed depends on the characteristics of the laser (energy, pulse duration, focusing condition, wavelength, and beam quality), on the

Chemometrics and Numerical Methods in LIBS, First Edition. Edited by Vincenzo Palleschi.
© 2023 John Wiley & Sons Ltd. Published 2023 by John Wiley & Sons Ltd.

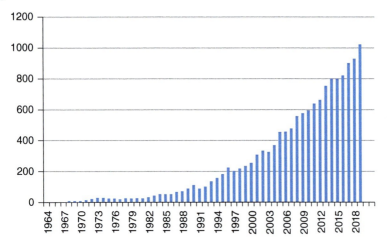

Figure 1 LIBS papers evolution according to Scopus database using specific key words.

characteristics of the sample (thermal conduction, melting and vaporization temperature, and so on), and on the ambient atmosphere (pressure, composition, and thermal conduction) where it is created. To extract the information from the light emitted, you need a spectrometer to diffract the light and a detector to convert photons to an electrical signal you can work with. It involves several fields of science, such as laser–matter interaction, plasma physics, atomic physics, plasma chemistry, spectroscopy, electro-optics, and signal processing. The LIBS plasma is transient (it is space- and time-dependent), unlike an inductively coupled plasma, arc plasma, or glow discharge plasma, which are all stationary. This characteristic dictates some restrictions on the ability to transfer tools used with other emission spectroscopy techniques to LIBS. Therefore, the development of LIBS over the years has been closely tied to the development of enabling tools (such as pulsed lasers, detectors, and spectrometers) and ongoing improvements in their performance.

We can distinguish four periods in the development and use of LIBS as technique over the last five decades. During the first period, prior to the 1990s, the plasma was generated by inadequate lasers, and the emission of the plasma was observed mostly time- and space-integrated, with the limited use of single channel photomultipliers (PMT) as detectors for time-resolved spectroscopy, so only limited analytical quantification was achievable.

During the second period, from 1990 to 2000, the arrival of the intensified charge-coupled device (ICCD) detector after the Cold War made it possible to observe time-resolved emission for several lines simultaneously in a given spectral window, rather than only one line as allowed by the single channel photomultiplier tube (PMT). This ability attracted some research groups to develop the understanding of the LIBS plasma and how it can be used for spectrochemistry. This development provided new capabilities for LIBS at the end of the 1990s and beginning of the 2000s, which allowed LIBS to address new emerging applications. In addition, the echelle spectrometer coupled with an intensified charge-coupled device (ICCD) camera allowed time-resolved broadband spectra and opened new ways to extract more information from the LIBS plasma. This capability was strengthened by the arrival of the Sony linear CCD array chip, which enabled the use of a low-cost gated

CCD camera. The combination of a gated CCD with low cost compact Czerny Turner spectrometers enabled a growth in the number of laboratories working on LIBS along with newcomers, and an increase of new applications that became feasible with the new capabilities. More importantly, it encouraged some LIBS spin-off companies to enter the market.

In the third period, from 2000 to 2010, the LIBS reached a milestone with the first conference devoted to LIBS organized in Pisa in 2000 by Vincenzo Palleschi's group. Since then, the series of LIBS International conferences has been organized every two years, alternating with the Euro-Mediterranean symposium conference (EMSLIBS), which was started in Cairo by Mohamed Abdel Harith's group in 2001. A similar LIBS symposium began in North America in 2007 and was organized by Jagdish Singh and Andrzej Miziolek. During that period, LIBS found its way across a variety of applications and disciplines in geology, metallurgy, planetary science, defense, food, environment, industry, mining, biology, automotive, materials science, aerospace, forensics, pharmaceuticals, security, and more. In addition, more companies entered the market to commercialize LIBS systems.

In the last 10 years, the miniaturization of LIBS equipment has opened new opportunities to perform real-time measurements and respond to emerging needs under conditions in which other spectroscopic techniques cannot be applied. In addition, the progress of laser technologies, such as the diode pumped laser and the fiber laser, with the improvement of the beam quality, led to better conditions for plasma generation and better analytical performance. Furthermore, the high repetition rate and the low cost of ownership of these devices have met the requirements of acceptance for several industrial applications in terms of speed of analysis and cost. Big players entered the market and now offer handheld LIBS systems. Nowadays, as an example, the operating lifetime of a fiber laser is around 100 000 hours, or 11 years, of 24/7 use without any consumables, which is better than the TV in our houses. We have seen some growth as well in R&D reflected by several regional symposium that has been organized in Asia (ASLIBS) and Latin America (LASLIBS).

To summarize, during the last three decades, extensive research has been carried out on the influence of the parameters affecting the analytical signal, to improve LIBS performance. Meanwhile, dynamic technological development in the field of solid-state lasers, electro-optical detectors, and signal processing was successfully harnessed for LIBS. The analytical performance of LIBS for a multielement analysis now achieves a level that is equal to, or even better than in some cases, that of classical methods. LIBS is currently considered one of the most active research areas in the field of analytical spectroscopy.

After the brief history and the introduction, this the first part of this book provides a brief explanation of the physics involved in plasma generation and the features of this plasma in LIBS (Chapter 1), then followed by a description of the basic components (Chapter 2), which compose a LIBS instrumentation. These devices are described associating their features with the properties of the laser-induced plasma. Finally, some key LIBS applications is described in Chapter 3.

This part will introduce the reader to the basic of LIBS, its fundamentals, instrumentation, and applications. It is not intended to be exhaustive survey of LIBS literature nor the state of art of the technique. It will bring generally to the reader a brief overview for the necessary ingredients needed to use the LIBS technique as analytical method for a given application and help understanding how to correlate spectra to composition and the factors affecting that correlation. It will provide a brief explanation of the physics involved in plasma generation and the features of this plasma in LIBS, followed by a description of the basic devices, which compose a LIBS instrumentation. These components will be described associating their features with the properties of the laser-induced plasma. Moreover, different kinds of parameters that affect the plasmas and its use for spectrochemistry will be highlighted in order to help the users understand different analysis algorithms and chemometrics method that go beyond the "spectral signatures" obtained with the technique. Finally, some key LIBS applications will be described, and the main research challenges that this approach face at the moment will be discussed.

Part I

Introduction to LIBS

1

LIBS Fundamentals

Mohamad Sabsabi

Conseil national de recherches Canada, Boucherville, Quebec, Canada

From laser ablation to nuclear fusion, from welding and drilling to surface treatment, and from laser deposition to selective chemical reactions in nanotechnologies, laser plasma physics has generated an enormous fallout of applications. By measuring the emission spectrum from the laser-induced plasma, qualitative and quantitative information about the sample's chemical composition can be obtained. Laser ablation means using laser light energy to remove a portion of a sample by melting, fusion, sublimation, ionization, erosion, and/or explosion. Laser ablation results in the formation of a gaseous vapor, luminous plasma, and in the production of fine particles. This is the base of laser-induced breakdown spectroscopy (LIBS) technique. LIBS is only a tiny fraction of all the application domains in which these physical mechanisms have found useful applications. Furthermore, the laser ablation process (a term that includes the processes of evaporation, ejection of atoms, ions, molecular species, and fragments; generation of shock waves; plasma initiation and expansion; plasma–solid interactions; etc.) influences the amount and composition of the ablated mass and must be understood and controlled in order to achieve accurate and sensitive quantitative analysis. The LIBS technique is at a crossroad of several fields of science, such as laser–matter interaction, plasma physics, atomic physics, plasma chemistry, spectroscopy, electro-optics, and signal processing. Readers with no substantial prior knowledge on this subject are invited to thoroughly read the review of Hahn and Omenetto [1–2] as well as the references of LIBS books and review papers contained herein. Also, expert readers can find more detailed information on those topics in the books and review paper contained herein [1–25]. This part of Chapter 1 is devoted to LIBS fundamentals. It will introduce the reader to the basics of LIBS plasmas and their fundamentals in order to understand their use for analytical chemistry. It will bring generally to the reader a brief overview of the necessary ingredients needed to use the LIBS technique as analytical method for a given application. We will discuss the fundamental principles of the laser–matter interaction, plasmas physics, atomic physics and optical emission spectroscopy involved in the LIBS technique and relevant for its use for spectrochemical analysis. By using optical means, the spectroscopist tries to correlate the LIBS signal to the chemical composition of the sample they would like to analyze. The spectral emission intensity in the plasma

Chemometrics and Numerical Methods in LIBS, First Edition. Edited by Vincenzo Palleschi.
© 2023 John Wiley & Sons Ltd. Published 2023 by John Wiley & Sons Ltd.

depends not only on the concentration of the element in the sample but it is also affected by the properties of the plasma itself. The latest depends on factors such as the characteristics of the laser excitation source (energy, pulse duration, focusing condition, wavelength, and beam quality), the sample characteristics (thermal conductivity, melting and evaporation temperature, density, and chemical composition), and the surrounding gas (nature, composition, thermal conduction, and ambient atmosphere). Furthermore, the laser ablation process (which includes the processes of evaporation, ejection of atoms, ions, molecular species, and fragments; generation of shock waves; plasma initiation and expansion; plasma–solid interactions; etc.) influences the amount and composition of the ablated mass and must be understood and controlled in order to achieve qualitative or accurate and sensitive quantitative analysis. The complexity of the phenomena involved can be assessed by resorting to a simple derivation of the dependence of the LIBS signal upon the various processes leading from the (solid) sample to the measured signal photons emitted from the (gas phase) atoms and ions excited in the plasma volume. The fundamental parameters governing the overall process can then be explicitly written (see Chapter 3 by Palleschi and Sabsabi Reference [4]):

$$S = A_{mn} f_{interaction} f_{excitation} f_{detection} \tag{1.1}$$

It can be concluded that the signal is influenced by three interrelated functions, describing the initial interaction between the sample and the laser, $f_{interaction}$ (leading to ablation/vaporization of solid material); the excitation/ionization mechanism leading to atomic (ionic) emission, $f_{excitation}$; and the characterization of the radiation environment, $f_{detection}$ (thin or thick plasmas), while A_{mn} is the transition probability of the transition chosen (photons/s). The $f_{interaction}$ and the $f_{excitation}$ will be developed in the following part of Chapter 1, while the $f_{detection}$ will be detailed in Chapter 2.

In summary, the quality of correlation between the LIBS analytical signal and the element concentration in the sample depends on the features of the laser-induced plasmas and their generation conditions. In addition, the plasmas' generation conditions should be reproducible and controlled in order to be useful for spectrochemistry.

1.1 Interaction of Laser Beam with Matter

The interaction of high-power laser light with a target material has been an active topic of research not only in plasma physics but also in the field of material science, chemical physics, and particularly in analytical chemistry [3–8]. The high intensity laser beam focused on a target (solid, liquid, or gas) may dissociate, excite, and/or ionize the constituent atomic species of the solid and produces plasma, which expands either in the vacuum or in the ambient gas depending on the experimental conditions. As a result of laser–matter interaction and depending on the laser characteristics (in particular its irradiance), various processes may occur such as ablation of material (the processes of evaporation, ejection of atoms, ions, molecular species and fragments; generation of shock waves; plasma initiation and expansion; plasma–solid interactions; etc.) that influences the amount and composition of the ablated mass, high-energy particle emission, generation of various parametric instabilities, as well as emission of radiation ranging from the infrared (IR) to X-rays. By measuring the emission spectrum from the laser-induced plasma, qualitative and quantitative information about the sample's chemical composition can be obtained. This is the base of the LIBS technique. These processes have many applications, but we are mainly interested in the one that is related to the study of optical emission from the plasma. For other applications of laser-produced plasma, the readers can find detailed information in a series of books [26–29]. Laser ablation is the first step in the LIBS process, and its influence will be reflected in the "figures of merit," temporal and spatial resolution, sensitivity, precision, and accuracy. The influence of laser ablation on LIBS has been studied extensively in several books, and we refer the reader particularly to Chapter 3 by Russo et al. in the book edited by Singh and Thakur [3].

1.2 Basics of Laser–Matter Interaction

When a high-power laser pulse is focused onto a material target (solid, liquid, gas, and aerosols), the intensity in the focal spot produces rapid local heating and intense evaporation followed by plasma formation. The interaction between a laser beam and a solid is dependent on many characteristics of both the laser and the solid material. It is a complicated process and not fully understood phenomenon, which is still under intensive investigation. Various factors affect ablation of material, which includes the laser pulse width, its wavelength, its spatial and temporal fluctuations, as well as its power fluctuations. The mechanical, physical, and chemical properties of the target material also play an important role in laser-induced ablation. The phenomena of laser–target interactions have been reviewed by several authors [30, 31], while the description of melting and evaporation at metal surfaces has been reported by Ready [30]. The ablated material compresses the surrounding atmosphere and leads to the formation of a shock wave. During this process, a wide variety of phenomena including rapid local heating, melting, and intense evaporation are involved. Then, the evaporated material expands as a plume above the sample surface. The plasma expands normal to the target surface at a supersonic speed in vacuum or in the ambient gas. The hot expanding plasma interacts with the surrounding gas mainly by two mechanisms: (i) the expansion of high-pressure plasma compresses the surrounding gas and drives a shock wave and (ii) during this expansion, energy is transferred to ambient gas by the combination of thermal conduction, radiative transfer, and heating by shock wave. The evolution of plasma depends on the intensity of laser, its wavelength, size of focal spot, target vapor composition, ambient gas composition, and its pressure. It has been found that the plasma parameters such as radiative transfer, surface pressure, plasma velocity, and plasma temperature are strongly influenced by the nature of the plasma. Since vaporization and ionization take place during the initial fraction of laser pulse duration, the rest of the laser pulse energy is absorbed in the vapor and expanding plasma plume. This laser absorption in the expanding vapor/plasma generates three different types of waves as a result of different mechanisms of propagation of absorbing front into the cool transparent gas atmosphere. These waves are (i) laser-supported combustion (LSC) waves, (ii) laser-supported detonation (LSD) waves, and (iii) laser-supported radiation (LSR) waves [26, 30]. Each wave is distinguished based on its velocity, pressure, and on the effect of its radial expansion during the subsequent plasma evolution, which is strongly dependent on the intensity of irradiation. See more details for low irradiance [31, 32] and intermediate and high irradiance [33, 34].

At low irradiation, LSC waves are produced, which comprise of a precursor shock that is separated from the absorption zone and the plasma. The shock wave results in an increase in the ambient gas density, pressure, and temperature, whereas the shock edges remain transparent to the laser radiation. The front edge of the expanding plasma and the laser absorption zone propagate into the shocked gas and give rise to LSC wave.

At intermediate irradiance, the precursor shock is sufficiently strong, and the shocked gas is hot enough to begin absorbing the laser radiation without requiring additional heating by energy transport from the plasma. The laser absorption zone follows directly behind the shock wave and moves at the same velocity.

At high irradiance, the plasma is so hot that, prior to the arrival of the shock wave, the ambient gas is heated to temperatures at which laser absorption begins. In the ideal condition, laser absorption is initiated without any density change, and the pressure profile results mainly from the strong local heating of the gas rather than a propagating shock wave.

1.3 Processes in Laser-Produced Plasma

As discussed in the previous section, the interaction of a high intensity laser light with solid target initially increases the surface temperature of the sample such that material transfer across the surface becomes significant (vaporization). As a result of material vaporization and plasma formation, target erosion appears in the form of craters on the sample surface. The theoretical considerations on plasma production and heating by means of laser beams have been proposed by several authors [32–35]. The initiation of plasma formation over a target surface begins in the hot target vapor. First of all, absorption of laser radiation takes place via electron-neutral inverse Bremsstrahlung, but when sufficient electrons are generated, the dominant laser absorption mechanism makes a transition to electron-ion inverse Bremsstrahlung. Photoionization of excited states can also contribute in the case of interactions with short wavelength radiations. The same absorption processes are responsible for the absorption by the ambient gas also. The laser-produced plasma expands into the vacuum or into the surrounding gas atmosphere, where the free electrons present in the plasma [25–29] modify propagation of laser light. The plasma formed by a high intensity or short time duration laser has a very steep density and temperature gradient in comparison to the plasma formed by the low intensity or long time duration laser. The density gradient in the plasma plays a very important role in the mechanism of light absorption and in the partition of absorbed energy between thermal and nonthermal particle distribution. There are three basic mechanisms through which intense laser light may interact with plasma [29]. The first mechanism is an inverse Bremsstrahlung, where electric field of the incident light rattles electrons, which then lose this energy in collision with ions. This mechanism is important with shallow density gradients in the plasma. The parametric processes also take part most efficiently, when the density gradient is shallow. There are three wave parametric interaction processes in which intense laser field drives one or more longitudinal plasma waves out of the noise and also parametric decay processes where laser light decays into a high-frequency electron acoustic wave and a low-frequency ion acoustic wave conserving energy and momentum. Another important short-pulse laser absorption mechanism is the resonance absorption. With a p-polarized light obliquely incident on plasma surface, the radial component of electric field resonates with plasma frequency and causes large transfer of energy to electrons near critical density N_c surface. Critical density for a given laser wavelength is

$$N_c = 10^{21}/\lambda^2 \text{ cm}^{-3} \tag{1.2}$$

where λ is in micron. Energy absorbed at or below the critical density in plasma is then conducted toward the target surface by various transport processes. The study of energy coupling to the target has many subareas such as laser light absorption, nonlinear interaction, electron energy transport, and ablation of material from the target surface. One of the important processes, in laser–plasma interaction, is emission of radiation from the plasma ranging from visible to hard X-rays [3], and it is very relevant for the understanding of LIBS. It has been found that X-rays are emitted from all parts of the absorption, interaction, and transport regime. At densities near and slightly above the critical, nearly 70% of the incident laser energy may be re-emitted as X-rays with energy ranging from 50 eV to 1 keV or above depending on the temperature of the plasma. However, as the plasma expands away from the target surface, its density as well as the temperature decreases. As the plasma temperature decreases, the wavelength of emission from the plasma increases, that is, emission shifts from X-rays to visible region.

1.4 Factors Affecting Laser Ablation and Laser-Induced Plasma Formation

1.4.1 Influence of Laser Parameters on the Laser-Induced Plasmas

The laser is the most important variable affecting the characteristics of the plasma since the effects of its parameters are twofold: first, during its interaction with the targeted sample and then with the plasma plume itself. In general, photons are coupled within the available electronic, or vibrational, states in the material depending on its wavelength. During this coupling, the material is heated to a particular temperature depending on the mechanism of interaction of the laser pulse with that, and the onset of ablation (either thermal or photochemical) occurs if the fluence is above a particular threshold. Once the plasma plume is generated, its density may obstruct ("plasma shielding" as explained in the previous section) partially or entirely laser radiation, depending on the laser wavelength and pulse length. Consequently, not the full energy is transferred from the laser pulse to the original material. With all these key aspects affecting the whole chain of possible events occurring during plasma formation, it seems obvious to the reader that different laser parameters may affect the physics of the plasma plume. In order to bring more clarification to the reader, an understandable framework devoted to the influences of laser parameters will be addressed as follows.

1.4.2 Laser Wavelength (λ)

The wavelength influence on LIBS can be explained from two points of view; the laser–material interaction (energy absorption) and the plasma development and properties (plasma–material interaction).

When photon energy is higher than bond energy, photon ionization occurs and nonthermal effects are more important. For this reason, the plasma behavior depends on wavelength in nanosecond LIBS setups. In the same way, the optical penetration is shorter for ultraviolet (UV) lasers, providing higher laser energy per volume unit of material. In general, the shorter the laser wavelength, the higher the ablation rate and the lower the elemental fractionation [36].

The plasma ignition and its properties depend on wavelength. The plasma initiation with nanosecond lasers is provoked by two processes; the first one is inverse Bremsstrahlung by which free electrons gain energy from the laser during collisions among atoms and ions. The second one is photoionization of excited species and excitation of ground atoms with high energies. Laser coupling is better with shorter wavelengths, but at the same time the threshold for plasma formation is higher. This is because inverse Bremsstrahlung is more favorable for IR wavelengths [37].

In contrast, for short wavelengths (between 266 and 157 nm), the photoionization mechanism is more important. For this reason, the shorter the wavelength in this range, the lower the fluence necessary (energy per unit area) to initiate ablation [38]. In addition, when inverse Bremsstrahlung occurs, part of the nanosecond laser beam reheats the plasma. This not only increases the plasma lifetime and intensity but also increases the background at the same time. Longer wavelengths increase inverse Bremsstrahlung plasma shielding but reduce the ablation rate and increase elemental fractionation (elemental fractionation is the redistribution of elements between solid and liquid phases that modifies plasma emission) [39].

1.4.3 Laser Pulse Duration (τ)

Irrespective of their duration, laser pulses usually reach the required conditions for ablation of targets since the rate of energy deposition greatly exceeds the rate of energy redistribution and dissipation, thus resulting in extremely high temperatures in those regions where energy absorption occurs. However, as a consequence of the different mechanisms of energy dissipation in the sample, differences in pulse duration result in fundamental differences of the ablation process. Indeed, interaction of nanosecond (ns) pulses with materials is substantially different from those of femtosecond (fs) pulses since the rate of energy deposition is significantly shorter in this last instance. Thus, for ns pulses, the material undergoes transient changes in the thermodynamic states from solid, through liquid, into a plasma state. Furthermore, the leading edge of the laser pulse creates plasma, and the remaining part of the pulse heats the plasma instead of interacting with the target. In the case of ultrashort laser pulses, at the end of the laser pulse, only a very hot electron gas and a practically undisturbed lattice are found.

If the selected laser is a femtosecond one, nonthermal processes will dominate the ionization. The pulse is too short to induce thermal effects; hence, other effects should ionize the atoms, depending on the kind of sample. The pulse has a huge amount of energy and effects like multiphoton absorption and ionization, tunneling, and avalanche ionization excite the sample. With this amount of energy, the electron–hole created will induce emission of X-rays, hot electrons, and photoemission. This will create highly charged ions through a process called Coulomb explosion [5]. The absence of thermal effects creates a crater with highly defined edges without melted or deposited materials.

In contrast, nanosecond lasers induce other effects. The electron-lattice heating time is around 10^{-12} second, much shorter than the pulse time. This causes thermal effects to dominate the ionization process. Briefly, the laser energy melts and vaporizes the sample, and the temperature increase ionizes the atoms. If the irradiance is high enough, nonthermal effects will be induced too and both will ionize the sample. Between 10^{-9} and 10^{-8} seconds, plasma becomes opaque for laser radiation, thus the last part of the laser pulse interacts with plasma surface and will be absorbed or reflected; hence, it will not ionize much more material. This effect is called plasma shielding and is strongly dependent on environmental conditions (surrounding gases or vacuum) and experimental conditions (laser irradiance and wavelength) [40, 41]. This shielding reduces the ablation rate because the radiation does not reach the sample surface. This induces a crater with melted and deposited material around it, but at the same time the plasma is reheated, and the lifetime and size of plasma is higher [40, 41].

1.4.4 Laser Energy (E)

The energy parameters related to laser material interaction are fluence (energy per unit area, J/cm^2) and irradiance (energy per unit area and time, W/cm^2). Ablation processes (melting, sublimation, erosion, explosion, etc.) have different fluence thresholds [5]. The effect of changes in the laser energy is related to laser wavelength and pulse time. Hence, it is difficult to analyze the energy effect alone. In general, the ablated mass and the ablation rate increase with laser energy. The typical threshold level for gases is around 10^{11} and 10^{10} W/cm^2 for liquids, solids, and aerosols [3]. These values are for guidance and depend strongly on laser pulse time and environmental conditions, reaching up to around 10^{15} W/cm^2 for nitrogen at 760 Torr with an Nd : YAG laser, 1064 nm, 7 ps [4].

1.4.5 Influence of Ambient Gas

The plasma size, propagation speed, energy, and emission properties are related to the ambient gas into which the plasma expands. The ambient gas can help or prevent the plasma shielding. For example, the gas can shield the sample from the laser beam if a gas breakdown occurs before sample vaporization [42]. These undesirable effects are less important for gases and aerosols, but they can be important for solid samples.

Gas pressure will influence plasma expansion. Low pressures increase energy losses and uniformity of the plasma energy distribution. In addition, different gases have different behaviors at different pressures [42]. Thermal conductivity of the gas can affect the cooling of the laser-induced plasma. For instance, it has been found from the literature that ambient gas having higher thermal conductivity may shrink the plasma and shorten its lifetime after the laser pulse due to the cooling of the plasma by thermal conduction. Ambient gas also plays an important role in the excitation and may increase the population in some excited levels of given atoms through favorable collisional excitation of the atoms of the group VII in particular fluorine and chlorine. It is known that helium gas can enhance also the excitation of the atoms in the group VII of the periodic table [42].

1.5 Plasma Properties and Plasma Emission Spectra

The plasma, induced by the interaction pulsed laser sample, emits light, which consists of discrete lines (bound–bound), bands, and an overlying continuum (free-bound and free–free). These discrete lines, which characterize the material, have three main features: wavelength, intensity, and shape. These parameters depend on both the structure of the emitting atoms and their environment. Each kind of atom has some different energy levels, which determine the wavelength of the line.

The main diagnostic technique for plasmas involves the relationship between plasma properties and spectral line characteristics. Line widths are related to plasma temperature and electron density. Line shapes and shifts can be a diagnostic for the principal broadening mechanism. There are other diagnostic techniques. Langmuir probes and Thompson scattering can be used to measure electron densities, Schlieren or other interferometric techniques can reveal refractive index changes, and Abel inversion can assist in unfolding properties in the plasma by layer. Even the plasma acoustic emission can be employed for diagnosis. Optical techniques other than the passive observations of emission lines, such as laser-induced fluorescence, can cast light on critical plasma parameters. On the other hand, the intensity and shape of the lines depend strongly on the environment of the emitting atom. For not too high plasma densities, both the natural broadening (due to Heisenberg's uncertain principle) and the Doppler broadening (Doppler broadening is due to the thermal motion of the emitters, the light emitted by each particle can be slightly red- or blue-shifted, and the final effect is a broadening of the line) dominate the linear shape [12]. For high plasma densities, atoms in the plasma are affected by electric fields due to fast-moving electrons and slow-moving ions, and these electric fields split and shift the atomic energy levels. As a consequence of these perturbations of the levels, the emission lines are broadened, and they change their intensity and shape. This effect is known as the Stark effect [12], and it dominates the line shape for dense plasmas. This broadening together with the different parameters of spectral lines (intensities and shapes) and even the continuum radiation features can be useful to determine plasma parameters, such as electron temperature, pressure, and electron number density [3–8]. These parameters are very important to characterize the plasma, giving information about the physical state of it.

A LIBS plasma is a local assembly of atoms, ions, and free electrons, over all electrically neutral, in which the charged species often act collectively. Plasmas are characterized by a variety of parameters, the most basic being the degree of ionization. A weakly ionized plasma is one in which the ratio of electrons to other species is less than 10%. At the other extreme, highly ionized plasmas may have atoms stripped of many of their electrons, resulting in very high electron to atom/ion ratios. LIBS plasmas typically fall in the category of weakly ionized plasmas. LIBS plasmas are highly dynamical systems, space and time dependent. Their physical characteristics (total density, electron number density, temperature) and behaviour resemble the ones of thermal plasmas like the ones induced by electrical arc or spark. At early times, ionization is high. As electron–ion recombination proceeds, neutral atoms and then molecules are formed. Throughout there is a background continuum that decays with time more quickly than the spectral lines. The continuum is primarily due to Bremsstrahlung (free–free) and recombination (free–bound) events. In the Bremsstrahlung process, photons are emitted by electrons accelerated or decelerated in collisions. Recombination occurs when a free electron is captured into an ionic or atomic energy level and gives up its excess kinetic energy in the form of a photon. Time resolution of the plasma light in LIBS allows for discrimination in favor of the region where the signals of interest predominate. The symbol t_d represents the delay from the initiation of the laser to the opening of the window during which signal will be accepted; t_b represents the length of that window. LIBS has also been performed in a double-pulse mode, where two laser pulses from the same or different lasers are incident on the target. For an overview of diagnostic techniques, the literature is very rich, and the reader can see any book devoted to atomic physics of plasmas and emission spectroscopy, for example, books [3–9] and the review paper of Hahn and Omenetto [2, 25] reported herein. The most powerful spectroscopic technique of determining the electron density Ne of discharge plasma comes from the measurement of the Stark broadening of spectral lines. In this method, absolute intensities of spectral lines are not required, merely line shapes and Full-Width at Half-Maximum (FWHM) are sufficient. Since broadening is quite appreciable for electron density $Ne \geq 10^{15}$ cm^{-3}, standard spectrometers often suffice to record the spectra for measurements of line shape. The electron density Ne is extracted by matching the line width (or the entire line shape) with the calculated one. Details of line shape calculations can be found in Griem's books [40, 41]. Plasma temperature can be derived from the Boltzman plot, Saha's equation, or from the value of the continuum that depends on the plasma temperature and electron density. More details can be found in the reference mentioned here [3–9].

References

1 D.W. Hahn, N. Omenetto, Laser-induced breakdown spectroscopy (LIBS), part I: review of basic diagnostics and plasma-particle interactions: still-challenging issues within the analytical plasma community, *Appl. Spectrosc.* 64 (2010), pp. 335–366.
2 D.W. Hahn, N. Omenetto, Laser-induced breakdown spectroscopy (LIBS), part II: review of instrumental and methodological approaches to material analysis and applications to different fields, *Appl. Spectrosc.* 66 (2012), pp. 347–419
3 D.A. Cremers and L.J. Radziemski, *Handbook of Laser-induced Breakdown Spectroscopy* (John Wiley & Sons Inc., Hoboken, NJ, (2013)).
4 A.W. Miziolek, V. Palleschi, and I. Schechter, Eds., *Laser Induced Breakdown Spectroscopy (LIBS): Fundamentals and Applications* (Cambridge University Press, Cambridge, UK, (2006)).
5 J.P. Singh and S.N. Thakur, Eds., *Laser-Induced Breakdown Spectroscopy* (2nd ed), (Elsevier, Amsterdam, The Netherlands, (2020)).

6 Y.-I. Lee and J. Sneddon, *Laser Induced Breakdown Spectroscopy: A Practical and Tutorial Approach* (Taylor & Francis, Abingdon, Oxfordshire, UK, (2016)).
7 R. Noll, *Laser Induced Breakdown Spectroscopy: Fundamentals and Applications* (Springer, Heidelberg, Germany, (2012)).
8 M. Baudelet, *Laser-Induced Breakdown Spectroscopy: A Fundamental Approach for Quantitative Analysis* (Momentum Press, Buchanan, NY, (2014)).
9 S. Musazzi and U. Perini, Eds., *Laser-Induced Breakdown Spectroscopy: Theory and Applications* (Springer, Heidelberg, Germany, (2012)).
10 D. Santos, L.C. Nunes, G.G. Arantes de Carvalho, M. da Silva Gomes, P.F. de Souza, F. de Oliveira Leme, L.G.C. dos Santos, F.J. Krug, Laser-induced breakdown spectroscopy for analysis of plant materials: a review, *Spectrochim. Acta Part B At. Spectrosc.*, 71–72 (2012), pp. 3–13.
11 R.S. Harmon, R.E. Russo, R.R. Hark, Applications of LIBS for geochemical and environmental analysis: a comprehensive review, *Spectrochim. Acta Part B At. Spectrosc.*, 87 (2013), p. 11.
12 G.S. Senesi, Laser-induced breakdown spectroscopy (LIBS) applied to terrestrial and extraterrestrial analogue geomaterials with emphasis to minerals and rocks, *Earth Sci. Rev.*, 139 (2014), pp. 231–267
13 C. Li, C.L. Feng, H.Y. Oderji, G.N. Luo, H.B. Ding, Review of LIBS application in nuclear fusion technology, *Front. Physiol.*, 11 (2016), p. 16.
14 R. Noll, C. Fricke-Begemann, S. Connemann, C. Meinhardt, V. Sturm, LIBS analyses for industrial applications—An overview of developments from 2014 to 2018, *J. Anal. At. Spectrom.*, 33 (2018), pp. 945–956.
15 A. Botto, B. Campanella, S. Legnaioli, M. Lezzerini, G. Lorenzetti, S. Pagnotta, F. Poggialini, V. Palleschi, Applications of laser-induced breakdown spectroscopy in cultural heritage and archaeology: a critical review, *J. Anal. At. Spectrom.*, 34 (2019), pp. 81–103.
16 R. Gaudiuso, N. Melikechi, Z.A. Abdel-Salam, M.A. Harith, V. Palleschi, V. Motto-Ros, B. Busser, Laser-induced breakdown spectroscopy for human and animal health: a review, *Spectrochim. Acta Part B At. Spectrosc.*, 152 (2019), pp. 123–148.
17 G. Nicolodelli, J. Cabral, C.R. Menegatti, B. Marangoni, G.S. Senesi, Recent advances and future trends in LIBS applications to agricultural materials and their food derivatives: an overview of developments in the last decade (2010–2019). Part I. Soils and fertilizers, *Trends Anal. Chem.*, 115 (2019), pp. 70–82.
18 G.S. Senesi, J. Cabral, C.R. Menegatti, B. Marangoni, G. Nicolodelli, Recent advances and future trends in LIBS applications to agricultural materials and their food derivatives: an overview of developments in the last decade (2010–2019). Part II. Crop plants and their food derivatives. *Trends Anal. Chem.*, 118 (2019), pp. 453–469.
19 G.S. Senesi, R.S. Harmon, R.R. Hark, Field-portable and handheld laser-induced breakdown spectroscopy: Historical review, current status and future prospects, *Spectrochim. Acta Part B At. Spectrosc.*, 175 (2021),106013.
20 S. Legnaioli, B. Campanella, F. Poggialini, S. Pagnotta, M. A. Harith, Z. A. Abdel-Salam and V. Palleschi, Industrial applications of laser-induced breakdown spectroscopy: a review, *Anal. Methods*, 12 (2020), pp. 1014–1029.
21 F.J. Fortes, J. Moros, P. Lucena, L.M. Cabalín, J.J. Laserna, Laser-induced breakdown spectroscopy, *Anal. Chem.* 85 (2013), pp. 640–669.

22 W. Li, X. Li, X. Li, Z. Hao, X. Zeng, A review of remote laser-induced breakdown spectroscopy, *Appl. Spectrosc. Rev.* 55 (2020), pp. 1–25.

23 L.J. Radziemski, D.A. Cremers, A brief history of laser-induced breakdown spectroscopy: from the concept of atoms to LIBS 2012, *Spectrochim. Acta Part B At. Spectrosc.*, 87 (2013), pp. 3–10.

24 P.A. Foster, Ed., *A Guide to Laser-Induced Breakdown Spectroscopy. Series: Physics Research and Technology* (Nova Science Publishers, Inc., New York, (2020).

25 L.J. Radziemski, D.A. Cremers, *Laser Induced Plasma and Applications* (Marcel Dekker, New York, (1989)).

26 G. Bekefi, *Principle of Laser Plasmas*, (John Wiley & Sons, New York, (1976)).

27 R.E. Cairn, J.J. Sanderson, *Laser Plasma Interaction*, (Institute of Physics Publishing, Edinberg, Scotland, (1980)).

28 C. E. Max, *Laser Plasma Interaction*, R. Balian, J. C. Adams, Eds., (North Holland Publishing Co., Amsterdam, The Netherlands (1982)).

29 W.L. Kruer, *The Physics of Laser Plasma Interaction*, (Addison-Wesley, New York (1988)).

30 J. F. Ready, *Effect of High Power Laser Radiation*, (Academic Press, New York (1971)).

31 N. H. Kemp and R. G. Root, Analytical study of laser-supported combustion waves in hydrogen, *J. Energy* 3 (1979), p. 40.

32 G. Weyl, A. Pirri and R. Root, Laser ignition of plasma off aluminum surfaces, *AIAA J.*, 19 (1981), p. 460.

33 W. E. Maher, R. B. Hall and R. R. Johnson, Experimental study of ignition and propagation of laser-supported detonation waves, *J. Appl. Phys.* 45 (1974), p. 2138.

34 J.B. Steverding, Ignition of laser detonation waves, *J. Appl. Phys.* 45 (1974), p. 3507.

35 N. Pirri, R. G. Root and P. K. S. Wu, Plasma energy transfer to metal surfaces irradiated by pulsed lasers, *AIAA J.* 16 (1978), p. 1296.

36 G. Abdellatif and H. Imam, A study of the laser plasma parameters at different laser wavelengths, *Spectrochim. Acta Part B At. Spectrosc.*, 57 (7) (2002) pp. 1155–1165.

37 L. M. Cabalin and J. J. Laserna, Experimental determination of laser induced breakdown thresholds of metals under nanosecond Q-switched laser operation, *Spectrochim. Acta Part B At. Spectrosc.*, 53(5), (1998), pp. 723–730.

38 R. E. Russo, X. L. Mao, O. V. Borisov, and L. Haichen, Influence of wavelength on fractionation in laser ablation ICP-MS, *J. Anal. At. Spectrom.*, 15(9), (2000), pp. 1115–1120.

39 X. Mao, W. T. Chan, M. Caetano, M. A. Shannon, and R. E. Russo, Preferential vaporization and plasma shielding during nano-second laser ablation, *Appl. Surf. Sci.*, (96–98) (1996), pp. 126–130.

40 J. M. Vadillo, J. M. Fernandez Romero, C. Rodrıguez, and J. J. Laserna, Effect of plasma shielding on laser ablation rate of pure metals at reduced pressure, *Surf. Interface Anal.*, 27 (11), (1999), pp. 1009–1015.

41 J. A. Aguilera, C. Aragon, and F. Penalba, Plasma shielding effect in laser ablation of metallic samples and its influence on LIBS analysis, *Appl. Surf. Sci.*, 127–129, (1998), pp. 309–314.

42 Y. Iida, Effects of atmosphere on laser vaporization and excitation processes of solid samples, *Spectrochim. Acta Part B At. Spectrosc.*, 45(12), (1990), pp. 1353–1367.

43 D.R. Keefer, H.L. Crowder, and R. E. Elkins, A two-dimensional model of the hydrogen plasma for a laser powered rocket, AIAA Meeting Paper, 20th Aerospace Sciences Meeting, 11 January 1982–14 January 1982, Orlando, FL, USA, 82-0404 (1982).

44 S.A. Ramsden and P. Sevic. A radiative detonation model for the development of a laser-induced spark in air. *Nature* 203, (1964), pp. 1217–1219.

45 Y. P. Raiser, Heating of a gas by a powerful light pulse, *Sov. Phys. JEPT* 21 (1965), p. 1009.

46 H. R. Griem, *Plasma Spectroscopy*, (McGraw-Hill, New York (1964)).

47 H. R. Griem, *Spectral Line Broadening by Plasmas*, (Academic Press, New York (1974)).

2

LIBS Instrumentations

Mohamad Sabsabi[1] and Vincenzo Palleschi[2]

[1] Conseil national de recherches Canada, Boucherville, Quebec, Canada
[2] Applied and Laser Spectroscopy Laboratory, Institute of Chemistry of Organometallic Compounds, Research Area of the National Research Council, Pisa, Italy

2.1 Basics of LIBS instrumentations

The purpose of this chapter is to provide a description of the basic components of a LIBS system (laser source, focusing and collection optics, and detection system) and how their technical features as well as their experimental arrangement may affect the measurements. New developments in laser sources, fiber optics technology, and silicon photon multiplier (SiPM) detectors are also highlighted. As seen in the first part of Chapter 1, the reader can find a description of LIBS setup in any LIBS-related literature. A LIBS setup needs a laser as source for plasma generation, optical components for focusing a light on the sample and collecting the light from the generated plasma, a spectrometer to diffract that collected light and finally a detector at the exit of the spectrometer to convert photons to electrons leading to a spectrum that will be treated by a computer for useful information. Unlike inductively coupled plasma atomic emission spectroscopy (ICP-AES) or other techniques based on optical emission spectroscopy (OES), the LIBS technique requires a time-resolved detection owing to its transient nature. LIBS is concerned with the radiation emitted by the microplasma induced by focusing a powerful laser on the sample. LIP emission is space- and time-dependent. In the initial moments, the plasma emission consists of an intense radiation continuum superimposed with much broadened lines. These lines are strongly broadened by the Stark effect, owing to the high electron density that exists in the plasma at this time. Thus, temporally gating off the earlier part of the plasma is essential for spectrochemical analysis. This dictates an important difference based on time-gated detection of the atomic emission between the detector requirements for LIBS and ICP-OES or other techniques based on OES. In addition, some other features are also of concerns such as the high repetition rate of the laser, which requires a fast read out of the data given by the detector.

Chemometrics and Numerical Methods in LIBS, First Edition. Edited by Vincenzo Palleschi.
© 2023 John Wiley & Sons Ltd. Published 2023 by John Wiley & Sons Ltd.

2.2 Lasers in LIBS Systems

There are a wide range of lasers sources used for LIBS and laser ablation studies [1–4]. As source for plasmas generation, the laser irradiance (also termed laser intensity) on the sample to be analyzed should exceed the threshold of ablation of the sample. The most wide spread types of lasers utilized in LIBS applications are flashlamp-pumped solid-state lasers and recently diode pumped solid laser diode-pumped solid-state (DPSS) with Nd:YAG as laser medium operated in the actively Q-switch mode to generate high-energy laser pulses with pulse durations in the nanosecond range. The Q-switching is a very well-known optical technique utilized for obtaining very narrow and intense laser pulses. In active Q-switching, the fast variation of the energy losses is obtained by modifying the polarization state of the intra-cavity radiation by means of an electro-optical device (like e.g. a Pockels cell) (Section 2.1). Table 2.1 (not exhaustive) gives an overview of the lasers and laser parameters used for LIBS together with the emitted wavelength and a rough indication of the pulse duration and the energy per pulse. For further details concerning characteristics, design, construction, and performance of solid-state lasers, see Sections 2.3 and 2.4. The main features of solid-state lasers and here especially the Nd : YAG lasers are presented, since these are of predominant importance for LIBS. The laser medium is a doped insulator; dopants are rare earths or transition metals. The laser medium is pumped optically with broadband noble gas or halogen lamps (emission range 200–1000 nm) or narrow band semiconductor lasers (806 nm). Figure 2.1 shows the principal setup of a flashlamp-pumped solid-state

Table 2.1 Laser sources utilized in LIBS experimental apparatuses.

Laser type	Wavelength (nm)	Pulse duration (ns)	Energy/pulse (mJ)
CO_2 Q-Switched	10.6×10^3	200	100
Er : YAG Q-switched	2.94×10^3	170	25
Nd : YAG	1.064×10^3	4–10	0.1–3
Nd : YAG second harmonic	532	4–8	0.05–2
Nd : YAG third harmonic	354.7	4–8	0.02–0.7
Nd : YAG fourth harmonic	256	3–5	0.01–0.3
Ruby Q-switched	694.3	5–30	1–50
Ruby ps pulse	694.3	10^{-2}	0.01–0.5
N_2 laser	337.1	3–6	0.1–0.6
XeCl excimer	308	20–30	$0.5 \times 10^3 - 10^3$
KrF excimer	248	25–35	$0.5 - 10^3 - 10^3$
ArF excimer	193	8–15	8–15
Fiber laser—Ytterbium doped	$1.03-1.08 \times 10^3$	$5 \times 10^{-5} - 10^3$	a
Fiber laser—Erbium doped	$1.53-1.62 \times 10^3$	$5 \times 10^{-5} - 10^3$	a
Ti: sapphire	800	$2 \times 10^{-5} - 2 \times 10^{-4}$	1–5

[a] Depends on both the pulse duration and the pulse repetition rate.

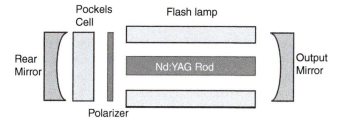

Figure 2.1 Schematic description of a flash-pumped actively Q-switched Nd : YAG laser.

laser. The laser medium is usually a rod cooled with water. The flashlamps are arranged parallel to this rod. The pump light is reflected and concentrated by the pump cavity. The inner surface of the pump cavity has a highly reflecting coating or a scattering surface to homogenize the illumination of the laser rod. For pulsed Nd : YAG lasers, Xe flashlamps are used with typical life times of 20–50 million flashes depending on the total energy dissipated during the flashlamp discharge. The emission spectrum of the Xe flashlamp only partly matches the excitation spectrum of Nd : YAG; hence, only a small fraction of the absorbed pump power is transferred to the upper laser level. As a consequence, less than 3% of the electrical input power is converted into laser radiation. Diode-pumped systems offer a better matching to the absorption band and reach 10–20% of overall efficiency (laser beam power in relation to electrical power input). Therefore, DPSS lasers (DPSSLs) systems save energy costs and secondary costs like cooling engines. The reduced heat load of the laser crystal leads to smaller internal stresses. The use of diode lasers is becoming more popular in spite of the higher cost. Diode pumped lasers, in fact, have reduced dimensions, better pulse to pulse reproducibility and better quality factor ($M^2 < 1.3$) and typical life times of few billions shots. As shown in Table 2.1, the fundamental wavelength of a Nd : YAG laser (1064 nm) can be down-converted to shorter wavelengths (532, 354.7, and 266nm) by means of passive harmonic generation techniques utilizing nonlinear crystals (like e.g. KDP or BBO). Nd : YAG lasers are commercially available in a wide range of sizes and output powers. Water-cooled models with up to 3 J/pulse at repetition rates between 10 and 50 Hz can be found as well as smaller air-cooled handheld versions with repetition rate of 1 Hz and pulse energy smaller than 20mJ.

Another important class of lasers, whose utilization in LIBS experiments has rapidly grown in the last years, is represented by the femtosecond (fs) lasers, i.e. pulsed laser sources with pulse duration ranging from tens to hundreds of femtoseconds (Section 2.4). Because of this very narrow pulse duration, the rates of energy deposition are extremely high, giving rise to interactions with the test materials that are substantially different from those of nanosecond laser pulses. Less damage to surrounding material and less and more reproducible ablation of the heated material, for example, have been observed in solid samples as well as an improvement in the spatial resolution and in the signal to background emission ratios [5–7]. The first femtosecond lasers (developed in 1980s) were colliding pulse mode-locked dye lasers, with pulse duration of the order of 30 fs. Dye lasers, however, have been replaced (in early 1990s) by self-mode-locked Ti: sapphire lasers, since these sources are easier to use and allow to obtain shorter pulse durations (down to few femtoseconds) [8, 9]. Main feature of these lasers is the use of the chirped pulse amplification (CPA) technique, which is a three-step process where the femtosecond pulses generated by a seeding laser (a mode-locked laser with an additional compression mechanism based on the use of a Kerr lens) are temporally stretched (to allow their amplification), then amplified with the aid of an additional pump laser and finally recompressed down to the femtosecond scale [10–13]. Examples of LIBS systems utilizing ultra-short pulsed lasers can be found in the literature [14–16].

A further fast growing laser technology that seems to be very promising for LIBS work (in particular in industrial applications) is represented by the high-power pulsed fiber lasers [17, 18]. Fiber lasers are a variation of the standard solid-state lasers where the active medium is an optical fiber rather than a rod, with the consequent advantage that the longer interaction length results in higher photon conversion efficiency solid-state lasers where the active medium is an optical fiber rather than a rod, with the consequent advantage that the longer interaction length results in higher photon conversion efficiency. In the typical dual-core fiber laser structure, an undoped outer core collects the pump light and guides it along the fiber, whereas the stimulated emission is generated in the inner doped core (typical dopants are ytterbium and erbium). Fiber lasers are usually pumped by diode lasers but in a few cases by other fiber lasers. Both end and side pumping mechanisms are used. In the end-pumping configuration, the light (coming from one or more pump lasers) is directly fired into the end of the fiber, while in the side pumping arrangement the light is coupled into the outer core via a fiber coupler. Both Q-switching and modelocking techniques are used to obtain very short pulses [19]. In the former case, pulse durations are in the nanosecond to microsecond range, while in the latter pulse durations as short as 50 fs can be achieved. In conclusion, we can state that the interest in fiber lasers is increasing not only for their compact size (which is an important requirement for reducing the overall dimensions of portable LIBS systems), but also for their characteristics in terms of energy per pulse (up to the mJ), beam quality factor (M2 close to 1), and pulse width (from femtoseconds to microseconds) [20, 21].

2.3 Desirable Requirements for Atomic Emission Spectrometers/Detectors

The combination of spectrometer and detector is an important factor to consider in OES for any plasma characterization or analytical spectrochemistry. The experimentalist must choose the type of measurements to be made. Based on that, the appropriate system can be designed. For example, using non-resonance lines of nonmetals such as F, Cl, Br, I, S, N, and O requires working in the near-infrared (NIR) region (<940 nm), while the resonance lines of these elements are located in the vacuum ultraviolet (VUV) (below 185 nm). The requirements for the ideal spectrometer and detector to provide for simultaneous determination of any combination of elements in the spectrum include a high resolution of 0.01–0.003 nm to resolve the lines of interest and avoid interferences. Second, wide wavelength coverage is needed, typically from 165 to 800–950 nm to be able to detect simultaneously several elements. Third, a large dynamic range is necessary to provide the optimum signal-to-noise ratio (SNR) for a large range of elemental intensities; the detector has to have a wide dynamic range, typically 6–7 orders of magnitude. The spectrometer/detector combination should have a high sensitivity and a linear response to radiation. The detector has to have high quantum efficiency (QE), particularly in the NIR and UV, and low noise characteristics. Furthermore, for rapid analysis, the readout and data acquisition time should be shorter, at least less than the time lap between the laser pulses. It is necessary to review briefly the characteristics of detectors and spectrometers used in LIBS applications to appreciate the impact of any combination of spectrometer/detectors has had on LIBS.

2.4 Spectrometers

A spectrometer separates light into its component wavelengths. Separation of light is achieved in all modern instruments by the use of a diffraction grating. A grating consists of a series of closely spaced lines, which are ruled or etched onto the surface of a mirror. When light strikes the grating, it is diffracted at a certain angle, according to the following:

$$\sin\alpha \pm \sin\beta = kn\lambda \tag{2.1}$$

where α and β are, respectively, the incidence and diffracted angles, n is the grating density (number of grooves per millimeter), λ is the wavelength, and k is diffraction order. Modern gratings are usually blazed, which means that the groove is shaped to concentrate a large fraction of the incident intensity into diffraction at a particular angle. Thus, it is possible to have a blaze diffraction grating that is more efficient for a specific wavelength region. In this case, the grooves are ruled at a specified angle (known as the blaze angle).

2.4.1 Czerny–Turner Optical Configuration

Figure 2.2 shows an optical layout incorporating the Czerny–Turner mounting, which is the most commonly used in LIBS systems. This spectrometer typically consists of entrance, exit optics, and grating. The grating groove density used is between 100 and 4800 grooves/mm.

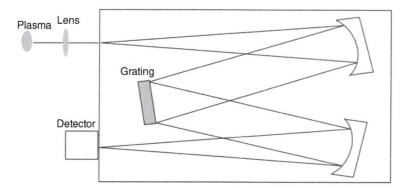

Figure 2.2 Czerny–Turner spectrometer system.

The Czerny–Turner optical configuration operates from the first order up to the fourth. The resolution or the ability of the spectrometer to separate wavelengths or lines is specified as the spectral bandpass. This is defined as one-half of the wavelength distribution passing through the exit slit. According to the Rayleigh criteria of resolution for diffraction line profiles, two lines are considered to be just resolved if the central maximum of the first-line profile coincides with the first minimum of the following line profile. In other words, to completely resolve two lines or wavelengths, the spectral bandpass must be no greater than one-half their difference in wavelength. For example, if two lines are separated by 0.2 nm, the spectral bandpass must be 0.1 nm or less. In this configuration, the selection of the desired wavelength is achieved by rotation of the grating within its mounting, which brings individual spectral lines into focus at a fixed position at the exit of the spectrometer. This approach allows any wavelength to be observed sequentially, one wavelength at a time if the spectrometer is equipped with a photomultiplier tube (PMT) at the exit. It also allows the measurement of a small section of the spectrum when it is attached to a linear diode array or a CCD detector. The spectral window obtained is only a few nanometers wide depending on the reciprocal linear dispersion of the spectrometer and the detector size.

2.4.2 Paschen–Runge Design

Polychromators generally use a Paschen–Runge optical configuration. The grating, entrance slit, and the multiple exit slits are fixed around the periphery of a circle, called the Rowland circle (Figure 2.3). In this configuration, the grating is concave and does not rotate. This approach allows simultaneously the acquisition of multiple wavelengths by using several PMT detectors. However, to cover the wavelength range available (for example from 160 to 770 nm) with 6 pm as spectral bandpass, 100 000 detectors would be necessary. The total length of the detector must be 200 cm with 14 µm pixel size. Thus, 40 interline CCD (ICCD) detectors (25 mm size) would be necessary to cover the whole

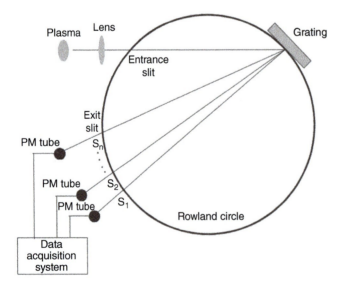

Figure 2.3 Paschen–Runge spectrometer system.

spectrum, and this would be very expensive. Another alternative simultaneously to collect several thousand independent channels of spectral information is the use of an echelle grating with a cross-disperser to form a two-dimensional spectrum. This spectrometer cuts a vertical spectrum into several segments horizontally. The format of the spectrum can therefore be made to match the rectangular format of the two-dimensional CCD detectors.

2.4.3 Echelle Spectrometer Configuration

Echelle spectrometers are orthogonal dual-dispersion devices usually employing a refractive element (prism) as an order sorter and a high-dispersion element (echelle grating) for wavelength resolution. The major characteristic of the echelle spectrometer in terms of components is the grating. In contrast to the blazed grating used in the Czerny–Turner Mounting, the design of the echelle grating utilizes the spectral order for maximum wavelength coverage [22]. The resolution R of a diffraction grating is related to the groove density n and the spectral order k ($R = kn$). In this case, instead of using a grating with a large number of grooves, resolution is improved by increasing both the blaze angle and the spectral order. For example, if we take $\sin \alpha + \sin \beta = 1$ and $\lambda = 333$ nm, the product kn will be 3000: for $k = 1$, $n = 3000$ grooves/mm, while, for $k = 50$, $n = 60$ grooves/mm. The problem with an echelle grating is the overlapping of spectra. To prevent the overlapping of spectral orders, a secondary dispersion stage or order sorter is required. This is accomplished by using a prism or another grating. If the prism is placed so that light separation occurs perpendicular to the diffraction grating (Figure 2.4), a two-dimensional spectral map is produced where the data are sorted into spectral order vertically and wavelength horizontally (the orthogonal dispersive element, either a prism or grating, provides the extended

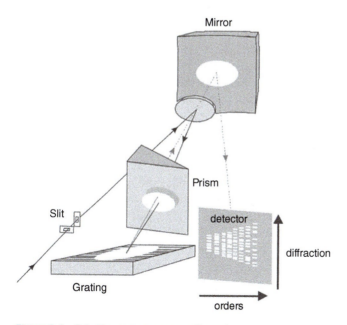

Figure 2.4 Echelle spectrometer configuration.

wavelength coverage by controlling the position at which wavelengths are imaged in space before being dispersed by the echelle grating). This gives the echelle spectrometer its two-dimensional wavelength dispersion character. Wavelength is dispersed horizontally within a given order and it decreases vertically as order number increases. The dimensional display of spectra is ideally suited for the use of solid-state detectors, in particular the ICCD which are of interest in this chapter. However, the intensifier cylindrical shape (18 or 25 mm diameter) adds some restrictions on the active pixels. For example, the pixels outside the intensifier tube will not be sensitive; this induces some gaps in the spectrum. The echelle grating can give excellent resolution and dispersion in a relatively compact spectrometer. One commercial echelle spectrometer provides an average reciprocal linear dispersion of 0.12 nm/mm and an average resolution of 0.003 nm in an instrument with 0.75 m focal length. For comparison, a conventional 0.75-m grating spectrometer with 2400 grooves/mm gives a reciprocal linear dispersion of 0.54 nm/mm and a resolution of 0.03 nm in the first order.

2.5 Detectors

2.5.1 Photomultiplier Detectors

The PMT has been the mainstay of low-light detectors for well over 50 years. Photomultipliers are extremely sensitive light detectors that provide a current output proportional to light intensity. They have a large area of light detection, wide range of spectral sensitivity (UV to visible), wide dynamic range (over six orders of magnitude), high QE (10–25% at the optimum response), fast temporal response (typically 10 ns to a few picoseconds), low detector cost, and ease of detectors use. The photomultiplier starts at the photocathode, followed by an assembly of dynodes (9–16 dynodes), each at a potential of about 100 V with respect to the last, and each emitting several secondary electrons for every one that hits it. The total gain of the multiplier depends on the number of stages and the total applied voltage. A typical PMT will have a gain of 10^6–10^7 for each incident photon. The electrical current measured at the anode is proportional to the amount of light that strikes the photocathode. This current is then converted into voltage signal, which is transferred via an analog-to-digital converter (ADC) to a computer for processing purposes. The noise from a PMT comes almost entirely from the photocathode dark current, and it can be reduced by cooling. The response of the photomultiplier to a delta pulse of light is governed by the electron trajectories within the tube and their transit time (typically 10 ns to a few picoseconds). For more details about PMTs, we refer the reader to the following references [23–26]. The major drawbacks of PMTs are that they are relatively bulky and are single channel, i.e. one PMT per analyte line. As explained earlier, to separate the analyte signal from the earlier part of the strong background LIP emission, some form of time gating is carried out with boxcar integration to acquire information about analytical lines. Transient events in LIP are difficult to monitor using a single-channel mode of detection. The spectral bandwidth of a PMT (typically 6–100 pm) in the range 170–800 nm corresponds to a small part and represents only 0.001% of the information available. As a consequence, single-channel methods of detection are not optimal for many plasma diagnostic and analytical spectrochemistry tools.

2.5.2 Solid-State Detectors

Solid-state array detectors consist of series of radiation-sensitive semiconductor picture elements or pixels that convert photons to a quantifiable charge. This charge is transferred to a readout amplifier for amplification before digitization by an ADC. Description, operation, and characteristics of these detectors, which include photodiode array (PDA), CCD, and CID, are beyond the scope of this chapter and can be found in the literature [27–28]. Because of its transient nature, the LIBS technique requires time-resolved detection. Therefore, the CTD and PDA detectors are often not adequate for transient plasma spectroscopy, in particular when signal integration time is ultra-short (nanosecond to microsecond scale). Because gating in LIBS is essential for acquiring temporal information or when discrimination of very weak lines superimposed on a very intense continuous background is needed, solid detectors need to be gated, that is rapidly turned on and off. To address these requirements, there are two options developed and commercially available to gate these detectors, by using an intensifier device coupled to a CCD or PDA or by an interline-transfer CCD structure. The IPDAs are slowly being displaced for the increased sensitivity, wider dynamic range, and better SNR that the CCDs have to offer, in particular the ICCDs' ability to provide two-dimensional information formats. A detailed comparison of these two detectors can be found in the literature [27]. Another subclass of the CTD detectors are the CIDs. The CID differs from the CCD primary by the mode of readout of the photogenerated charge [27–28]. To our knowledge, these detectors have not been used in LIBS applications and are not available commercially with intensifiers. As a consequence, they will not be studied here. This section focuses on highlighting some characteristics of the ICCD and the interline CCD without giving details, to allow the reader to realize the advantages of the ICCD detectors or of the interline CCD for their specific application. In particular, these detectors are generally two-dimensional and can be coupled with appropriate dispersing systems such as the echelle spectrometer to take advantage of this characteristic.

2.5.3 The Interline CCD Detectors

A CCD detects light through the creation of electron–hole pairs in silicon. The electrons are trapped by an imposed electric field and read out as charge by the camera electronics. CCDs measure the photogenerated charge accumulated in pixels by sequentially shifting the charge to an on-chip preamplifier located at the periphery of the device, and the signal is converted to computer-readable form by an ADC. Depending on the architecture of the device (full frame, frame transfer or interline transfer), the specifics of how this is done vary. Interline charge-coupled device architecture is designed to compensate for many of the shortcomings of frame-transfer CCDs. The interline CCD is a hybrid sensor with photosensitive diodes on one part of the pixel that are electrically coupled to a CCD-type storage region which resides under a mask structure. The masks are long structures running along the vertical axis of the CCD alternating with the open regions, hence the name interline CCD. In this approach, the sensor is subdivided into interdigitized lines of photoactive regions and masked registers.

The photoactive regions integrate light and then pass the charge to the adjacent masked registers. Like the full-frame and frame-transfer architectures, interline-transfer CCDs undergo readout by shifting rows of image information in a parallel fashion, one row at a time, to the serial shift register. The serial register then sequentially shifts each row of image information to an output amplifier as a serial data stream. The entire process is repeated until all rows of image data are transferred to the output amplifier and off the chip to an ADC-integrated circuit.

During the period in which the parallel storage array is being read, the image array is busy integrating charge for the next image frame, similar to the operation of the frame-transfer CCD. A major advantage of this architecture is the ability of the interline-transfer device to operate without a shutter or synchronized strobe, allowing for an increase in device speed and faster frame rates. Image "smear," a common problem with frame-transfer CCDs, is also reduced with interline CCD architecture because of the rapid speed (only one or a few microseconds) in which image transfer occurs. Drawbacks include a higher unit cost to produce chips with the more complex architecture, and a lower sensitivity due to a decrease in photosensitive area present at each pixel site since only one-half of the array is photoactive.

Also, this affects the spatial resolution on one axis owing to the gaps between photoactive regions. The advantage of the interline-transfer approach is that there is virtually no optical smearing during transfer and possibility of gating.

2.5.3.1 The Image Intensifier

The image intensifier in front of the CCD chip acts as a superfast shutter capable of operating on nanosecond timescales. An image intensifier is a cylindrical vacuum device typically 18 or 25 mm in diameter. It consists of three major components: photocathode, microchannel plate (MCP), and the phosphor screen [27, 29]. The photocathode transduces the photon image to an electron image. The MCP amplifies the electrons and the phosphor screen converts the amplified electron image to a corresponding photon image. The MCP is a disk of special glass with many individual hollow microchannels, each acting as an electron multiplier. When the photocathode is struck with incident photons, this causes the release of free electrons by photoelectric effects. A gate voltage (typically 200 V) accelerates the emitted electrons to the MCP. After leaving the MCP, the resulting electrons are further accelerated into a phosphor screen by high voltage (kilovolts) then converted to photons and detected by the CCD. The photons detected by the CCD are always monochromatic, independent of the incident photon wavelength received by the photocathode of the intensifier. This means the ICCD sensitivity is dependent on the photocathode characteristics.

The ability rapidly to gate the potential applied to the photocathode allows electronic shuttering of the camera. The result is the ability to image and resolve short-lived events. The shortest gate times are limited by the speed of switching the gate potential. Intensifier manufacturers have addressed this issue by lowering the impedance of the photocathode by using a conductive nickel undercoating on one of the cathode surfaces. However, this approach lowers the effective QE in the UV because of absorption of the light by nickel [27, 28]. Presently, the shortest gate is in the range 1–3 ns, while it is 20–40 ns for the slower gate.

References

1. R.A. Hann, D. Bloor (eds.), *Organic Materials for Nonlinear Optics* (Royal Society of Chemistry, London), (1989).
2. A.E. Siegman, *Lasers* (University Science Book, Sansalito, CA), (1986).
3. O. Svelto, *Principles of Lasers*, 5th edn. (Springer, New York), (2010).
4. E.L. Guerevich, R. Hergenroder, *Appl. Spectrosc.* **61** (2007), 233A.
5. A. Semerok, C. Chaleard, V. Detalle, J.L. Lacour, P. Mauchien, P. Meynadier, C. Nouvellon, B. Sallé, P. Palianov, M. Perdrix, G. Petite, *Appl. Surf. Sci.* **311** (1999), 138–139.
6. S. Guizard, A. Semerok, J. Gaudin, M. Hashida, P. Martin, F. Quéré, *Appl. Surf. Sci.* **186** (2002), 364.
7. M.R. Leahy-Hoppa, J. Miragliotta, R. Osiander, J. Burnett, Y. Dikmelik, C. McEnnis, J.B. Spicer, *Sensors* **10** (2010), 4342.
8. D.E. Spence, P.N. Kean, W. Sibbett, *Opt. Lett.* **16** (1991), 42.
9. K. Yamakawa, C.P.J. Barty, *IEEE J. Sel. Top. Quant.* **6** (2000), 658.
10. D. Strickland, G. Mourou, *Opt. Commun.* **55** (1985), 447.
11. D. Strickland, G. Mourou, *Opt. Commun.* **56** (1985), 219.
12. T. Brabec, F. Krausz, *Rev. Mod. Phys.* **72** (2000), 545.
13. O. Barthelemy, J. Margot, M. Chaker, M. Sabsabi, F. Vidal, T.W. Johnston, S. Laville, B. Le Drogoff, *Spectrochim. Acta Part B* **60** (2005), 905.
14. X. Zeng, X. Mao, R. Greif, R.E. Russo, *Proc. SPIE* **5448** (2004), 1150.
15. I.V. Cravetchi, M.T. Taschuk, Y.Y. Tsui, *Anal. Bioanal. Chem.* **385** (2006), 287.
16. J.F.Y. Gravel, F.R. Doucet, P. Bouchard, M. Sabsabi, *J. Anal. At. Spectrom.* **26** (2011), 1354.
17. F. He, J.H.V. Price, K.T. Vu, A. Malinowski, J.K. Sahu, D.J. Richardson, *Opt. Exp.* **14** (2006), 12846.
18. B. Shiner, *Ind. Laser Solut. Manuf.* **21** (2006), 24.
19. Y. Wang, *J. Lightwave Technol.* **23** (2005), 2139.
20. G.M. Weyl, in *Laser Induced Plasma and Application*, edited by L.J. Radziemski, D.A. Cremers (Marcel Dekker, New York), (1989), pp. 302–305
21. R.E. Russo, X.L. Mao, H. Liu, J. Gonzales, S.S. Mao, *Talanta* **57** (2002), 425.
22. J. Olesik, *Spectrosc.*, **14** (1999), 36.
23. A.P. Thorne, *Spectrophysics* (Chapman and Hall, New York), (1988).
24. W. Demtroder, *Laser Spectroscopy: Basic Concepts and Instrumentation*, 2nd edn. (Springer-Verlag, Berlin), (1981).
25. K.W. Busch and A.M. Busch, *Multielement Detection Systems for Spectrochemical Analysis* (Wiley, New York), (1990).
26. J. Mika and T. Torok, *Analytical Emission Spectroscopy: Fundamentals*, translated by A. P. Floyd (Crane, Russak & Company, New York), (1974), pp. 473–496.
27. J.V. Sweedler, K. Ratzlaff and M.B. Denton, *Charge Transfer Device in Spectroscopy* (VCH Publishers, New York), (1994).
28. F.M. Pennebaker, R.H. Williams, J.A. Norris and M.B. Denton, Developments in detectors in atomic spectroscopy. In *Advances in Atomic Spectroscopy*, edited by J. Sneddon, volume **5**, chapter 3 (JAI Press, Stamford, CT), (1999).
29. C. Earle, *Laser Focus World*, **10** (1999), 69.

3

Applications of LIBS

Vincenzo Palleschi[1] and Mohamad Sabsabi[2]

[1] *Applied and Laser Spectroscopy Laboratory, Institute of Chemistry of Organometallic Compounds, Research Area of the National Research Council, Pisa, Italy*
[2] *Conseil national de recherches Canada, Boucherville, Quebec, Canada*

The acronym LIBS for Laser-induced breakdown spectroscopy was firstly introduced in 1981 by Radziemski and Loree [1, 2]. Several excellent reviews have been recently published on industrial applications of LIBS [3], biomedical applications [4], geological applications [5], environmental diagnostics [6], and cultural heritage conservation and study [7]. It would be impossible trying to describe all the applications of LIBS in these and other fields, even limiting the attention to the relevant key papers. In the spirit of this book, we will focus our discussion on the most important applications where chemometrics has demonstrated to be important for the success of the LIBS technique.

3.1 Industrial Applications

3.1.1 Metal Industry

Among the industrial applications reviewed in the paper published in 2020 by Legnaioli et al. [3], the applications in metal industry are predominant. The reason for the success of LIBS in this field is related to its intrinsic capability of performing remote analysis in short times, with a throughput that can reach thousands of measurements in a second. The combined advantages of performing a fast compositional analysis without the need to be in close proximity of the target makes LIBS the ideal technique when the sample is not closely accessible, as in the case of high-temperature metals in foundries and/or when it moves at high speed, as in metal sorting for recycling. In this kind of applications, the number of spectra that must be treated can easily reach numbers that would make a conventional analysis unpractical, if not impossible.

For this reason, most of the recent application of LIBS in industry, and in particular in metal industry, are based on real-time chemometric analysis of the data. A few examples would make this clearer. The world two most active groups in LIBS development and research, the Pisa group led by Palleschi and the Malaga group of Laserna, were recently involved in a joint project funded by the European Commission (Laser-based continuous monitoring and resolution of steel grades in sequence casting machines – LACOMORE), for the optimization of the continuous casting process of steel [8–10]. In this industrial process, the change between different steel grades occurs gradually, and this produces a transition zone in the final product (steel slabs or billets) which must be removed because its composition does correspond neither to the first nor to the second steel grade. The traditional way of dealing with this problem is based on simple models of the mixing of molten metal in the tundish and its transportation down to the cutting zone. However, since the mathematical models cannot predict exactly the cutting point, corresponding to the reach of the second steel grade composition, some safety margin is added for being sure to have reached the wanted composition of the second grade within the established tolerance.

The extra material discarded represents a cost both for the material wasted and for the costs of its disposal. To optimize the cutting point, it is necessary to monitor the steel composition while it travels, in the form of slabs or billets, a few meters before the cutting zone, to activate the cutter at the right moment, i.e. when the composition of the steel corresponds to the required one. The steel, at a temperature of about 900 °C, typically travels at casting speed around 1.5 m/min; therefore, the use of more conventional techniques, (such as X-ray fluorescence [XRF], which is competitive with LIBS for laboratory analysis) is ruled out by the need of performing fast and remote analysis of the steel. Moreover, the decision on the cutting point must be taken in a few seconds, since the logistics of the plant requires the point of analysis to be relatively close to the cutting zone. Given that constraints, the use of chemometric tools for LIBS spectra analysis is mandatory.

The two groups involved in the LACOMORE research proposed two different chemometric approaches: the Pisa group demonstrated (in off-line test of steel billets coming from the factory) that a nonlinear calibration-based approach was able to precisely determine the size of the transition zone [8] (see Chapter 13 on nonlinear quantification methods). Moreover, the quantitative analysis evidenced that even when the AVERAGE steel composition reaches the specification of the second grade, the local fluctuations of the composition may be strong. This phenomenon was explained in terms of an inhomogeneous mixing of the second and first grade raw materials in the tundish [10], an effect which was not considered in previous mathematical models [11].

The Malaga group, on the other hand, proposed to use discriminant analysis (see Chapter 7 – Distance-based classification) for detecting the point where the steel composition stabilizes (i.e. the point from which the LIBS spectra are similar one to the other above a given threshold) and defining that point as the end of the transition zone [9]. This approach has the advantage, with respect to the one proposed in Pisa, of not requiring a calibration/training of the algorithm; on the other hand, the setting of the discrimination threshold for the determination of the end of the transition zone is critical, and the applications at the factory were made more complex by the presence of an oxidized layer at the surface of the steel billets. In any case, the LACOMORE project has demonstrated the concrete possibility of setting up a permanent LIBS system, operating remotely and potentially capable to detect the optimum cutting point to minimize the waste of material in continuous casting steel manufacturing.

Another interesting industrial application of LIBS in metal analysis is the classification of metallic scraps for recycling. Among the many proofs of principle published in recent papers, it is worth mentioning again a project funded by the European Commission which directly involved the Pisa group. The SHREDDERSORT (Selective Recovery of non-Ferrous Metal Automotive Shredder by Combined Electromagnetic Tensor Spectroscopy and Laser-Induced Plasma Spectroscopy) project was aimed to the automatic classification of nonferrous scraps at the conveyor belt [12, 13]. With respect to the application on steel, in a recycling plant the scraps travel at speeds of a few meters per second; the decision on the class to sort the scraps (in the case of the SHREDDERSORT project, cast vs. wrought aluminum) must thus be taken in a fraction of second. Moreover, the typical dimensions of the scraps do not allow to take more than two or three points of measurement, depending on the repetition rate of the laser and, of course, not more than one laser shot per point. The outcome of the project has been the demonstration of a successful fast and automatic sorting of the aluminum scraps using a neural network having as input the intensity of the elemental lines discriminating between the two kinds of aluminum and as the output the probability of the sample to be a cast or wrought aluminum scrap. Despite the large variability in the physical appearance of the scraps, which are not planar and can be covered by surface dirt or paint, the use of the double-pulse LIBS approach described in Chapter 2 allowed to minimize the number of the scraps of uncertain nature that, according to the SHREDDERSORT strategy, should be analyzed off-line with a second LIBS instrument.

In the last few years, several portable or handheld LIBS instruments have been put on the market. One of the principal targets of this technology are metallurgy applications, as possible alternative to XRF instrumentation for rapid analysis of metallic scraps. LIBS has the advantage, with respect to its XRF competitor, of operating without the need for ionizing radiations, is faster, sensitive to low-Z elements (aluminum, in particular) and can perform in-depth analysis. All these positive characteristics, however, are made less appealing by the limited analytical performances of the portable instrumentation, which are further reduced by the need of compacting in a portable instrument the laser, the spectrometer, and the processor for controlling the instrument, providing the user interface and calculating the sample composition from the LIBS spectra.

The most challenging samples are the iron-based alloys, because of the presence of the predominant emission line of iron which, acquired with a low spectral resolution spectrometer, might overlap with the other elements in the alloy, making difficult the quantitative analysis. A chemometric approach is thus necessary for the operation of HH-LIBS instrumentation; moreover, the algorithms used must be not only fast but also capable of running on the low computational power processors controlling the instrument. Afgan et al. [14] wrote the software for the B&W Tech HH-LIBS instrument, basing the procedure on the so-called "standardization" of the LIBS signal (in practice, a simplified application of the Calibration-Free LIBS (CF-LIBS) method, proposed by the Palleschi group in 1999 [15]) and a modification of the partial least squares (PLS) algorithm [16–18]. More recently, Poggialini et al. [19] compared the internal calibration of the Bruker HH-LIBS with an improved calibration based on two combined artificial neural networks, demonstrating a definite improvement of the analytical performances of the instrument for the elements considered, including carbon, which was not quantifiable with the factory built-in software.

Many other applications have been proposed involving LIBS and chemometrics in metal analysis. Some of them will be discussed in the relevant chapters of this book.

3.1.2 Energy Production

Another important industrial application that has attracted a considerable interest in the last decade is the use of LIBS for quality assessment of coal quality. Many papers have been published – mostly by the Chinese LIBS community [20, 21] – demonstrating the possibility of determining the calorific power and ash content, i.e. the main quality characteristics of industrial coal, using chemometric methods. However, most of the works in the literature describe measurements on pelletized coal samples, which are not extendable to the industrial environment. Only a few contributions remain that can be relevant for real industrial applications, involving single-shot analysis on solid carbon moving at high speed on a conveyor belt.

One of the most successful approaches for LIBS online coal analysis has been reported by Gaft et al. [22, 23], who used an ultrasonic autofocusing strategy for compensating the height variation of coal on the conveyor belt. In the paper reporting the results of a field test on a coal plant in South Africa, the authors described the procedure of detection and removal of outliers in the group of LIBS spectra acquired but did not reveal the details of the analytical procedure used for quantifying the elemental composition of coal "because of property issues" [23]. However, their general description of the method used ("Analytical algorithms were developed to transform the spectra to concentration measurements. This was developed using a trial-and-error method to minimize the absolute deviation of predicted versus actual analysis for the data."[23]) can be interpreted as the description of a rudimentary artificial neural network.

A feed-forward back propagation artificial neural network was later used by De Saro et al. for the determination of coal composition in off-line testing at a coal power plant [24, 25].

In the previous examples, the concentration of the main elements in the coal samples was determined for calculating, using empirical formulas, the corresponding ash content and/or calorific values of the coal.

However, in 2019, the Palleschi group demonstrated that the determination of the elemental composition of the coal might be plagued by a relatively large experimental uncertainty, which would be amplified by the approximated empirical formulas used for the calculation of ash content or upper calorific value (UCV) [26]. The authors demonstrated, however, the possibility of determining the ash content in untreated solid coal using artificial neural networks and linking the experimental intensity of the LIBS emission lines directly to the intended output (ash content or UCV). The experimental configuration proposed by Redoglio et al. [27] for remote analysis of coal blocks transported at high speed on a conveyor belt was used. The choice of the emission lines to use as inputs was driven by physical/chemical considerations. For example, the UCV of coal was determined using only the intensity of the carbon and hydrogen lines since, according to the PPM (Palleschi, Paganini, and Masci) formula [3], the concentration of these two elements was predominant in determining the UCV.

3.2 Biomedical Applications

In the last decade, many applications of LIBS for the study of biological systems and medical diagnostics have been proposed, most of them based on chemometric approaches [4]. Having resort to chemometric methods is almost mandatory when dealing with complex systems; since LIBS is essentially an elemental technique, it is not easy to link the information obtained to the properties of the biological system under study. Results have been reported on the successful identification by LIBS of pathogenic organisms [28, 29], using chemometric classification methods, and several works have been published on the discrimination of healthy and cancerous tissues by LIBS [30, 31]. It should be considered, however, that a complete clinical study on the diagnostic capabilities of LIBS is still missing, and the results obtained by some authors must be interpreted with caution. In fact, the classification algorithms can differentiate, for examples, between spectra obtained on different bacterial strains, but it is not obvious that the difference detected by the chemometric algorithms in the LIBS spectra would be associated with a difference between bacterial strains, because the changes can derive from a slightly different growth medium, or to minimal differences in the experimental conditions between the measurements on the first and second strain. In other words, the chemometric classification algorithms are designed to recognize the differences between the LIBS spectra but cannot give any information about the biological mechanism, if any, which produces these differences.

Similar problems may occur when interpreting the differences in LIBS analysis of blood of healthy and cancer patients [32–34], when the effect of the medical treatment underwent by the cancer patients could reflect in changes which LIBS (and chemometrics) would recognize. A definitive assessment on the capability of LIBS in this field would be the demonstration of its effectiveness for early diagnosis, after a careful, statistically significant clinical study which, at the present, is still missing.

Although the perspectives of using LIBS for direct analysis of tissues are promising, but still to be fully assessed, the applications based on the study of specific biomarkers, suitably tagged with micro- or nanoparticles, have been already demonstrated by the group of Melikechi, who introduced the so-called Tag-LIBS method [35, 36]. In Tag-LIBS, the biomarker is detected through the detection of the specific chemical element carried by the nanoparticle bound to it. The method developed by Melikechi is very selective because the process of binding the nanoparticle to a specific protein has been studied since long time in medical diagnostics, and in this applications LIBS is used at the best of its possibilities, i.e. for detecting and quantifying the characteristic chemical elements of the nanoparticles.

Moving along this line, the group of Motto-Ros presented an excellent study on the accumulation and elimination of gadolinium-based nanoparticles in murine kidney [37], obtaining by LIBS high-resolution microscopic multielemental images of kidney sections at different times after the injection of the nanoparticles. The same authors also demonstrated the possibility of recovering 3D compositional information by analyzing several sections of the kidney and stacking together the resulting 2D images in a full 3D model [38]. The authors evaluated that about 0.4 ng of the sample were ablated in a single pulse, with a diameter of the laser-induced crater of about 20 mm. The quantitative analysis on the Gd concentration in the kidney was performed by the authors of [37] using univariate calibration methods. The limits of detection (LOD) for Gd, Si, Na, and Fe resulted of the order of 0.1–0.2 mmol, corresponding to absolute LOD, considering the dimensions of the laser crater, of the orders of tens of attomoles per pulse.

3.3 Geological and Environmental Applications

The ability of LIBS of performing rapid analysis, at a distance as well as in close proximity of the sample, has made it interesting since the beginning of its use for environmental analysis and geological research. In most environmental applications, however, the technique suffers of its poor sensitivity, and the difficulties in obtaining information about organic pollutants, because of its nature of elemental technique. Traditionally, the LIBS analysis of soil was performed using univariate calibration methods [39], since the concentration of the pollutants in most of the cases is too low for affecting the soil matrix, and the emission of the different trace elements can be considered as independent of the one on the other. However, in recent years, several proposals of using LIBS for precision agriculture [40, 41] have made interesting the analysis of minor and major components of soil, and in this case the precise quantification of the soil composition requires the use of multivariate methods [42–44].

In geology, the use of the technique has experienced a great spur of interest after the arrival on Mars of the first extraterrestrial LIBS instrument. It has been commented in more than one occasion that the best results of the Mars LIBS missions have been obtained on Earth, with the space technology pushing the development of portable and handheld LIBS instruments that have found their natural field of application in metallurgy, as we have seen, but also in geology [5]. Senesi et al. [45] wrote an extensive historical review on the development of handheld LIBS and field-portable LIBS instruments and treated specifically the geological applications of these instruments in [46].

Besides the important in situ applications of LIBS, a lot of work has also been devoted to improving the analytical performances of the technique in laboratory analysis of geological samples. As an elemental technique, for laboratory applications in geology, LIBS competes with the wavelength dispersive XRF technique. LIBS instruments are considerably less expensive than Wavelength Dispersive X-Ray Fluorescence (WD-XRF) and easier to maintain; moreover, they do not involve the use of ionizing radiations. On the other hand, the LIBS quantitative analysis is still considered problematic. For demonstrating the potential of LIBS in laboratory analysis of geological materials, in 2020, the Palleschi group adapted a multivariate nonlinear algorithm used for the analysis of WD-XRF data to LIBS, showing that the trueness of the quantitative results obtained was very good and close to the one obtained using an artificial neural network, with the advantage of using linear algorithms (PLS analysis) for the solution of the nonlinear relations between line intensities and concentrations [47].

Geologic samples are typically heterogeneous. With the recent development of fast LIBS imaging systems, elemental micro-mapping has become one of the most interesting applications of the technique for the analysis of geological samples, with the possibility of performing microscopic analysis on real or "virtual" thin sections, as shown by Senesi et al. in [48]. The high spatial resolution obtained with the current instruments reflects in a large number of LIBS spectra acquired that should be processed for inferring the mineral species from the measured elemental composition. Pagnotta et al. [49] discussed the chemometric methods that can be used for fast analysis of highly inhomogeneous materials by LIBS, exploiting the method of self-organizing maps clustering, introduced for the first time in LIBS by the Pisa group [50], coupled with the already mentioned calibration-free LIBS approach [15].

An important step toward the assessment of the feasibility of LIBS in geology application has been done in 2019 at the X Euro-Mediterranean Symposium on LIBS (EMSLIBS) held in Brno, Czech Republic. The organizing group, led by Prof. Kaiser, set up an open contest aimed to the classification of 12 different types of mineral samples from their LIBS spectra. The data were provided to the participants as a training set of labeled spectra belonging to the 12 classes, and a set of test spectra to classify. The number of spectra per class in the test set was not specified (although this essential information was easily obtainable exploiting the automatic feedback provided by the data server). The participants were free to choose the most appropriate chemometric approach for the classification.

The results of the contest, reported in ref. [51], were extremely good: the first five most proficient groups obtained a classification accuracy of 90.3, 83.1, 82.6, 77.9, and 77.0%, respectively. However, these results must be interpreted with some caution. When asked to apply their optimized algorithm to another unknown set of samples (without feedback, this time), the winning group declined to participate to the verification test, and the other groups who accepted the further challenge, all obtained poorer results. Particularly noticeable was the drop in the performances of the fourth and fifth classified whose proficiency dropped from about 80 to 50%.

In any case, the accuracy of the classification obtained by the other groups with validated models remained around 80%, which can be considered as a realistic estimation of the performance of the LIBS technique for this kind of applications.

3.4 Cultural Heritage and Archaeology Applications

Despite its intrinsic micro-destructive nature, LIBS has been widely applied to the analysis of the materials of interest in Cultural Heritage and Archaeology applications [7]. Many applications of LIBS refer to the analysis of metal alloys, mostly copper-based as bronze [52–54] and brass [50, 55, 56] (thus exploiting all the advantageous characteristics of the technique described in the previous sections) and of geological materials (stones, mostly marble [57], ceramics [58–60], and mortars [61]). Several applications have been proposed for the analysis of pigments (mainly inorganic pigments [62–64]); among these, it is worth mentioning the work of Duchene et al. [65] which used Soft Independent Modelling of Class Analogy (SIMCA) and Partial Least-Squares Discriminant Analysis (PLS-DA) for the automatic identification of pigments from their LIBS spectra.

Several key methods in LIBS have been initially developed for Cultural Heritage applications. The first (rudimentary) LIBS elemental map was realized by the Pisa group on a fragment of a Roman Fresco [66], the first application of Self-Organized Maps clustering was proposed for the discrimination of two medieval brass alloys [50], and the first application of Graph clustering in LIBS also involved the analysis of brass samples [67].

3.5 Other Applications

Many other applications have been proposed in over 40 years of LIBS, since the first papers published by Loree and Radziemski in 1981 [1, 2]. Of special interest are the applications in building industry; Omenetto et al. [68] performed an extensive study on the best approaches for the detection of chlorine in concrete and mortars, obtaining LOD of the order of 0.05 wt%.

Also important are the LIBS applications to nuclear industry proposed by Horsfall et al. [69] who used chemometric methods for discriminating nuclear grade graphite from other carbon-based materials in nuclear waste, and the stratigraphic study of nuclear contaminated steel performed by Lang et al. [70]. This kind of studies is important for decommissioning and management of the nuclear waste.

Finally, it must be stressed that the LIBS technique has been applied to the study of many different materials, in most of the cases because of the interest of the application but also, sometimes, just because a LIBS instrument was available to the purpose. These studies are often associated with a chemometric approach, even when the specific application would not require it.

One of the purposes of this book is promoting a responsible use of the chemometric algorithms that should always be applied when, and only when, they are required for improving the results of LIBS analysis and always backed up by a solid physical/chemical model of the system under study.

References

1 L.J. Radziemski, T.R. Loree, Laser-induced breakdown spectroscopy: Time-resolved spectrochemical applications, *Plasma Chem. Plasma Process.* **1** (1981), 281–293. doi:10.1007/BF00568836.
2 T.R. Loree, L.J. Radziemski, Laser-induced breakdown spectroscopy: Time-integrated applications, *Plasma Chem. Plasma Process.* **1** (1981), 271–279. doi:10.1007/BF00568835.
3 S. Legnaioli, B. Campanella, F. Poggialini, S. Pagnotta, M.A. Harith, Z.A. Abdel-Salam, V. Palleschi, Industrial applications of laser-induced breakdown spectroscopy: A review, *Anal. Methods.* **12** (2020), 1014–1029. doi:10.1039/c9ay02728a.
4 R. Gaudiuso, N. Melikechi, Z.A. Abdel-Salam, M.A. Harith, V. Palleschi, V. Motto-Ros, B. Busser, Laser-induced breakdown spectroscopy for human and animal health: A review, *Spectrochim. Acta Part B At. Spectrosc.* **152** (2019), 123–148. doi:10.1016/j.sab.2018.11.006.
5 R.S. Harmon, G.S. Senesi, Laser-induced breakdown spectroscopy – A geochemical tool for the 21st century, *Appl. Geochem.* **128** (2021), 104929. doi:10.1016/J.APGEOCHEM.2021.104929.
6 D.A. Gonçalves, G.S. Senesi, G. Nicolodelli, Laser-induced breakdown spectroscopy applied to environmental systems and their potential contaminants. An overview of advances achieved in the last few years, *Anal. Chem.* **30** (2021), e00121. doi:10.1016/J.TEAC.2021.E00121.
7 A. Botto, B. Campanella, S. Legnaioli, M. Lezzerini, G. Lorenzetti, S. Pagnotta, F. Poggialini, V. Palleschi, Applications of laser-induced breakdown spectroscopy in cultural heritage and archaeology: A critical review, *J. Anal. At. Spectrom.* **34** (2019), 81–103. doi:10.1039/C8JA00319J.
8 G. Lorenzetti, S. Legnaioli, E. Grifoni, S. Pagnotta, V. Palleschi, Laser-based continuous monitoring and resolution of steel grades in sequence casting machines, *Spectrochim. ActaPart B At. Spectrosc.* **112** (2015), 1–5. doi:10.1016/j.sab.2015.07.006.

9 J. Ruiz, T. Delgado, L.M. Cabalín, J.J. Laserna, At-line monitoring of continuous casting sequences of steel using discriminant function analysis and dual-pulse laser-induced breakdown spectroscopy, *J. Anal. At. Spectrom.* **32** (2017), 1119–1128. doi:10.1039/C7JA00093F.

10 D. Mier, P. Nazim Jalali, P. Ramirez Lopez, J. Gurel, A. Strondl, L.M.M. Cabalín, T. Delgado, J. Ruiz, J. Laserna, B. Campanella, S. Legnaioli, G. Lorenzetti, S. Pagnotta, F. Poggialini, V. Palleschi, A stochastic model of the process of sequence casting of steel, taking into account imperfect mixing, *Appl. Phys. B Lasers Opt.* **125** (2019), 65. doi:10.1007/s00340-019-7175-2.

11 D. Mier Vasallo, J. Ciriza Corcuera, J.J. Laraudogoitia Elortegui, Procedimiento de Optimización de la Longitud de corte ce Palanquillas de Mezcla en Coladas Secuenciales de Aceros de Diferente Calidad, ES2445466, 2004.

12 B. Campanella, E. Grifoni, S. Legnaioli, G. Lorenzetti, S. Pagnotta, F. Sorrentino, V. Palleschi, Classification of wrought aluminum alloys by ANN evaluation of LIBS spectra from aluminum scrap samples, *Spectrochim. ActaPart B At. Spectrosc.* **134** (2017), 52–57. doi:10.1016/j.sab.2017.06.003.

13 SHREDDERSORT: Libs system video demo – YouTube, (n.d.). https://www.youtube.com/watch?v=OSCpntZcvUw (accessed March 5, 2019).

14 M.S. Afgan, Z. Hou, Z. Wang, Quantitative analysis of common elements in steel using a handheld μ-LIBS instrument, *J. Anal. At. Spectrom.* **32** (2017), 1905–1915. http://xlink.rsc.org/?DOI=C7JA00219J (accessed August 22, 2018).

15 A. Ciucci, M. Corsi, V. Palleschi, S. Rastelli, A. Salvetti, E. Tognoni, New procedure for quantitative elemental analysis by laser-induced plasma spectroscopy, *Appl. Spectrosc.* **53** (1999), 960–964. doi:10.1366/0003702991947612.

16 J. Feng, Z. Wang, L. Li, Z. Li, W. Ni, A nonlinearized multivariate dominant factor-based partial least squares (PLS) model for coal analysis by using laser-induced breakdown spectroscopy, *Appl. Spectrosc.* **67** (2013), 291–300. doi:10.1366/11-06393.

17 X. Li, Z. Wang, Y. Fu, Z. Li, W. Ni, A model combining spectrum standardization and dominant factor based partial least square method for carbon analysis in coal using laser-induced breakdown spectroscopy, *Spectrochim. Acta Part B At. Spectrosc.* **99** (2014), 82–86. doi:10.1016/J.SAB.2014.06.017.

18 V. Palleschi, Comment on "a multivariate model based on dominant factor for laser-induced breakdown spectroscopy measurements" by Zhe Wang, Jie Feng, Lizhi Li, Weidou Ni and Zheng Li, *J. Anal. At. Spectrom.*, 2011, doi:10.1039/c1ja10041f, *J. Anal. At. Spectrom.* **26** (2011), doi:10.1039/c1ja10197h.

19 F. Poggialini, B. Campanella, S. Legnaioli, S. Pagnotta, S. Raneri, V. Palleschi, Improvement of the performances of a commercial hand-held laser-induced breakdown spectroscopy instrument for steel analysis using multiple artificial neural networks, *Rev. Sci. Instrum.* **91** (2020), 073111. doi:10.1063/5.0012669.

20 Y. Zhao, L. Zhang, S.-X. Zhao, Y.-F. Li, Y. Gong, L. Dong, W.-G. Ma, W.-B. Yin, S.-C. Yao, J.-D. Lu, L.-T. Xiao, S.-T. Jia, Review of methodological and experimental LIBS techniques for coal analysis and their application in power plants in China, *Front. Phys.* **11** (2016), 114211. doi:10.1007/s11467-016-0600-7.

21 S. Sheta, M.S. Afgan, Z. Hou, S.-C. Yao, L. Zhang, Z. Li, Z. Wang, Coal analysis by laser-induced breakdown spectroscopy: A tutorial review, *J. Anal. At. Spectrom.* (2019), doi:10.1039/C9JA00016J.

22 M. Gaft, I. Sapir-Sofer, H. Modiano, R. Stana, Laser induced breakdown spectroscopy for bulk minerals online analyses, *Spectrochim. Acta Part B At. Spectrosc.* **62** (2007), 1496–1503. doi:10.1016/j.sab.2007.10.041.

23 M. Gaft, E. Dvir, H. Modiano, U. Schone, Laser induced breakdown spectroscopy machine for online ash analyses in coal, *Spectrochim. Acta Part B At. Spectrosc.* **63** (2008), 1177–1182. doi:10.1016/j.sab.2008.06.007.

24 C. E. Romero, R. De Saro, LIBS analysis for coal, in: S. Musazzi, U. Perini (Eds.), *Laser-Induced Breakdown Spectroscopy*, Springer Series in Optical Sciences, vol **182**, Springer, Berlin, Heidelberg, (2014), pp. 511–529. doi:10.1007/978-3-642-45085-3_19.

25 C.E. Romero, R. De Saro, J. Craparo, A. Weisberg, R. Moreno, Z. Yao, Laser-induced breakdown spectroscopy for coal characterization and assessing slagging propensity, *Energy Fuels* **24** (2010), 510–517. doi:10.1021/ef900873w.

26 S. Legnaioli, B. Campanella, S. Pagnotta, F. Poggialini, V. Palleschi, Determination of ash content of coal by laser-induced breakdownspectroscopy*Spectrochim. Acta Part B At. Spectrosc.*, **155** (2019), 123–126. doi:10.1016/j.sab.2019.03.012.

27 D. Redoglio, E. Golinelli, S. Musazzi, U. Perini, F. Barberis, A large depth of field LIBS measuring system for elemental analysis of moving samples of raw coal, *Spectrochim. Acta Part B At. Spectrosc.* **116** (2016), 46–50. https://biblioproxy.cnr.it:2114/science/article/pii/S0584854715002797?via%3Dihub (accessed November 1, 2018).

28 S. Manzoor, S. Moncayo, F. Navarro-Villoslada, J.A. Ayala, R. Izquierdo-Hornillos, F.J.M. De Villena, J.O. Caceres, Rapid identification and discrimination of bacterial strains by laser induced breakdown spectroscopy and neural networks, *Talanta* **121** (2014), 65–70. doi:10.1016/j.talanta.2013.12.057.

29 W.A. Farooq, M. Atif, W. Tawfik, M.S. Alsalhi, Z.A. Alahmed, M. Sarfraz, J.P. Singh, Study of bacterial samples using laser induced breakdown spectroscopy, *Plasma Sci. Technol.* **16** (2014),1141–1146. doi:10.1088/1009-0630/16/12/10.

30 D. Pokrajac, T. Vance, A. Lazarević, A. Marcano, Y. Markushin, N. Melikechi, N. Reljin, Performance of multilayer perceptrons for classification of LIBS protein spectra, in: *Proceedings of the 10th Symposium on Neural Network Applications in Electrical Engineering (NEUREL 2010)*, Serbia, 23–25 September 2010 (2010), pp. 171–174. doi:10.1109/NEUREL.2010.5644078.

31 T. Vance, N. Reljin, A. Lazarevic, D. Pokrajac, V. Kecman, N. Melikechi, A. Marcano, Y. Markushin, S. McDaniel, Classification of LIBS protein spectra using support vector machines and adaptive local hyperplanes, in: *Proceedings of the International Joint Conference on Neural Networks*, 18–23 July 2010, Spain, Barcelona, IEEE, (2010), pp. 1–7. doi:10.1109/IJCNN.2010.5596575.

32 Y. Chu, F. Chen, Z. Sheng, D. Zhang, S. Zhang, W. Wang, H. Jin, J. Qi, L. Guo, Blood cancer diagnosis using ensemble learning based on a random subspace method in laser-induced breakdown spectroscopy, *Biomed. Opt. Express.* **11** (2020), 4191. doi:10.1364/BOE.395332.

33 X. Chen,X. Li,S. Yang,X. Yu,A. Liu, Discrimination of lymphoma using laser-induced breakdown spectroscopy conducted on whole blood samples, *Biomed. Opt. Express.* **9** (2018), 1057. doi:10.1364/BOE.9.001057.

34 Z. Yue, C. Sun, F. Chen, Y. Zhang, W. Xu, S. Shabbir, L. Zou, W. Lu, W. Wang, Z. Xie, L. Zhou, Y. Lu, J. Yu, Machine learning-based LIBS spectrum analysis of human blood plasma allows ovarian cancer diagnosis, *Biomed. Opt. Express.* **12** (2021), 2559. doi:10.1364/BOE.421961.

35 Y. Markushin, N. Melikechi, Sensitive detection of epithelial ovarian cancer biomarkers using tag-laser induced breakdown spectroscopy, in: S.A. Farghaly (Ed.), *Ovarian Cancer – Basic Science Perspective*, InTech, Rijeka, Croatia, (2012).

36 Y. Markushin, N. Melikechi, A. Marcano, O. S. Rock, E. Henderson, D. Connolly, LIBS-based multi-element coded assay for ovarian cancer application, in: S. Achilefu, R. Raghavachari (Eds.), *Proceeding of the SPIE 7190, Reporters, Markers, Dyes, Nanoparticles, and Molecular Probes for Biomedical Applications, 719015*, 20 February 2009 (2009), p. 719015. doi:10.1117/12.810247.

37 L. Sancey, V. Motto-Ros, B. Busser, S. Kotb, J.M. Benoit, A. Piednoir, F. Lux, O. Tillement, G. Panczer, J. Yu, Laser spectrometry for multi-elemental imaging of biological tissues, *Sci. Rep.* **4** (n.d.) 6065. doi:10.1038/srep06065.

38 Y. Gimenez, B. Busser, F. Trichard, A. Kulesza, J.M. Laurent, V. Zaun, F. Lux, J.M. Benoit, G. Panczer, P. Dugourd, O. Tillement, F. Pelascini, L. Sancey, V. Motto-Ros, 3D imaging of nanoparticle distribution in biological tissue by laser-induced breakdown spectroscopy, *Sci. Rep.* **6** (2016), doi:10.1038/srep29936.

39 A. Ciucci, V. Palleschi, S. Rastelli, R. Barbini, F. Colao, R. Fantoni, A. Palucci, S. Ribezzo, H.J. L. Van Der Steen, Trace pollutants analysis in soil by a time-resolved laser-induced breakdown spectroscopy technique, *Appl. Phys. B Lasers Opt.* **63** (1996), doi:10.1007/BF01095271.

40 P.R. Villas-Boas, M.A. Franco, L. Martin-Neto, H.T. Gollany, D.M.B.P. Milori, Applications of laser-induced breakdown spectroscopy for soil characterization, part II: Review of elemental analysis and soil classification, *Eur. J. Soil Sci.* **71** (2020), 805–818. doi:10.1111/EJSS.12889.

41 P.R. Villas-Boas, M.A. Franco, L. Martin-Neto, H.T. Gollany, D.M.B.P. Milori, Applications of laser-induced breakdown spectroscopy for soil analysis, part I: Review of fundamentals and chemical and physical properties, *Eur. J. Soil Sci.* **71** (2020), 789–804. doi:10.1111/EJSS.12888.

42 A. Erler, D. Riebe, T. Beitz, H.G. Löhmannsröben, R. Gebbers, Soil nutrient detection for precision agriculture using handheld laser-induced breakdown spectroscopy (LIBS) and multivariate regression methods (PLSR, Lasso and GPR), *Sensors* **20** (2020), 418. doi:10.3390/s20020418.

43 J.M. Anzano, A. Cruz-Conesa, R.J. Lasheras, C. Marina-Montes, L.V. Pérez-Arribas, J.O. Cáceres, A.I. Velásquez, V. Palleschi, Multielemental analysis of Antarctic soils using calibration free laser-induced breakdown spectroscopy, *Spectrochim. Acta Part B At. Spectrosc.* **180** (2021), 106191. doi:10.1016/J.SAB.2021.106191.

44 M. Corsi, G. Cristoforetti, M. Hidalgo, S. Legnaioli, V. Palleschi, A. Salvetti, E. Tognoni, C. Vallebona, Double pulse, calibration-free laser-induced breakdown spectroscopy: A new technique for in situ standard-less analysis of polluted soils, *Appl. Geochem.* **21** (2006), doi:10.1016/j.apgeochem.2006.02.004.

45 G.S. Senesi, R.S. Harmon, R.R. Hark, Field-portable and handheld laser-induced breakdown spectroscopy: Historical review, current status and future prospects,*Spectrochim.Acta Part B At. Spectrosc.* **175** (2021), doi:10.1016/J.SAB.2020.106013.

46 G.S. Senesi, Portable hand held laser-induced breakdown spectroscopy (LIBS) instrumentation for in-field elemental analysis of geological samples, *Int. J. Earth Environ. Sci.* **2017** (2017), 146. doi:10.15344/2456-351X/2017/146.

47 S. Pagnotta, M. Lezzerini, B. Campanella, S. Legnaioli, F. Poggialini, V. Palleschi, A new approach to non-linear multivariate calibration in laser-induced breakdown spectroscopy analysis of silicate rocks, *Spectrochim. Acta Part B At. Spectrosc.* **166** (2020), doi:10.1016/j.sab.2020.105804.

48 G.S. Senesi, B. Campanella, E. Grifoni, S. Legnaioli, G. Lorenzetti, S. Pagnotta, F. Poggialini, V. Palleschi, O. De Pascale, Elemental and mineralogical imaging of a weathered limestone rock by double-pulse micro-laser-induced breakdown spectroscopy, *Spectrochim. Acta Part B At. Spectrosc.* **143** (2018), 91–97. doi:10.1016/J.SAB.2018.02.018.

49 S. Pagnotta, M. Lezzerini, B. Campanella, G. Gallello, E. Grifoni, S. Legnaioli, G. Lorenzetti, F. Poggialini, S. Raneri, A. Safi, V. Palleschi, Fast quantitative elemental mapping of highly inhomogeneous materials by micro-laser-induced breakdown spectroscopy, *Spectrochim. Acta Part B At. Spectrosc.* **146** (2018), 9–15. doi:10.1016/j.sab.2018.04.018.

50 S. Pagnotta, E. Grifoni, S. Legnaioli, M. Lezzerini, G. Lorenzetti, V. Palleschi, Comparison of brass alloys composition by laser-induced breakdown spectroscopy and self-organizing maps, *Spectrochim. Acta Part B At. Spectrosc.* **103–104** (2015), 70–75. doi:10.1016/j.sab.2014.11.008.

51 J. Vrábel, E. Képeš, L. Duponchel, V. Motto-Ros, C. Fabre, S. Connemann, F. Schreckenberg, P. Prasse, D. Riebe, R. Junjuri, M.K. Gundawar, X. Tan, P. Pořízka, J. Kaiser, Classification of challenging laser-induced breakdown spectroscopy soil sample data – EMSLIBS contest, *Spectrochim. Acta Part B At. Spectrosc.* **169** (2020), doi:10.1016/j.sab.2020.105872.

52 S. Legnaioli, G. Lorenzetti, L. Pardini, G.H. Cavalcanti, V. Palleschi, Applications of LIBS to the analysis of metals, in: S. Musazzi, U. Perini (Eds.), *Laser-Induced Breakdown Spectroscopy*, Springer Series in Optical Sciences, vol **182**, Springer, Berlin, Heidelberg, (2014), doi:10.1007/978-3-642-45085-3_7.

53 A. Foresta, F. Anabitarte García, S. Legnaioli, G. Lorenzetti, D. Díaz Pace, L. Pardini, V. Palleschi, LIBS analysis of twelve bronze statues displayed in the National Archaeological Museum of Crotone, *Opt. Pura y Apl.* **45** (2012), doi:10.7149/OPA.45.3.277.

54 M. Ferretti, G. Cristoforetti, S. Legnaioli, V. Palleschi, A. Salvetti, E. Tognoni, E. Console, P. Palaia, In situ study of the Porticello Bronzes by portable X-ray fluorescence and laser-induced breakdown spectroscopy, *Spectrochim. Acta Part B At. Spectrosc.* **62** (2007), 1512–1518. doi:10.1016/j.sab.2007.09.004.

55 N. Ahmed, M. Abdullah, R. Ahmed, N.K. Piracha, M.A. Baig, Quantitative analysis of a brass alloy using CF-LIBS and a laser ablation time-of-flight mass spectrometer, *Laser Phys.* **28** (2018), 016002. doi:10.1088/1555-6611/aa962b.

56 J.M. Andrade, G. Cristoforetti, S. Legnaioli, G. Lorenzetti, V. Palleschi, A.A. Shaltout, Classical univariate calibration and partial least squares for quantitative analysis of brass samples by laser-induced breakdown spectroscopy, *Spectrochim. Acta Part B At. Spectrosc.* **65** (2010), 658–663. doi:10.1016/j.sab.2010.04.008.

57 F. Colao, R. Fantoni, V. Lazic, A. Morone, A. Santagata, A. Giardini, LIBS used as a diagnostic tool during the laser cleaning of ancient marble from Mediterranean areas, *Appl. Phys. A.* **79** (2004), 213–219.

58 R.-J. Lasheras, J. Anzano, C. Bello-Gálvez, M. Escudero, J. Cáceres, Quantitative analysis of roman archeological ceramics by laser-induced breakdown spectroscopy, *Anal. Lett.* **50** (2017), 1325–1334. doi:10.1080/00032719.2016.1217000.

59 V. Lazic, F. Colao, R. Fantoni, A. Palucci, V. Spizzichino, I. Borgia, B.G. Brunetti, A. Sgamellotti, Characterisation of lustre and pigment composition in ancient pottery by laser induced fluorescence and breakdown spectroscopy, *J. Cult. Herit.* **4** (2003), 303–308.

60 A. Ramil, A.J. López, M.P. Mateo, A. Yáñez, Classification of archaeological ceramics by means of laser induced breakdown spectroscopy (LIBS) and artificial neural networks, in: *Proceedings of the International Conference on Lasers in the. Conservation of Artworks, LACONA 7*, Madrid, Spain, 17–21 September 2007 (2008), pp. 121–125. doi:10.1201/9780203882085.ch19.

61 S. Pagnotta, M. Lezzerini, L. Ripoll-Seguer, M. Hidalgo, E. Grifoni, S. Legnaioli, G. Lorenzetti, F. Poggialini, V. Palleschi, Micro-laser-induced breakdown spectroscopy (micro-LIBS) study on ancient roman mortars, *Appl. Spectrosc.* **71** (2017), 721–727. doi:10.1177/0003702817695289.

62 L. Angeli, C. Arias, G. Cristoforetti, C. Fabbri, S. Legnaioli, V. Palleschi, G. Radi, A. Salvetti, E. Tognoni, Spectroscopic techniques applied to the study of Italian painted neolithic potteries, *Laser Chem.*2006 (**2006**) 1–7. doi:10.1155/2006/61607.

63 E.A. Kaszewska, M. Sylwestrzak, J. Marczak, W. Skrzeczanowski, M. Iwanicka, E. Szmit-Naud, D. Anglos, P. Targowski, Depth-resolved multilayer pigment identification in paintings: Combined use of laser-induced breakdown spectroscopy (LIBS) and optical coherence tomography (OCT), *Appl. Spectrosc.* **67** (2013), 960–972. doi:10.1366/12-06703.

64 M. Hoehse, A. Paul, I. Gornushkin, U. Panne, Multivariate classification of pigments and inks using combined Raman spectroscopy and LIBS, *Anal. Bioanal. Chem.* **402** (2012), 1443–1450. doi:10.1007/s00216-011-5287-6.

65 S. Duchene, V. Detalle, R. Bruder, J. Sirven, Chemometrics and laser induced breakdown spectroscopy (LIBS) analyses for identification of wall paintings pigments, *Curr. Anal. Chem.* **6** (2010), 60–65. doi:10.2174/157341110790069600.

66 M. Corsi, G. Cristoforetti, V. Palleschi, A. Salvetti, E. Tognoni, Surface compositional mapping of pigments on a roman fresco by CF-LIBS, in: *Proceedings of the First International Conference on Laser-Induced Breaking Application*, Tirrenia, Naples, Italy, 8–12 October, 2000 (2000), p. 74.

67 E. Grifoni, S. Legnaioli, G. Lorenzetti, S. Pagnotta, V. Palleschi, Application of graph theory to unsupervised classification of materials by laser-induced breakdown spectroscopy, *Spectrochim. Acta Part B At. Spectrosc.* **118** (2016), doi:10.1016/j.sab.2016.02.003.

68 N. Omenetto, W.B. Jones, B.W. Smith, T. Guenther, E. Ewusi-Annan, University of Florida, Feasibility of atomic and molecular laser induced breakdown spectroscopy (LIBS) to in-situ determination of chlorine in concrete: Final report, (2016), https://rosap.ntl.bts.gov/view/dot/31477 (accessed February 17, 2019).

69 J.P.O. Horsfall, D. Trivedi, N.T. Smith, P.A. Martin, P. Coffey, S. Tournier, A. Banford, L. Li, D. Whitehead, A. Lang, G.T.W. Law, A new analysis workflow for discrimination of nuclear grade graphite using laser-induced breakdown spectroscopy, *J. Environ. Radioact.* **199–200** (2019), 45–57. doi:10.1016/j.jenvrad.2019.01.004.

70 A. Lang, D. Engelberg, N.T. Smith, D. Trivedi, O. Horsfall, A. Banford, P.A. Martin, P. Coffey, W.R. Bower, C. Walther, M. Weiß, H. Bosco, A. Jenkins, G.T.W. Law, Analysis of contaminated nuclear plant steel by laser-induced breakdown spectroscopy, *J. Hazard. Mater.* **345** (2018), 114–122. doi:10.1016/j.jhazmat.2017.10.064.

Part II

Simplications of LIBS Information

4

LIBS Spectral Treatment

Sabrina Messaoud Aberkane[1], Noureddine Melikechi[2], and Kenza Yahiaoui[1]

[1] *Ionized Media and Lasers Division, Center for Development of Advanced Technologies, Algiers, Algeria*
[2] *Department of Physics and Applied Physics, Kennedy College of Sciences, University of Massachusetts Lowell, MA, USA*

4.1 Introduction

Analysis of laser-induced breakdown spectroscopy (LIBS) spectra requires correction of the acquired data from signals that are not necessarily useful to extract qualitative and quantitative information of a sample of interest. These signals include background and electron continuum emission caused by free–free (bremsstrahlung) and bound–free (recombination radiation) transitions, characteristics of the laser pulses used to generate the plasma, potential wavelength shifts due to environmental conditions, dark current, and other potential instrument contributions. In many situations, these signals may deteriorate the signal-to-noise ratio (SNR) and, as a result, limit the potential of the analysis of the sample investigated. Identifying and removing from raw LIBS spectra the contributions of these effects are typically referred to as preprocessing. Below, we describe various preprocessing techniques that have been developed to address and isolate the main contributing signals to the LIBS spectra.

4.2 Baseline Correction

The baseline correction, also called background subtraction, is a key step in LIBS data preprocessing. It consists of removing signal variations originating from the spectral background without causing a loss of information. Recently, with the growth of robust numerical algorithms, a great deal of attention has been paid to rapid baseline correction. Several researchers have proposed approaches that address baseline correction; in the next section, we will provide a brief description of some of the key developments in this area.

Chemometrics and Numerical Methods in LIBS, First Edition. Edited by Vincenzo Palleschi.
© 2023 John Wiley & Sons Ltd. Published 2023 by John Wiley & Sons Ltd.

4.2.1 Polynomial Algorithm

In 2003, Gornushkin et al. [1] proposed a background correction algorithm based on high-order polynomial functions. This algorithm was developed for a 2048-pixel one-dimensional CCD array but could be for detectors with an arbitrary number of arrays. The algorithm is based on dividing the LIBS spectrum into N groups, identifying the local minima for each group, as well as a predefined number of minor minima that correspond to the pixels with minimal intensities.

Local minima are considered minor minima if they are within three standard deviations of the major minima. Following this step, a polynomial is drawn using a least square fit through the "major minima" of all groups and a standard deviation, σ_N, is determined. This is expected to produce the smallest difference between the minor minima and the generated polynomial baseline. Note that the algorithm is optimized by iteratively identifying the optical parameters, which include the polynomial power order, number of groups, and the number of major minima in each group. The polynomial function drawn from both major and minor minima represents the continuum background.

Later, Sun and Yu [2] developed an algorithm that starts by identifying all minima of the spectrum. To do this, Sun and Yu select a threshold based on signal variance ratio. This, they argue, tends to exclude all "inappropriate minima" and results in a reduction in computational time. Thus, estimation of the threshold of the variance ratios of two neighboring minima plays a critical role in this approach. An overestimation of the threshold results in an overestimation of the background, and vice versa [3]. Following this step, a linear interpolation is performed between adjacent points, and one or more polynomial functions with optimal power order are used to approximate the continuous baseline. This method, evaluated using several spectra with different levels of complexity, can estimate relatively accurately the continuum background, see Figure 4.1.

Spline interpolation is another process of continuous background detection and correction. It was used by Tan et al. [4] to check the reliability of the quantitative and qualitative analyses of the LIBS technique. The results of the simulation show an improvement in the signal-to-background ratio (SBR), in the linear correlation coefficient between Cu concentration and spectral intensity, and by the way, in the baseline Cu quantification correction after spline interpolation has been used.

In 2018, Iu et al. [5] proposed another variance of polynomial function-based automated background correction. It consists of looking for third-order minima and is designed without intensified CCD for low-cost LIBS. This method does not require user parameters to be predefined, thus avoiding the adverse effects of artificial selection parameters for the baseline estimation. The number of third-order minima points is defined by the number of segments in the spectrum, while the size of the segment is determined by the interval between adjacent minimum points, and the threshold is defined by the mean value of the second-order minima in the segment. In addition, all the effective points of the spectrum can be achieved by crossing all segments only once, which helps reduce the number of iterations and computational time required.

The performance of the proposed method [5] was evaluated on simulated and real LIBS spectra with different complexities. The accuracy of the algorithm is evaluated by experimentally acquiring the calibration curve of the different elements in cast iron alloy samples.

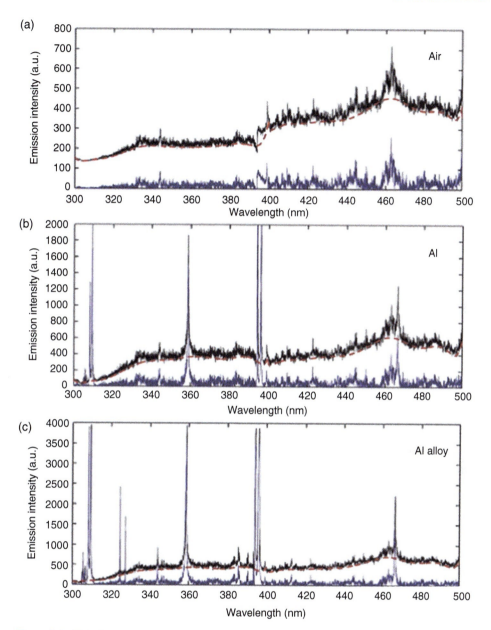

Figure 4.1 This figure represents examples of the estimation of baseline corrections for simple LIBS spectra: (a) air, (b) Al, and (c) Al alloy containing Cu and Mn elements. In each subfigure, the upper solid line is the original spectrum, the middle dashed line is the estimated background, and the lower solid line is the corrected spectrum. *Source:* From Sun and Yu [2] / With permission of Elsevier.

4.2.2 Model-free Algorithm

Yaroshchyk and Eberhardt [6] proposed, for the first time in the LIBS data baseline correction, the use of an updated version of the model-free algorithm version designed by Friedrichs [7] for the baseline removal of NMR data. By exploiting the ubiquity of noise in experimental spectrums, this algorithm circumvents the problem of discriminating between NMR and noise peaks [7]. Friedrichs's approach is based on the concept that the baseline for a noise spectrum may be determined by the criterion that the area above the baseline equals the area below it, i.e. the average value of the spectrum, except in the presence of NMR resonances, when the area under these peaks is greater than the area under the noise peaks. In this case, the number of local maxima and minima is used to approximate the area, and the baseline is determined by the criterion of having an equal number of local extrema above and below the line. Using Friedrichs's nomenclature, the background structure is therefore controlled for each pixel "i" in the spectrum by determining the median value of the extrema in a small window, W, centered on the pixel. An extremum is a point "i" whose value of $I(i)$ is either strictly greater or less than the intensities of $I(i-1)$ and $I(i+1)$. Then, the median of extrema, $Med(i)$, is calculated for "i." The preliminary distorted baseline is formed by a collection of median values named Med. Then, to smooth all strong discontinuities, a Gaussian function G (centered on 0 and normalized to 1) is combined with Med, and the final estimated baseline at "i," $B(i)$, is provided by:

$$B(i) = \sum_{j=i-\frac{W}{2}+1}^{j=i+\frac{W}{2}} Med(j)*G(i-j) \tag{4.1}$$

Friedrichs's approach performs well with LIBS spectra, especially when the spectra include a continuum background and isolated or slightly interfering spectral lines compared to the window size. However, when the algorithm is used to rich-line spectra, strongly overlapped spectra, or unresolved spectra, the measured background can appear distorted. This distortion is related to the fact that many of the extrema are generated by peaks in overlapping spectral areas rather than by the background.

By expanding the size of the window, W, the precision of the Friedrichs's algorithm can be optimized, but the baseline results at the cost of visually "real" LIBS background. The precision of Friedrichs's approach can be improved by increasing the size of the window, W, but the baseline results does not much the visually "true" LIBS background.

To make the algorithm more efficient, Friedrichs uses elastic windows that differ in width depending on the complexity of the field. In this case, when computing Med, the window size for each pixel is automatically selected to contain X extrema, where X is the maximum amount of extrema for the given window size chosen by the user.

The approach of Yaroshchyk and Eberhardt [6] is based on the assumption that the baseline for a spectrum that has no analytical lines and consists only of white noise can be computed by monitoring minimum values using only the median of the noise magnitude. The initial distorted baseline is calculated using a moving minimum for a window size W because most of the spectra minima are unaffected by the presence or absence of analytical peaks.

Furthermore, a normalized boxcar is convoluted with W-point moving minima to smooth off any sharp discontinuities and produce the final estimated baseline. The modified baseline is then defined by the following Eq. (4.2):

$$B(i) = \sum_{j=i-\frac{W}{2}+1}^{j=i+\frac{W}{2}} Min(j)*rect(i-j) \tag{4.2}$$

where $Min(j)$ is W-point moving minima, and $rect(i-j)$ is a rectangular function (boxcar) that is constant, $1/W$, within the bounds of the W-point domain, and zero outside.

The simulated spectrum fragment is shown in Figure 4.2 (right) [6], along with the corresponding moving minimum (a minimum interval value $[I - W/2 + 1, I + W/2]$ for each

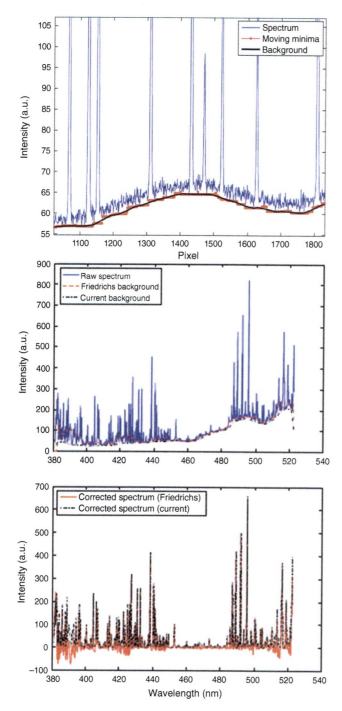

Figure 4.2 This figure represents (a): A region in spectrum (narrow solid line), moving minima (computed for $W = 50$), (dots with narrow solid line), and resulting background (wide solid line) (b): High-grade iron ore spectrum and the two background profiles computed with Friedrichs's (dashed line) and current (dashed-dotted line) algorithms. Corresponding background-corrected spectra resulting from using Friedrichs's (solid line) and proposed (dash-dotted line) algorithms (c). *Source:* From Yaroshchyk and Eberhardt [6] / With permission of Elsevier.

pixel i) and the final baseline function determined following Eq. (4.2). It is shown that the computed background tracks the true background, except in the presence of a relatively small near-constant offset that is approximately equal to the noise magnitude. We note that the presence of such an offset does not pose a problem for linear calibration when a procedure such as the trapezoid rule (that, in turn, removes the offset from the analytical peak) is used to calculate the peak area.

To correct for the impact of the LIBS baseline, Yaroshchyk and Eberhardt [6] demonstrated the robustness of this new version of Friedrichs's algorithm [7]. The method can be easily automated since no input variables are required. Several tests were performed on experimental LIBS spectra of different complexity, poor and rich-line spectra, and SNR. Figure 4.2b offers an example of an unscaled spectrum associated with a high-grade ore (66.9% Fe) plotted with the baseline corrections using the two algorithms mentioned earlier. It was noted that high Fe emission intensity, rich-line spectrum, and high SNR characterize the spectrum. The baseline-corrected spectrum calculated using the Friedrichs's algorithm (solid line) and its modified version (dash-dotted line) is presented at the left of Figure 4.2.

We can see that the modified version of the free model reliably monitors the baseline over the entire wavelength range in comparison to the Friedrichs's algorithm, which overestimated the baseline in many wavelength intervals. That makes this modified algorithm potentially a good candidate for use in automated data processing processes.

4.2.3 Wavelet Transform Model

The wavelet transform (WT) is a mathematical transformation used to analyze a signal in frequency space. It is built on the convolution of signal with a special function called the mother wavelet and is based on its ability to parse a signal into fast, medium, and slow components. This technique has benefited from the significant contributions of Ingrid Daubechies [8] and Stephane Mallat [9]. It has been used in various applied spectroscopy as well as in image compression applications.

WT has grown significantly in the last two decades. Compared to Fourier transform (FT), WT has the characteristics of time-frequency localization and multi-resolution analysis of signals [10]. Using WT, it is possible to decompose the original signal into localized contributions defined by parameters of scale, and each contribution represents a part of the signal at different frequencies. Generally, a signal contains three main parts depending on the frequency-type information: high-frequency noise, medium-frequency peaks, and low-frequency background.

WT is defined as the dilation and translation of the basis function $\psi(t)$, and the so-called "mother function" is given by:

$$\psi_{a,b}(t) = \frac{1}{\sqrt{|a|}} \psi\left(\frac{t-b}{a}\right), \quad (a, b \in R, \ a \neq 0) \tag{4.3}$$

where a and b are the scale (dilation) and position (translation) parameters, respectively.

A signal function $f(t)$ in any space $L^2(R)$, can be expanded in a series of wavelet bases, so that

$$WT_f(a,b) = \langle f(t), \psi_{a,b}(t) \rangle = \frac{1}{\sqrt{a}} \int_{-\infty}^{+\infty} f(t) \psi^* \left(\frac{t-b}{a} \right) dt \qquad (4.4)$$

$WT_f(a, b)$ is the wavelet coefficient that represents the projection of signal function $f(t)$ on the wavelet basis.

For the LIBS application, the WT can transform the spectral LIBS signal into a two-dimensional space-frequency plane, which can be useful for extracting the substitutive characteristics of the signals. As defined in the following equation, the energy is conserved during the WT process:

$$\int_{-\infty}^{+\infty} |f(t)|^2 dt = \frac{1}{C_\psi} \int_0^{+\infty} \int_{-\infty}^{+\infty} |WT_f(a,b)|^2 \frac{dadb}{a^2} \qquad (4.5)$$

where

$$C_\psi = \int_{-\infty}^{+\infty} \frac{|\psi(\omega)|^2}{|\omega|} d\omega < \infty \qquad (4.6)$$

The last equation, Eq. (4.6), represents the conditions of validity of the wavelet function $\psi_{a,b}(t)$.

In general, the spectral line intensity reflects the signal energy. Regarding the energy conservation condition of the WT, the wavelet coefficients can also represent the energy of the signal, in that the wavelet coefficients contain full-line intensity information.

To eliminate the redundancy of continuous wavelet transform (CWT) in real implementations, the wavelet function is discretized. Two algorithms are commonly used for this, "à trous" and the multi-resolution decomposition of Mallat intended for non-orthogonal and orthogonal wavelets [11], respectively. Apart from the difference in their filter types, the two algorithms are similar.

Figure 4.3 shows an illustration of the standard discrete wavelet transform (DWT) decomposition scheme based on the Mallat algorithm:

DWT consists of a series of successive signal decomposition (with length 2^n) into two components "Detail coefficients" and "Approximation coefficients," both with reduced size of 2^{n-j}, where j is the decomposition level (DL). At each level, the signal is decomposed by high-pass and low-pass filters to extract high frequency and low frequency, respectively,

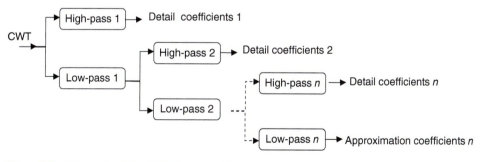

Figure 4.3 Schematic of the DWT decomposition.

for the next scale. This procedure permits dividing, iteratively, the signal into scales of different details until level j is reached, $j \leq n$.

Using the filter bank, lower wavelet coefficients are determined from higher resolution ones. It is possible to view the scaling and wavelet functions as low-pass and high-pass filters, respectively, which iteratively divide the signal into scales of different details.

In its standard mode, DWT conserves the amount of signal data; the total number of coefficients is equal to the number of data points of the initial signal. However, it is relatively simple to compress the transformed data, since the most important information is stored in few coefficients, which makes DWT suitable (appropriate for) in the data compression field. Note also that DWT is sensitive to the frequency shift of inputs signals. Therefore, a small shift in the input signal causes large changes in the wavelet coefficients and significant variations in the distribution of energy at different scales. To address this problem, and at the cost of additional computational complexity, Beylkin [12] and Shensa [11] have proposed a shift-invariant DWT approach based on the scheme of "à trous" algorithm [13] and is called redundant or stationary discrete wavelet transform (SWT). This algorithm differs from the "standard" DWT decomposition scheme, based on Mallat's algorithm, in varying the low-pass and high-pass filters at each consecutive level. Compared to DWT, the SWT of the signal includes redundant data, so the number of coefficients is conserved that makes this approach less suitable for data compression.

The background noise (baseline) represents the lower-frequency component of the spectrum, which typically has a continuous shape. The baseline correction consists of isolating the lower-frequency component through a series of decomposition with an "adequate" level.

Ma et al. [14] and Chen et al. [15] have shown that both the wavelet function and the DL play a crucial role in the baseline removal in inductively coupled plasma atomic emission spectrometry (ICP-AES) and near-infrared spectra (NIR). The optimization parameters of the calibration model are the R^2 coefficient, the root mean square error of prediction (RMSEP), and the average relative error (ARE). To correct the baseline for a LIBS application designed to investigate carbon content in coal, Yuan et al. [16] implemented the DWT method. Considering LIBS and ICP-AES as a type of atomic emission spectrometry, these researchers adopted the Daubechies family of orthogonal wavelets function and selected db6 for baseline correction. This function, db6, has been optimized by Ma et al. [14] based on RMSEP criterion. To determine the concentration of trace elements, Zou et al. [17] proposed an updated CWT algorithm that automatically subtracts the continuum baseline. This work is based on the assumption that, in the low-frequency part of the wavelet decomposition, the measured spectrum consists of the continuum background and the spectral peaks representative of the sample. The corresponding wavelet approximation coefficients should be multiplied by scaling factor γ, which varies in the range 0–1. The baseline correction depends on wavelet function first, DL in the second order, and scale factor γ. For a significant improvement in baseline correction, Zou et al. [17] determined the algorithm optimal values of db8, 10 DLs, and $\gamma = 0.89$. Muchao [18] employed SWT decomposition to correct the LIBS signal baseline. It was found that the corrected signal spectra can improve the accuracy of the element content measurement and reduce the stimulating times of laser (the laser usage time).

Kepes et al. [19] investigated the effect of baseline corrections on LIBS spectra in terms of detection limits (LOD), signal to background ratio (SBR) and relative standard deviation

(RSD) of several aluminum alloy-based elements. In addition to manual background subtraction, they used three baseline correction methods: free model [6], WT [20], and Gornushkin's polynomial algorithm [1]. It was found that the algorithms of Yaroshchyk (free model) and Gornushkin (polynomial) provide the highest robustness and the algorithm of Yaroshchyk can be the best overall choice considering ease of implementation. Recently, Dyar et al. [21] investigated the impact and causes of baseline removal using different methods of background correction used in other LIBS data techniques and showed its complexity.

4.3 Noise Filtering

Noise inevitably contributes to LIBS signals and impacts systematically the limit of detection (LOD) determination and the qualitative and quantitative spectral analysis. Generally, it includes dark noise and background noise.

Dark noise consists of photon noise (generated in instruments using photomultiplier detectors), detector noise (related to solid-state photodiode detector instrumentation), and flicker noise (caused by laser instability, vibration, sample turbulence, suspended particle light dispersion, dust, etc.) [16]. This form of noise is well distributed and has relatively low energy at all wavelengths. Following optimization of the hardware conditions, environmental noise can be minimized, but its complete elimination is impossible.

Background noise is characterized by a continuous spectral shape due to the bremsstrahlung effect of electrons, ions, and electrons recombination, and the stray lights in the plasma radiation process. This type of noise is also difficult to eliminate but can be reduced using adequate delay and gate time in the experiment.

Noise filtering plays a critical role in increasing the SNR. Often the noise is reduced by smoothing. Developing effective methods to filter the LIBS information from the overall signal that contains noise is a major initial step in the analysis. The most common de-noising methods include Savitzky–Golay (smoothing procedure), fast FT (mathematical filtering of the signal), boxcar, and wavelet-based filters. FT assumes that the signal is stationary, but the LIBS signal is always nonstationary [22]. Compared with traditional FT, WT is a time-frequency localization analysis method. It is an important tool for LIBS spectrum de-noising that has attracted interest in recent years due to its simplicity and effectiveness in realization.

4.3.1 Wavelet Threshold De-noising (WTD)

As shown in Section 4.2.3, the WT discretizes the noisy signal into wavelet coefficients. De-noising using WT consists of modifying its coefficients. Since the CWT and SWT are representatives of compaction of signal intensity, the smallest and the largest coefficients are related to the noise effects and significant features of the LIBS data, respectively [23]. Then, de-noising consists of removing or modifying the smallest coefficients without influencing the relevant signal features. In this context, wavelet thresholding (WTD) is used to smooth or eliminate some of the associated noise WT coefficients from the recorded signal.

WTD consists of three steps:

1) Extraction of the wavelets coefficients by decomposition of the signal using orthogonal WT with appropriate wavelet function and DL.

2) Selection of the threshold for each DL and processing of wavelet coefficient as a function of the adopted threshold function.
3) Reconstruction of the signal based on approximated coefficients via inverse wavelet.

In general, the WTD application is based on three types of thresholding functions, hard thresholding, soft thresholding, and semi-soft thresholding, implemented by Donoho and Johnstone [24]. The hard thresholding and soft thresholding modify the CWT and the SWT coefficients as follows (Eqs 4.7 and 4.8):

$$WT_h(d, \lambda) = \begin{cases} d, |d| \geq \lambda \\ 0, |d| < \lambda \end{cases} \quad (4.7)$$

$$T_s(d, \lambda) = \begin{cases} sgn(d).(|d| - \lambda), |d| \geq \lambda \\ 0, \quad |d| < \lambda \end{cases} \quad (4.8)$$

Using the hard thresholding, all wavelet coefficients below a defined threshold value become zero, while the soft thresholding approach reduces the values of wavelet coefficients by the given threshold. The estimation of the optimal threshold value is one of the critical components of the WTD. Donoho and Johnstone [24] have proposed the universal threshold as an asymptotically optimal solution:

$$\lambda = \sigma.\sqrt{2\ln N_c} \quad \text{and} \quad \sigma = \frac{median\left(d_j(k)\right)}{06745}, \quad d_j(k) \in \text{finess scale} \quad (4.9)$$

with σ is the noise standard deviation and N_c is the total number of wavelet coefficients $d_j(k)$.

When the threshold value is too low, the noise reduction is inefficient, while if it is too high, the detailed signal information would be lost. To achieve a compromise between the two modes, the semi-soft thresholding is proposed. It is based on the introduction of two threshold values, λ_1 and λ_2, and they were $\lambda_1 < \lambda_2$. Using this threshold function, the wavelet coefficients lower than λ_1 (the low threshold) turn to zero, while the wavelet coefficients higher than λ_2 remain unaffected and the coefficients within the range $[\lambda_1, \lambda_2]$ are modified as shown in Eq (4.10):

$$WT(d, \lambda) = \begin{cases} d, & |d| > \lambda_2 \\ sgn(d).\left[\frac{(|d| - \lambda_1).\lambda_2}{\lambda_2 - \lambda_1}\right], \lambda_1 < |d| \leq \lambda_2 \\ 0, & |d| \leq \lambda_1 \end{cases} \quad (4.10)$$

Thus, taking into account the adapted computational threshold selection procedure, it is not surprising to note that threshold selection is one of the hardest tasks in WTD. Generally, it is necessary to optimize the higher threshold λ_2, which is defined by Donoho's threshold and is given by Eq. (4.9), and the lower threshold λ_1, which is obtained by the minimal threshold and is given by [25].

WTD has been used, for the first time, for removing noise from LIBS spectra by Schlenke et al. [23]. These researches evaluated the hard and soft thresholding approaches on DWT and SWT. It was found that hard thresholding of SWT induces the best results in terms of MSE as well as corrected signal smoothness. Computational complexity, which leads to time-consuming processing and increased memory space requirements, was the only drawback of the SWT approach.

Wiens et al. [26] used the WTD de-noising method in ChemCam LIBS data preprocessing and got a satisfactory de-noising effect. The decomposition of the WT was performed by an iterative application of a pair of quadrature mirror filters (QMFs).

SNR improvements of carbon lines in different ambient gases were obtained by Yuan et al. [16] using the WT method. These authors adopted in the DWT, the db4 (Daubechies function), three DLs based on RMSEP, a threshold that is 0.5 times the universal threshold, and a hard thresholding function.

Zhang et al. [27] proposed a new double threshold de-noising method and confirmed its validity and usefulness for noisy LIBS spectrum, see Figure 4.4. This method is based on the

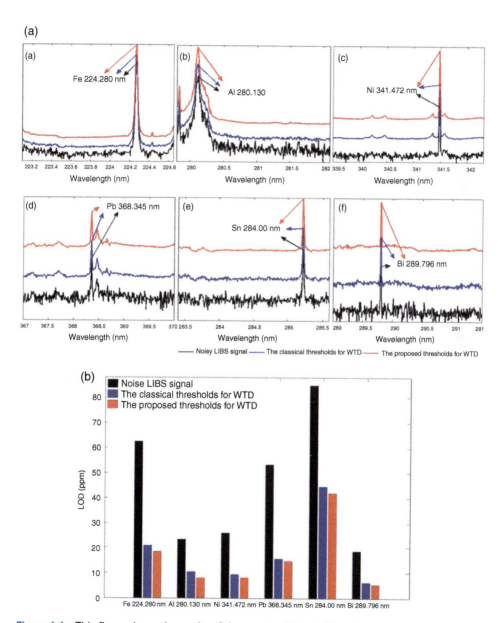

Figure 4.4 This figure shows the merits of the proposed threshold method compared to soft threshold one (a) and the impact of applying thresholding methods on the LOD of some copper sample elements (b). *Source:* From Zhang et al. [27] / With permission of Royal Society of Chemistry.

introduction of a new upper threshold defined as $\lambda_{2-new} = \gamma_j \cdot \lambda_2$, where γ_j is the correlation factor that could be calculated on each scale j. The DL was defined using the white noise testing method. In another contribution, the same authors, Zhang et al. [22], investigated the role of the DL on de-noising results by WTD with modified wavelet function multiplied by a scaling factor j. To optimize the DL, they proposed the use of an entropy analysis of the wavelet coefficients way. These researchers improved the elements noise ratios and consequently the LOD by more than 50%.

Xie et al. [28] reconstructed a new threshold function in WTD de-noising of LIBS spectra based on the modification of the hard and soft threshold method. The wavelet coefficients modification is defined by:

$$WT(d, \lambda) = \begin{cases} sgn(d) \cdot \left(|d| - \alpha * \dfrac{\lambda}{\exp(d - \lambda/N)}\right), & |d| \geq \lambda \\ 0, 0 \ll \alpha \ll 1 & |d| < \lambda \end{cases} \qquad (4.11)$$

where α is an adjustment parameter between 0 and 1, defined as a flexible choice between the soft and the hard threshold. Using this method, authors [28] eliminate the white noise and minimize the error of prediction in aluminum alloys analysis.

Another de-noising method based on improved wavelet dual-threshold function (IWDTF) was developed for LIBS data preprocessing for quantitative modeling of Cu and Zn in Chinese animal mature composts [29]. In Eq. (4.12), we present the related coefficients calculation equations using the IWDTF function:

$$WT(d, \lambda) = \begin{cases} d; & d \geq \lambda_2 \\ -(a\sin\theta + b\sin\theta - a) + \sqrt{b^2 - [d - (\lambda + a\cos\theta + b\cos\theta)]^2}; & \lambda_1 + a\cos\theta \leq d \leq \lambda_2 \\ a - \sqrt{a^2 - (d-\lambda)^2}; & \lambda_1 \leq d \leq \lambda_1 + a\cos\theta \\ 0; & d < \lambda_1 \\ -a + \sqrt{a^2 - (d-\lambda)^2}; & -\lambda_1 - a\cos\theta \leq d \leq -\lambda_1 \\ (a\sin\theta + b\sin\theta - a) - \sqrt{b^2 - [d - (\lambda + a\cos\theta + b\cos\theta)]^2}; & -\lambda_2 \leq d \leq -\lambda_1 - a\cos\theta \\ d; & d \leq -\lambda_2 \end{cases}$$

(4.12)

where λ_1 and λ_2 are the low and high threshold, respectively. θ represents the complementary angle of slope angle for the tangent point of the two arcs; a and b are the radii of the first and second arcs, respectively. Using this method and following adjustment of all parameters, results showed that optimum spectral de-noising effects for the Cu and Zn contents were achieved. The performances of the model were improved and yielded an R^2 of 0.9807 and RMSEP of 0.9177.

Chu Zhang et al. [30] used the LIBS technique coupled with chemometric methods for the identification of coffee varieties. The results show that LIBS preprocessing data using WTD reduces noise that can lead to the development of a qualitative online process.

In 2019, Huang et al. [31] demonstrated that the use of a hybrid model based on WTD and K-fold support vector machine recursive feature elimination (K-SVM-RFE) can improve the estimation of the indictors aging grade and hardness grade of steel. It was found that WTD can effectively de-noise the LIBS spectra, and the de-noising process benefited the followed feature selection step. This improved the accuracy of the classification of steel aging grade and hardness grade from 0.90 to 0.96.

Fu et al. [32] investigated the quantification of soils LIBS data using chemometric models. Spectral de-noising by WTD methodology based on soft threshold function has been used for the data preprocessing. Instead of a Daubechies decomposition function, another family of orthogonal wavelets function has been employed named "Symlet" function. The Symlet wavelet is an improvement of the Daubechies function, which has better symmetry for the reduction of the reconstruction phase shift [33]. Wavelet parameters, Sym function, and DL were identified corresponding to each element on basis of root mean square error of calibration (RMSEC), for example, Sym7 and DL4 were the optimum values for Cu.

Yongsheng Zhang et al. [33] investigated the effect of WTD de-noise on the quantitative study of coal. The soft thresholding function has been used for the WTD. Based on the root mean squared cross-validation error (RMSECV), the optimum number of decomposition was found to be five. In a further contribution to improving the accuracy of the coal quantitative analysis, Lu et al. [34] suggested a revised WTD approach based on the soft threshold function. Their contribution is related to the method of setting the threshold. The threshold value λ estimate is based on the wavelet coefficients of the DL1, assuming that the first DL contains more noisy signals than other DLs. This can lead to an overestimation of the noise level and over-filtering. For this reason, the modified threshold given by Lu et al. [34] is

$$\lambda_{new} = \alpha * \lambda = \alpha * \sigma \sqrt{2 \ln N_c}, \quad \alpha = \frac{E_{1,k}}{E_{j,k}}, \tag{4.13}$$

$$E_{j,k} = \sum_{K=1}^{K}(W_{j,k}), \tag{4.14}$$

where $E_{j,k}$ and $E_{1,k}$ are the wavelet coefficients of the energy associated with the j and the first DL. ($W_{j,k}$) is the wavelet coefficients at the j level. Then, we can see that using (Eq. 4.13), the current threshold can be modified for different noise levels at different DLs. Using this modified WTD process, the noisy coal signal of the LIBS spectrum was found to be effectively eliminated, see Figure 4.5.

WTD has recently become a strong de-noise tool, but its effectiveness depends on several WT-related parameters such as wavelet function, DL, and threshold estimation function. De-noising, as an important preprocessing process in the treatment of LIBS data employing chemometric methods, is still a challenge as can be seen by the continuous growth of the number of publications and the revised methodology.

4.3.2 Baseline Correction and Noise Filtering

In preprocessed spectra, the baseline and the noise correction are the ones that require the most time due to the iterative number involved. For this reason, Ewusi-Annan et al. [26] suggested a hybrid approach to correct background and noise spectra in a single step in order to minimize processing time. The approach is based on partial least square (PLS)

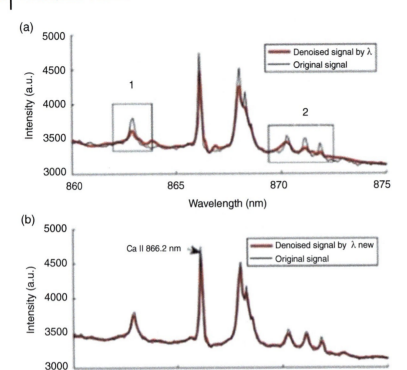

Figure 4.5 This figure depicts on a comparison of spectra de-noising by using universal threshold λ (a) and modified threshold λ_{new} (b) [5]. *Source:* Adapted from Iu et al. [5].

regression and artificial neural network (ANN) machine learning tools. To validate their method, the authors choose raw LIBS spectra of two data: one of the geochemical samples used to develop the calibration model for ChemCam (refer to Earth data) and the other Martian data from the curiosity rover (refer to Martian data). The method allowed for Earth and Martian spectra to be corrected for dark signal, continuum, noise, wavelength change, and instrument response. For both data sets, the implementation of the PLS was used from the Scikit-learn Library, where the cross-validation used in the loop and the training-test-split approaches were indeed selected. The use of PLS and ANN produced a projected preprocessed spectra with high accuracy.

4.4 Overlapping Peak Resolution

The overlapping is a common problem occurring in Raman spectroscopy, UV-Vis absorption spectroscopy, Fourier transform infrared spectroscopy (FTIR), and LIBS. The spectra exhibit the emission of peak's interference. It is essential due to the resolution limitation of detection material. When the improvement of the performance of the experimental equipment is limited, then data processing methods are used to overcome the overlapping of spectral lines. To resolve the overlapped lines [35], four methods are identified: curve fitting [36–40], WT [40–43], FT [44], and neutral network methods [45].

For the quantitative analysis of samples by LIBS, overlapping is considered to be one of the major issues to be discussed. Overlapping is not only related to the limitation of detection system resolution, but also non-appropriate gate delay may produce peaks overlapping, in LIBS [46]. Two methods are used to solve overlapping peaks: the curve fitting method [37–40] and the WT method [42, 43].

4.4.1 Curve Fitting Method

The basic approach of curve fitting is to decompose overlapping peaks by a multitude of functions that match with the Levenberg–Marquart (L–M) algorithm. Lorentzian and Voigt are the most fitting functions used for deconvolution. Some authors coupled the curve fitting to other models in order to achieve the best deconvolution. Zhang et al. [37] used the fractional differential theory to evaluate the required curve fitting initial values. They used the Lorentzian fit function for convenience but recommended a more effective use of the Voigt function. They consider that the spectral signal containing the overlapping lines is the contribution of the individual Lorentz function characterized by three parameters: peak height, location, and distance. The overlapping peaks are expressed as follows [37]:

$$F(x) = \sum_{j=1}^{M} L_j(x; A_j, \mu_j, \sigma_j) = \sum_{j=1}^{M} A_j \frac{\sigma_j^2}{(x-\mu_j)^2 + \sigma_j^2} \quad (4.15)$$

where $L_j(x; A_j, \mu_j, \sigma_j)$ is the individual peak with the wavelength variable; A_j, μ_j, and σ_j are the peak height, peak position, and full-width half-maximum (FWHM), respectively. M and j are the numbers of individual peaks and the serial number of the individual peak. The least square solution is optimized under the condition of the best agreement between the data and the model according to the equation below [37]:

$$RSS = \left(\sum_{i=1}^{N} y(x_i) - F(x)\right)^2 \quad (4.16)$$

where $y(x_i)$ is the experimental data expressed as function of x, N is the number of experimental data, and $F(x)$ is the function of fitting data.

The best fitting is heavily dependent on the initial values selected to prevent divergence or more iterations.

Zhang et al. [37] use a continuous spline wavelet transform (CSWT) to identify the number and position of each peak. For heights and FWHM, authors [37] used a procedure based on Leibniz's fractional differential theory, which provides the information of interpolation between integer order differential. For this system, three definitions are most commonly used: the definition of Grunwald–Letnikov (GL), the definition of Riemann–Liouville, and the definition of Caputo [47–49]. Authors [37] used the GL description for a signal differential order, where initial values for curve fitting are determined. Higher accuracy of both simulated and experimental results was obtained by the proposed procedure [37].

Tan et al. [38] used an algorithm called "error compensation algorithm" to correct and decompose overlapping peaks. This algorithm takes into account the residual fittings omitted from the curve fitting algorithm. The procedure is based on restoring the error to overlapping peaks several times and performing multiple fitting processes in order to obtain a lower residual fitting result. Tan et al. [38] report that Voigt profile can describe the spectral line profile better than the Lorentzian. In this case, the intensity of two superposed peaks can be expressed as:

$$I_\lambda = I_{1\lambda} + I_{2\lambda} = \left[a_1 \frac{i_1 \omega_1^2}{(\lambda - \lambda_1)^2 + \omega_1^2} + (1 - a_1) i_1 e^{((\lambda - \lambda_1)^2 / 2\omega_1^2)} \right] + \left[a_2 \frac{i_2 \omega_2^2}{(\lambda - \lambda_2)^2 + \omega_2^2} + (1 - a_2) i_2 e^{((\lambda - \lambda_2)^2 / 2\omega_2^2)} \right] \quad (4.17)$$

I_λ, $I_{1\lambda}$, and $I_{2\lambda}$ are the associate spectral intensities of overlapping peaks and of two spectral lines, respectively. λ is the wavelength of the overlapped peaks, λ_1, i_1, ω_1 and λ_2, i_2, ω_2 are central wavelength intensity and half-width of two spectral lines, respectively. a_1 and a_2 are constants. All these parameters are obtained after the curve fitting process. The residual error correction approach was investigated and implemented to correct overlapping peaks. The spectral intensity of the overlapping peaks is expressed as:

$$RI_\lambda = I_{1\lambda} + I_{2\lambda} + Sr_\lambda \quad (4.18)$$

where Sr_λ is the residual value related to the regression residual result noted $Sr1_\lambda$ and to the error noted err_λ by:

$$Sr_\lambda = Sr1_\lambda + err_\lambda \quad (4.19)$$

The $Sr1_\lambda$ will be compensated to the peaks P_1 and P_2. P_1 and P_2 are the decomposition results of the spectral lines with intensity values of $I_{1\lambda}$ and $I_{2\lambda}$, respectively. The intensity values of each point of the two peaks are updated to the following equations:

$$I_{1\lambda} = I_{1\lambda} + \frac{I_{1\lambda}}{I_{1\lambda} + I_{2\lambda}} Sr1_\lambda$$
$$I_{2\lambda} = I_{2\lambda} + \frac{I_{2\lambda}}{I_{1\lambda} + I_{2\lambda}} Sr1_\lambda \quad (4.20)$$

A new residual noted $Sr1'_\lambda$ is defined, and the peaks intensities are rewritten as:

$$I_{1\lambda} = I_{1\lambda} + \frac{I_{1\lambda}}{I_{1\lambda} + I_{2\lambda}} Sr1_\lambda + \frac{I_{1\lambda}}{I_{1\lambda} + I_{2\lambda}} Sr1'_\lambda$$
$$I_{2\lambda} = I_{2\lambda} + \frac{I_{2\lambda}}{I_{1\lambda} + I_{2\lambda}} Sr1_\lambda + \frac{I_{2\lambda}}{I_{1\lambda} + I_{2\lambda}} Sr1'_\lambda \quad (4.21)$$

The operation of curve fitting and error correction is repeated until a certain number of repetitions or error values that are less than or no longer change a given threshold are fulfilled. Using this approach, the authors resolved the overlapping of the Cu and Fe lines and improved the calibration curve between the concentration and intensity of Cu with a correlation coefficient R^2 of 0.9956 compared to Lorentz ($R^2 = 0.9836$) and Voigt fitting ($R^2 = 0.9898$).

Guezenoc et al. [39] developed a new method based on PLS-assisted variable selection to overcome overlapping lithium doublet and calcium line peaks. The authors compared the results of lithium quantification using the new method and the traditional multipoint fitting with several Voigt functions. The PLS regression models were estimated using the Rstudio software and the online "pls" package. This approach is based on the general advantages of multivariate analysis. In order to solve the problems of fitting and uncertainty, PLS models were built to evaluate the optimal number of components to be considered. The authors validated their procedure for 30 calibration samples from Martian data based on two different parameters, namely the RMSECV and the percentage of the explained concentration variance.

A first PLS model was applied for 1713 variables related to the spectral range 492.7–856.8 nm to take into account potential matrix effects. Since the first PLS showed a strong effect of the lithium doublet, a second PLS model was constructed to focus on the spectral interference between the lithium doublet and the calcium line (671.769 nm) in a reduced spectral range of 669.07–672.49 nm. Here, an interesting alternative to the traditional multi-peak fitting solution may be the PLS method.

An automatic curve fitting method based on a CWT was given by Yang et al. [40], which also extracted the major peaks in the LIBS spectrum. The approach used resolved overlapping peaks with a low degree of separation. The method is to calculate the second derivative of the LIBS spectrum by a CWT. The number and positions of individual peaks of the LIBS spectrum are obtained from the local minimum of the second derivative.

Validation of the method is accomplished by comparing the simulated spectrum to the experimental one. The number and positions of the individual LIBS spectrum peaks are derived from the second derivative's local minimum. From the separation of the two maximum sides of the local minimum, the FWHM of the individual peaks is calculated. In order to extract the major peaks in the LIBS spectrum, the threshold is set such that low-intensity peaks are eliminated. A Trust-Region algorithm is used to ensure the convergence of the method. With the convolution of several Voigt functions describing the emission of lines, the theoretical spectrum with filtered noise and corrected baseline is written as follows:

$$I_{th}(\lambda) = \sum_{i=1}^{M} I_i V(\lambda; \lambda_{0,i}, \gamma_i, \beta_i) + I_b \tag{4.22}$$

where $V(\lambda; \lambda_{0,i}, \gamma_i, \beta_i)$ is the individual Voigt peak with wavelength. I_i, λ, $\lambda_{0,i}$, γ_i, and β_i are the peak height, peak position, and FWHM, and β_i is a fractional parameter satisfying $0 \leq \beta_i \leq 1$. M and i are the number of the individual peaks and the serial number of the individual peaks, respectively. I_b is a constant background representing a residual background signal due to incomplete background correction.

The optimized parameters are obtained by the simultaneous determination of the fitting parameters when the square of the difference between the measured spectrum $I_m(\lambda)$ and the theoretical one $I_{th}(\lambda)$ is minimum. Adjusted R-squared (R_{adj}) and root mean squared error (RMSE) are used to test the efficiency of multiparameter minimization. Figure 4.6 shows the local minimum, delimited by two local maxima, in the case of a single peak.

Yang et al. [40] conducted a LIBS study of five artifacts samples. To apply their proposed procedure, seven steps have been taken. First, when the wavelet scale is chosen according to

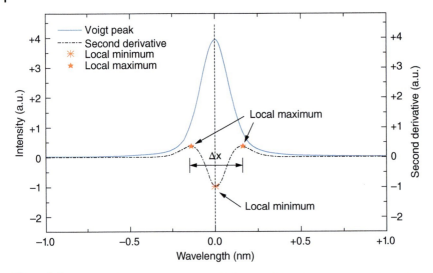

Figure 4.6 Voigt peak and corresponding CWT results [41]. *Source:* Adapted from Shao et al. [41].

the SNR, the CWT is performed on the original signal to obtain the second derivative. Second, the number and positions of individual peaks from the second derivative's local minimum are obtained and the interferences of artifacts are eliminated. Third, a threshold is set and only the individual peaks are extracted in such a way that the minimum values exceed the threshold. Fourth, the FWHM of individual peaks from the separation of the two maxima is calculated and the height of individual peaks is determined. The Trust-Region algorithm is minimized to have the individual peak parameters at the same time. This yields the theoretical LIBS spectrum, which is subsequently compared to the spectrum measured. Due to the good resolution of the overlapping peaks, an improvement in the results of the composition of N2/SF6 gas mixtures with different concentrations is achieved using the proposed curve fitting method proposed by Yang et al. [40].

Multi-peak fittings tend to be the ideal approach for addressing spectral interference from all of the works cited.

4.4.2 The Wavelet Transform

In LIBS, the shape of the spectral line is an important indicator of spectral interference. Under typical electron temperatures and electron densities, the broadening of the spectral line is primarily due to the Stark effect. Subsequently, this produces a Lorentz profile $L(\lambda)$. If the instrumental broadening is ignored with respect to the Stark broadening and the same experimental conditions are applied, the Stark coefficient can be assumed to be equal to the FWHM. In this case, if the self-absorption effect can be ignored, the integrated line intensity is proportional to the concentration of the element. A representative Figure 4.7 [42] is used to display the regression curves for the univariate calibration method based on LIBS data with two overlapping peaks (1) and (2). The spectral intensities S_1 and S_2 are described in this figure as:

Figure 4.7 This figure shows synthetic spectra with the absence of background. (a) With spectral interference. (b) Without spectral interference. *Source:* From Guo et al. [42] / With permission of Royal Society of Chemistry.

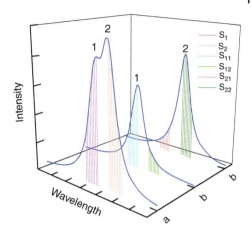

$$S_1 = S_{11} + S_{21} = k_{11}C_1 + k_{21}C_2 \quad (4.23)$$

$$S_2 = S_{22} + S_{12} = k_{22}C_2 + k_{12}C_1 \quad (4.24)$$

where the integrated intensities of the element of interest (1) are S_{11} and S_{12}. The integrated intensities of the element of interest (2) are S_{22} and S_{21}. k_{11}, k_{12}, k_{22}, and k_{21} are the change rates of the integrated intensities versus the concentrations, and C_1 is the concentration of the element of interest; C_2 is the interference element concentration.

LIBS spectra are typically composed of line interference and continuum background. For the simultaneous correction of spectral interference and continuum background, Guo et al. [42] suggested the use of DWT to describe the continuum background. Then the integrated intensities, Eqs (4.23) and (4.24) become

$$S_1 = S_{11} + S_{21} = k_{11}C_1 + k_{21}C_2 + \sum_{m_1 \in Z} c_{J,m_1} \varphi_{J,m_1}(t) \quad (4.25)$$

$$S_2 = S_{22} + S_{12} = k_{22}C_2 + k_{12}C_1 + \sum_{m_1 \in Z} c_{J,m_1} \varphi_{J,m_1}(t) \quad (4.26)$$

where $\varphi_{J,m_1}(t)$, J, m_1, and c_{J,m_1} are the scaling function, the scale of the highest DL, the scale index, and wavelet approximation coefficients, respectively.

Taking into account that the scaling factor is defined as: $\alpha = (k_{21}/k_{22})$.

The regression curve of the concerned element can be obtained from the following equation:

$$S_1 - \alpha S_2 = (k_{11} - \gamma k_{12})C_1 + \sum_{m_1 \in Z} c_{J,m_1} \varphi_{J,m_1}(t) - \alpha \sum_{m_1 \in Z} c_{J,m_1} \varphi_{J,m_1}(t) \quad (4.27)$$

To correct for continuum background and spectral interferences simultaneously according to Eq. (4.27), the three parameters, the wavelet function, the DL, and the scaling factor α, need to be determined. The DL is the dominant parameter in background interference correction and must be determined in the first position. Authors [42] selected the wavelet function from the Daubechies wavelet series, and used the RMSEC as an optimization criterion for the determination of the three parameters.

In the first step, Daubechies 1 (db1) wavelet function is used to decompose the spectrum by varying the DL from 4 to 24. The coefficients of approximation were set to zero and the spectrum was reconstructed. The univariate regression model is then built with integrated intensities. The DL and the scaling factor are determined where RMSEC is minimal. In this step, according to the optimized DL, and the scaling factor α, the spectrum is decomposed, and the wavelet function was optimized from the db1 to db20 function series. In the last step, the LIBS spectrum is processed with the optimum wavelet function, the DL, and the scaling factor to obtain the interference-corrected data. The optimum decomposition was found using db19 wavelet function, DL 8, and α of 0.18.

A couple of overlapping peaks were deconvoluted using the stated method, such as Cr I 461.61 nm/Ti I 461.73 nm, Si I 288.16 nm/Fe II 288.08 nm, Ti II 368.52 nm/Fe I 368.60 nm, and Mn I 401.81 nm/Fe I 401.71 nm. Quantitative analysis results [42] improved significantly when curve fitting, WT, and the proposed interference correction approaches were used. Later, Liu et al. [43] developed an algorithm based on iterative discrete wavelet transform (IDWT) and Richardson–Lucy deconvolution (RLD) to reduce the effect of spectral interference. RLD is a nonlinear spectral deconvolution method based on the Bayes formula and the maximum similarity estimation [50, 51]. The algorithm for discrete spectral data is given by:

$$I^{n+1}(i) = I^{n}(i) \sum_{j=0}^{N-1} h(j,i) \frac{I^{0}(j)}{\sum_{k=0}^{M-1} h(j,k) I^{n}(k)} \qquad (4.28)$$

where $I^{(n)}$ is the spectrum within a number of iterations n, and $I^{(0)}$ is the original spectrum. h is the convolution kernel which is Gaussian line shape and can be calculated by curve fitting of calibration light source spectrum. i, j, and k are the data number of the spectrum and the convolution kernel. N and M are the total data amount of the convolution kernel.

Before performing deconvolution, the spectra background was removed using an IDWT. The optimized parameters for IDWT were achieved using the sample of iron alloy spectra, and (db) 7 and DL 7 were found. As for IDWT, the iron alloy spectra were used for the optimization of the RLD. Spectra using the different data processing algorithms, IDWT, RLD, and coupled IDWT-RLD [43], see Figure 4.8, were compared to the original spectrum. These authors show that the combination IDW and RLD obviously reduced background and interferences lines and improved the quantitative analysis of elements with severe spectral interference.

4.5 Features Selection

Feature selection is the process of selecting from a relatively large set of data-relevant features that can describe the problem with a minimum loss of information. This process relies on reducing the dimensionality of the data, identifying, and selecting a subset of features from the initial ones. These extracted features are used in the model construction. This process offers several advantages in that it allows

- The identification of the key features that may be used to analyze the data.
- The improvement and simplification of the model's performance.

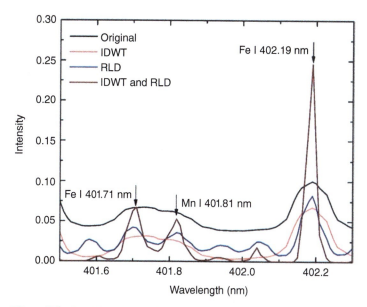

Figure 4.8 The figure shows the spectra after different data processing algorithms done by Liu et al. [43]. *Source:* From Liu et al. [43] / With permission of Royal Society of Chemistry.

- The reduction of data yields, shorter training times, and cost savings.
- The potential for the generalization and the use of simpler models.

Depending on the subset generation process, the feature extraction can be divided into three patterns, i.e. filter, wrapper, and embedded [52]. Each of the three strategies is distinguished by advantages and disadvantages. The filter method relies on the correlation between dependent and independent variables, but it did not interact with any learning algorithms. Therefore, the filter approach has the benefit of low computational costs, but its results are not always satisfactory. The embedded and wrapper approaches take account of the model's efficiency when selecting features compared to filter approaches [52].

Due to its rapid online integration, LIBS is widely used in various applications as described in several review papers [53–63]. Typically, LIBS measurements yield a large amount of data that contain information on the sample investigated and the experimental conditions used to generate such data. Performing LIBS analysis depends on a large number of variables buried in the spectra, usually higher than hundreds or thousands. Thus, selecting the key variables from a wide array of potential ones is a critical step in the analysis of LIBS data sets for quantitative or qualitative purposes using chemometric methods [11].

Researchers have proposed different procedures that can be used to select the key variables (or features) and improve chemometrics classifications and quantifications. These include predefined or manual spectral lines related to the content elements [64–72] and wavelength ranges [64, 73–75]. Due to the matrix effects and self-absorption, identification of emission lines that can be used in the model construction is not a trivial task and it is therefore important to perform a features extraction (variables extraction).

Several features selection methods have been used in LIBS such as the successive projection algorithm (SPA) [76, 77], the stepwise formulation (SW) [76], the ridge regression and RFE [78], the analysis of variance (ANOVA), the logistic regression (LGR) filter [79], and the PLS regression model [80]. Recently, hybrid feature selection models based on combining several models types as V-WSP-PSO [81] and iPLS-mIPW-PLS [32] have been used.

In the next section, we present the most used (or the most cited in LIBS literature) statistical feature selection algorithms proposed to reduce the possibility of losing meaningful information in LIBS data sets.

4.5.1 Principal Component Analysis

Principal component analysis (PCA) is a widely adopted statistical method for the analysis of LIBS spectra. An excellent review of the state of PCA is provided by Porizka et al. [82]. It has many implementations for low-dimensional visualization, clustering, filtering outliers, quantification, classification, and multivariate nonconventional mapping feature selection [82]. PCA is a statistical procedure that uses an orthogonal transformation to convert a set of observations of possibly correlated variables into a set of values of linearly uncorrelated variables called principal components (PCs). These PCs can be used as input features for the chemometrics algorithm. As shown in Table 4.1, this approach has been used widely as a feature selection in many contributions.

4.5.2 Genetic Algorithm (GA)

The genetic algorithm (GA) is a versatile technique applied for artificial intelligence developed by Holland in 1970 [101]. GAs methods progress is linked to the ideas developed in evolutionary biology, where different potential solutions are viewed as competing individuals in an evolving population. They have been used for many applications including variable extraction for multivariate calibration, molecular modeling, regression analysis, model identification, curve fitting, and classification [102]. When applied to feature extraction, the GA encodes subsets of features in the form of strings of binary (0/1) values called "chromosomes"[103]. Each location (or "gene") of the binary string corresponds to one particular variable, coded as "1" if the variable is selected in the model and "0" if not. The GA algorithm creates the extracted features named "offspring generations" created from crossover and mutation operation. The probability of selecting a given chromosome is proportional to its fitness, which is correlated with the evaluation test. Population size is maintained constant in each generation, and new individuals take the place of old ones, while the best ones are automatically transferred to the next stage to avoid the loss of solutions. The iterative process is repeated until an acceptable predefined error is reached. On Table 4.2, we summarize the investigated classification and quantification models based on GA features selection in LIBS.

4.5.3 Wavelet Transformation (WT)

Wavelet transform (WT) has been proven to be a powerful analytical data compression method [121–123]. Using WT, and as reported in Section 4.2.3, the LIBS data can be decomposed into several components of different scales and frequencies, and the relative spectral

Table 4.1 This table summarizes an overview of literature-related chemometric methods based on the principal component analysis features selection model.

PCA-chemometric methods	Applications/aim of research	References
PCA^a-SIMCA^b	• Classification of 31 powder geo-standards and 195 rock slab samples from the Mars analog • Classification of human bone • Classification of plastic samples	[83] [84] [85]
PCA-PLS-DA^c	• Classification of human bone • Classification of oil samples contaminated with heavy metals and oil residues • Discriminating power of the elements in edible sea salts	[84] [86] [87]
PCA-ANN^d	• Classification of human bone • Qualitative and quantitative investigation of chromium-polluted soils • Development of LIBS sensor to mine detection	[84] [88] [89]
PCA-k-NN^e	• Classification of explosive data sheet • Classification of iron ore data	[90] [91]
PCA-SVM^f	• Classification of human bone • Detection of protective coating traces on boron steel with aluminum–silicon covering	[84] [92]
PCA-LDA^g-ANN	• Fast identification of bio-minerals	[93]
PCA-CCA^h	• Estimation of the mechanical properties of steel	[94]
PCA-LDA	• Classification of similar metals • Classification of brick samples	[95] [96]
PCA-CART^i	• Classification of human bone	[84]
PCA-BLR^j	• Classification of human bone	[84]
PCA-PLS-DA-ANN-HCA^k-PLS	• Tissue classification (chicken brain, lung, spleen, liver, kidney, and skeletal muscle)	[97]
PCA-LDA-HCA	• Classification of bacterial spores	[98]
PCA-k-NN-SVM PCA-WNN^l	• Discrimination of soft tissues using • Quantitative analysis of coal ash	[99, 100]

^{a} Principal component analysis.
^{b} Soft independent modeling of class analogy.
^{c} Partial least squares-discriminant analysis.
^{d} Artificial neural network.
^{e} k-Nearest neighbors.
^{f} Support vector machines.
^{g} Linear discriminant analysis.
^{h} Canonical correlation analysis.
^{i} Classification and regression tree.
^{j} Binary logistic regression.
^{k} Hierarchical cluster analysis.
^{l} Wavelet neural network.

Table 4.2 This table summarizes an overview of literature-related chemometric methods based on the genetic algorithm features selection model.

GA-chemometric methods	Applications/aim of research	References
GA-PLS[a] GA-mIPW-PLS[b] GA-siPLS[c] GA-PLS-DA[d]	• Classification of recycled thermoplasts from consumer electronics • Improvement of boron by molecular analysis • Detection and quantification of adulterants in honey • Quantitative analysis of soil samples • Analysis of the calorific value of pulverized coal particle flow • Explosives classification	[104] [105] [106] [77] [107] [108]
GA-ANN[e] GA-BPNN[f]	• LIBS operational conditions for plant materials analysis • LIBS operational conditions for the determination of macro and micronutrients in sugarcane leaves • Quantitative analysis of Ni, Zr, and Ba in soil • Multielemental analysis of heavy metals in soil • Quantitative analysis of steel samples • Online measurement of gross calorific value • Rapid coal analyzer • Carbon in metal-smelting and classification processes	[109] [110] [111] [112] [113] [114] [115] [116]
GA-LDA[g]	• Classification model for classification of Brazilian soils • Explosives discrimination	[76] [117]
GA-SVM[h]	• Quantitative analysis of multielements in steels • Ash content, volatile matter, and calorific value in coal	[75] [118]
GA-Fuzzy logic[i]	• Identification and classification of meteorites by a handheld LIBS	[119]
GA-KELM[j]	• Prediction of the alloy elements of steel and iron coupled with LIBS	[120]

[a] Partial least squares.
[b] Modified iterative predictor weighting-partial least squares.
[c] Synergy interval partial least squares.
[d] Partial least squares-discriminant analysis.
[e] Neural network model.
[f] Genetic algorithm – backpropagation neural network model.
[g] Linear discriminant analysis.
[h] Support vector machine.
[i] Fuzzy logic.
[j] Kernel extreme learning machine.

information and their details could be found in the wavelet coefficients. Due to the property of WT, the wavelet coefficients with small amplitude can be considered uninformative; these can therefore be eliminated without significantly affecting the useful details or information. Consequently, a compressed number of wavelet coefficients, representing the original spectra, can be used in the model construction. Furthermore, WT can be used for selecting spectral windows based on the CWT) that can incorporate most information of spectral peaks and reduce the influence of noisy variables on the classification results.

Table 4.3 This table summarizes an overview of literature-related chemometric methods based on the wavelet transform features selection model.

WT-chemometric methods	Applications/aim of research	References
WT-PLS	• Carbon content in coal using LIBS.	[16]
WT-SVM	• Classification of suspect powders based on Bacillus anthracis spores content • Automatic classification for wheat flour grades	[124] [125]
WT-PLS-DA	• Plastic bottles classification	[126]
WT-MIV-KELM	• Calorific value determination of coal using LIBS	[123]
WTD-RFECV-SVM	• Quantitative analysis of coal properties	[34]
WT-RF WTD-K-SVM-RFE	• Quantification of K content in potassic salt ore by LIBS • Indictors aging grade and hardness grade of steel via LIBS technique	[127] [31]

Chemometrics methods based on WT *features selection, for LIBS classification and* quantification, are listed in Table 4.3.

Various papers on LIBS investigations show the benefit of the feature selection process. With the availability of powerful computers, feature selection is a fast-moving field. New methods are constantly being proposed including approaches that combine several approaches, usually filters and wrappers, to ensure a better resolution of data processing.

References

1 Gornushkin, I.B., Eagan, P.E., Novikov, A.B., Ith, B.W.S.M., and Winefordner, J.D. (2003), Automatic correction of continuum background in laser-induced breakdown and Raman spectrometry. *Appl. Spectrosc.*, **57** (2), 197–207.
2 Sun, L., and Yu, H. (2009), Spectrochimica acta part B automatic estimation of varying continuum background emission in laser-induced breakdown spectroscopy. *Spectrochim. Acta Part B At. Spectrosc.*, **64** (3), 278–287.
3 Zhang, B., Sun, L., and Yu, H. (2015), An improving method for background correction in laser induced breakdown spectroscopy. *Appl. Mech. Mater.*, **751**, 86–91.
4 Tan, B., Huang, M., Zhu, Q., Guo, Y., and Qin, J. (2017), Spectrochimica Acta Part B Detection and correction of laser induced breakdown spectroscopy spectral background based on spline interpolation method. *Spectrochim. Acta Part B At. Spectrosc.*, **138**, 64–71.
5 Iu, J.I.L., Hang, R.U.I.Z., Iaotian, X.L.I., Hen, J.I.C., Ianan, J., Iu, L., Iu, J.U.N.Q., Ao, X.U.N.G., Ui, J.I.C., and Eshig, B.A.H. (2018), Continuous background correction using effective points selected in third-order minima segments in low-cost laser-induced breakdown spectroscopy without intensified CCD. *Opt. Express*, **26** (13), 145–151.

6 Yaroshchyk, P., and Eberhardt, J.E. (2014), Automatic correction of continuum background in Laser-induced Breakdown Spectroscopy using a model-free algorithm. *Spectrochim. Acta Part B At. Spectrosc.*, **99**, 138–149.

7 Friedrichs, M.S. (1995), A model-free algorithm for the removal of baseline artifacts. *J. Biomol. NMR*, **5**, 147–153.

8 Daubechies, I. (1992), *Ten Lectures on Wavelets*, SIAM, Philadelphia, PA.

9 Mallat, S.G. (1989), A theory for multiresolution signal decomposition. *IEEE Trans. Pattern Anal. Mach. Intell.*, **11** (7), 674–693.

10 Shao, X.G., Leung, A.K.M., and Chau, F.T. (2003), Wavelet: a new trend in chemistry. *Acc. Chem. Res.*, **36** (4), 276–283.

11 Shensa, M.J. (1992), The discrete wavelet transform: wedding the À Trous and Mallat algorithms. *IEEE Trans. Signal Process.*, **40** (10), 2464–2482.

12 Beylkin, G. (1992), On the representation of operators in bases of compactly supported wavelets. *Siam J. Numer. Anal.*, **6** (6), 70–83.

13 Holschneider, M., Kronland-Martinet, R., Morlet, J., Tchamitchian, P., Combes, J.M., and Grossman, A. (1989), *Wavelets, Time-Frequency Methods and Phase Space. A Real-Time Algorithm Signal Analysis with the Help of the Wavelet Transform*. Springer, Berlin, 289–297.

14 Ma, X.-G., and Zhang, Z.-X. (2003), Application of wavelet transform to background correction in inductively coupled plasma atomic emission spectrometry. *Anal. Chim. Acta*, **485**, 233–239.

15 Chen, D., Shao, X., Hu, B., and Su, Q. (2004), A background and noise elimination method for quantitative calibration of near infrared spectra. *Anal. Chim. Acta*, **511** (1), 37–45.

16 Yuan, T., Wang, Z., Li, Z., Ni, W., and Liu, J. (2014), A partial least squares and wavelet-transform hybrid model to analyze carbon content in coal using laser-induced breakdown spectroscopy. *Anal. Chim. Acta*, **807**, 29–35.

17 Zou, X.H., Guo, L.B., Shen, M., Li, X.Y., Hao, Z.Q., Zeng, Q.D., Lu, Y.F., Wang, Z.M., and Zeng, X.Y. (2014), Accuracy improvement of quantitative analysis in laser-induced breakdown spectroscopy using modified wavelet transform. *Opt. Express*, **22** (9), 10233.

18 Muchao, L. (2014), Laser induced breakdown spectroscopy data processing method based on wavelet analysis. *Adv. Intell. Syst. Comput.*, **297**, 5.

19 Kepes, E., Porizka, P., Klus, J., Modlitbova, P., and Kaiser, J. (2018), Influence of baseline subtraction on laser-induced breakdown spectroscopic data. *J. Anal. At. Spectrom.*, **33**, 2107–2115.

20 Galloway, C.M., LeRu, E.C., and Etchegoin, P.G. (2009), An iterative algorithm for background removal in spectroscopy by wavelet transforms. *Appl. Spectrosc.*, **63** (12), 1370–1376.

21 Dyar, M.D., Giguere, S., Carey, C.J., and Boucher, T. (2016), Comparison of baseline removal methods for laser-induced breakdown spectroscopy of geological samples. *Spectrochim. Acta Part B At. Spectrosc.*, **126**, 53–64.

22 Zhang, B., Sun, L., Yu, H., Xin, Y., and Cong, Z. (2015), A method for improving wavelet threshold denoising in laser-induced breakdown spectroscopy. *Spectrochim. Acta Part B At. Spectrosc.*, **107**, 32–44.

23 Schlenke, J., Hildebrand, L., Moros, J., and Laserna, J.J. (2012), Adaptive approach for variable noise suppression on laser-induced breakdown spectroscopy responses using stationary wavelet transform. *Anal. Chim. Acta*, **754**, 8–19.

24 Donoho, D.L., and Johnstone, I.M. (1994), Ideal spatial adaptation via wavelet shrinkage. *Biometrika*, **81**, 425–455.
25 Donoho, D.L., and Johnstone, I.M. (1995), Adapting to unknown smoothness via wavelet shrinkage. *J. Am. Stat. Assoc.*, **90** (432), 1200–1224.
26 Ewusi-Annan, E., Delapp, D.M., Wiens, R.C., and Melikechi, N. (2020), Automatic preprocessing of laser-induced breakdown spectra using partial least squares regression and feed-forward artificial neural network: applications to Earth and Mars data. *Spectrochim. Acta Part B At. Spectrosc.*, **171** (July), 105930.
27 Zhang, B., Sun, L., Yu, H., Xin, Y. and Cong, Z. (2013), Wavelet denoising method for laser-induced breakdown spectroscopy. *J. Anal. At. Spectrom*, **3** (207890), 10715–10722.
28 Xie, S., Xu, T., Han, X., Lin, Q., and Duan, Y. (2017), Accuracy improvement of quantitative LIBS analysis using wavelet threshold de-noising. *J. Anal. At. Spectrom.*, **32** (3), 629–637.
29 Duan, H., Ma, S., Han, L., and Huang, G. (2017), A novel denoising method for laser-induced breakdown spectroscopy: improved wavelet dual threshold function method and its application to quantitative modeling of Cu and Zn in Chinese animal manure composts. *Microchem. J.*, **134**, 262–269.
30 Zhang, C., Shen, T., Liu, F., and He, Y. (2018), Identification of coffee varieties using laser-induced breakdown spectroscopy and chemometrics. *Sensors*, **18** (1).
31 Huang, J., Dong, M., Lu, S., Yu, Y., Liu, C., Yoo, J.H., and Lu, J. (2019), A hybrid model combining wavelet transform and recursive feature elimination for running state evaluation of heat-resistant steel using laser-induced breakdown spectroscopy. *Analyst*, **144** (12), 3736–3745.
32 Fu, X., Duan, F.J., Huang, T.T., Ma, L., Jiang, J.J., and Li, Y.C. (2017), A fast variable selection method for quantitative analysis of soils using laser-induced breakdown spectroscopy. *J. Anal. At. Spectrom.*, **32** (6), 1166–1176.
33 Zhang, Y., Dong, M., Cheng, L., Wei, L., Cai, J., and Lu, J. (2020), Improved measurement in quantitative analysis of coal properties using laser induced breakdown spectroscopy. *J. Anal. At. Spectrom.*, **35** (4), 810–818.
34 Lu, P., Zhuo, Z., Zhang, W., Tang, J., Tang, H., and Lu, J. (2020), Accuracy improvement of quantitative LIBS analysis of coal properties using a hybrid model based on a wavelet threshold de-noising and feature selection method. *Appl. Opt.*, **59** (22), 6443.
35 Zhang, T., Tang, H., and Li, H. (2018), Chemometrics in laser-induced breakdown spectroscopy. *J. Chemom.*, **32** (11), 1–18.
36 Hu, Y., Li, W., and Hu, J. (2005), Resolving overlapped spectra with curve fitting. *Spectrochim. Acta Part A Mol. Biomol. Spectrosc.*, **62** (1–3), 16–21.
37 Zhang, B., Yu, H., Sun, L., Xin, Y., and Conga, Z. (2013), A method for resolving overlapped peaks in laser-induced breakdown spectroscopy (LIBS). *Appl. Spectrosc.*, **67** (9), 1087–1097.
38 Tan, B., Huang, M., Zhu, Q., Guo, Y., and Qin, J. (2017), Decomposition and correction overlapping peaks of LIBS using an error compensation method combined with curve fitting. *Appl. Opt.*, **56** (25), 7116.
39 Guezenoc, J., Payré, V., Fabre, C., Syvilay, D., Cousin, A., Gallet-Budynek, A., and Bousquet, B. (2019), Variable selection in laser-induced breakdown spectroscopy assisted by multivariate analysis: an alternative to multi-peak fitting. *Spectrochim. Acta Part B At. Spectrosc.*, **152** (December 2018), 6–13.

40 Yang, W., Li, B., Zhou, J., Han, Y., and Wang, Q. (2018), Continuous-wavelet-transform-based automatic curve fitting method for laser-induced breakdown spectroscopy. *Appl. Opt.*, **57** (26), 7526.

41 Shao, L., Lin, X., and Shao, X. (2002), A wavelet transform and its application to spectroscopic analysis. *Appl. Spectrosc. Rev.*, **37** (4), 429–450.

42 Guo, Y.M., Deng, L.M., Yang, X.Y., Li, J.M., Li, K.H., Zhu, Z.H., Guo, L.B., Li, X.Y., Lu, Y.F., and Zeng, X.Y. (2017), Wavelet-based interference correction for laser-induced breakdown spectroscopy. *J. Anal. At. Spectrom.*, **32** (12), 2401–2406.

43 Liu, K., Zhou, R., Zhang, W., Tang, Z., Yan, J., Lv, M., Li, X., Lu, Y., and Zeng, X. (2020), Interference correction for laser-induced breakdown spectroscopy using a deconvolution algorithm. *J. Anal. At. Spectrom.*, **35** (4), 762–766.

44 Kauppinen, J.K., Moffatt, D.J., Cameron, D.G., and Mantsch, H.H. (1981), Noise in Fourier self-deconvolution. *Appl. Opt.*, **20** (10), 1866.

45 Gallant, S.R., Fraleigh, S.P., and Cramer, S.M. (1993), Deconvolution of overlapping chromatographic peaks using a cerebellar model arithmetic computer neural network. *Chemom. Intell. Lab. Syst.*, **18** (1), 41–57.

46 Yao, S.C., Chen, J.C., Lu, J.D., Shen, Y.L., and Pan, G. (2015), Influence of C—Fe lines interference correction on laser-induced breakdown spectroscopy measurement of unburned carbon in fly ash. *Guang pu xue yu Guang pu fen xi= Guang pu*, **35** (6), 1719–1723.

47 Almeida, R., and Torres, D.F.M. (2009), Calculus of variations with fractional derivatives and fractional integrals. *Appl. Math. Lett.*, **22** (12), 1816–1820.

48 Baleanu, D., and Trujillo, J.I. (2010), A new method of finding the fractional Euler-Lagrange and Hamilton equations within Caputo fractional derivatives. *Commun. Nonlinear Sci. Numer. Simul.*, **15** (5), 1111–1115.

49 Jumarie, G. (2010), Cauchy's integral formula via the modified Riemann-Liouville derivative for analytic functions of fractional order. *Appl. Math. Lett.*, **23** (12), 1444–1450.

50 Morh, M., and Matouek, V. (2011), High-resolution boosted deconvolution of spectroscopic data. *J. Comput. Appl. Math.*, **235** (6), 1629–1640.

51 Eichstädt, S., Schmähling, F., Wübbeler, G., Anhalt, K., Bünger, L., Krüger, U., and Elster, C. (2013), Comparison of the Richardson-Lucy method and a classical approach for spectrometer bandpass correction. *Metrologia*, **50** (2), 107–118.

52 Sánchez-Maroño, N. (2016), Feature selection for high-dimensional data. *Prog. Artif. Intell.*, **5**, 65–75.

53 Hahn, D.W., and Omenetto, N. (2012), Laser-induced breakdown spectroscopy (LIBS), part II: review of instrumental and methodological approaches to material analysis and applications to different fields. *Appl. Spectrosc.*, **66** (4), 347–419.

54 Harmon, R.S., Russo, R.E., and Hark, R.R. (2013), Applications of laser-induced breakdown spectroscopy for geochemical and environmental analysis: a comprehensive review. *Spectrochim. Acta Part B At. Spectrosc.*, **87**, 11–26.

55 Botto, A., Campanella, B., Legnaioli, S., Lezzerini, M., Lorenzetti, G., Pagnotta, S., Poggialini, F., and Palleschi, V. (2019), Applications of laser-induced breakdown spectroscopy in cultural heritage and archaeology: a critical review. *J. Anal. At. Spectrom.*, **34** (1), 81–103.

56 Yu, X., Li, Y., Gu, X., Bao, J., Yang, H., and Sun, L. (2014), Laser-induced breakdown spectroscopy application in environmental monitoring of water quality: a review. *Environ. Monit. Assess.*, **186** (12), 8969–8980.

57 Spizzichino, V., and Fantoni, R. (2014), Laser induced breakdown spectroscopy in archeometry: a review of its application and future perspectives. *Spectrochim. Acta Part B At. Spectrosc.*, **99**, 201–209.

58 Zhao, Y., Zhang, L., Zhao, S.-X., Li, Y.-F., Gong, Y., Dong, L., Ma, W.-G., Yin, W.-B., Yao, S.-C., and Lu, J.-D. (2016), Review of methodological and experimental LIBS techniques for coal analysis and their application in power plants in China. *Front. Phys.*, **11** (6), 114211.

59 Li, C., Feng, C.-L., Oderji, H.Y., Luo, G.-N., and Ding, H.-B. (2016), Review of LIBS application in nuclear fusion technology. *Front. Phys.*, **11** (6), 114214.

60 Markiewicz-Keszycka, M., Cama-Moncunill, X., Casado-Gavalda, M.P., Dixit, Y., Cama-Moncunill, R., Cullen, P.J., and Sullivan, C. (2017), Laser-induced breakdown spectroscopy (LIBS) for food analysis: a review. *Trends Food Sci. Technol.*, **65**, 80–93.

61 Noll, R., Fricke-Begemann, C., Connemann, S., Meinhardt, C., and Sturm, V. (2018), LIBS analyses for industrial applications – an overview of developments from 2014 to 2018. *J. Anal. At. Spectrom.*, **33** (6), 945–956.

62 Gaudiuso, R., Melikechi, N., Abdel-Salam, Z.A., Harith, M.A., Palleschi, V., Motto-Ros, V., and Busser, B. (2019), Laser-induced breakdown spectroscopy for human and animal health: a review. *Spectrochim. Acta Part B At. Spectrosc.*, **152**, 123–148.

63 Jolivet, L., Leprince, M., Moncayo, S., Sorbier, L., Lienemann, C.-P., and Motto-Ros, V. (2019), Review of the recent advances and applications of LIBS-based imaging. *Spectrochim. Acta Part B At. Spectrosc.*, **151**, 41–53.

64 Messaoud Aberkane, S., Abdelhamid, M., Mokdad, F., Yahiaoui, K., Abdelli-Messaci, S., and Harith, M.A. (2017), Sorting zamak alloys: via chemometric analysis of their LIBS spectra. *Anal. Methods*, **9** (24), 3696–3703.

65 Bousquet, B., Sirven, J.B., and Canioni, L. (2007), Towards quantitative laser-induced breakdown spectroscopy analysis of soil samples. *Spectrochim. Acta Part B At. Spectrosc.*, **62** (12), 1582–1589.

66 Serrano, J., Moros, J., Sánchez, C., Macías, J., and Laserna, J.J. (2014), Advanced recognition of explosives in traces on polymer surfaces using LIBS and supervised learning classifiers. *Anal. Chim. Acta*, **806**, 107–116.

67 Dong, M., Wei, L., Lu, J., Li, W., Lu, S., Li, S., Liu, C., and Yoo, J.H. (2019), A comparative model combining carbon atomic and molecular emissions based on partial least squares and support vector regression correction for carbon analysis in coal using LIBS. *J. Anal. At. Spectrom.*, **34** (3), 480–488.

68 Guo, G., Niu, G., Shi, Q., Lin, Q., Tian, D., and Duan, Y. (2019), Multi-element quantitative analysis of soils by laser induced breakdown spectroscopy (LIBS) coupled with univariate and multivariate regression methods. *Anal. Methods*, **11** (23), 3006–3013.

69 El Haddad, J., Canioni, L., and Bousquet, B. (2014), Good practices in LIBS analysis: review and advices. *Spectrochim. Acta Part B At. Spectrosc.*, **101**, 171–182.

70 Sanghapi, H.K., Jain, J., Bol'Shakov, A., Lopano, C., McIntyre, D., and Russo, R. (2016), Determination of elemental composition of shale rocks by laser induced breakdown spectroscopy. *Spectrochim. Acta Part B At. Spectrosc.*, **122**, 9–14.

71 Guezenoc, J., Bassel, L., Gallet-Budynek, A., and Bousquet, B. (2017), Variables selection: a critical issue for quantitative laser-induced breakdown spectroscopy. *Spectrochim. Acta Part B At. Spectrosc.*, **134**, 6–10.

72 De Lucia, F.C., and Gottfried, J.L. (2011), Influence of variable selection on partial least squares discriminant analysis models for explosive residue classification. *Spectrochim. Acta Part B At. Spectrosc.*, **66** (2), 122–128.

73 Godoi, Q., Leme, F.O., Trevizan, L.C., Pereira-Filho, E.R., Rufini, I.A., Santos, D., and Krug, F.J. (2011), Laser-induced breakdown spectroscopy and chemometrics for classification of toys relying on toxic elements. *Spectrochim. Acta Part B At. Spectrosc.*, **66** (2), 138–143.

74 Moncayo, S., Rosales, J.D., Izquierdo-Hornillos, R., Anzano, J., and Caceres, J.O. (2016), Classification of red wine based on its protected designation of origin (PDO) using Laser-induced breakdown spectroscopy (LIBS). *Talanta*, **158**, 185–191.

75 Zhang, T., Liang, L., Wang, K., Tang, H., Yang, X., Duan, Y., and Li, H. (2014), A novel approach for the quantitative analysis of multiple elements in steel based on laser-induced breakdown spectroscopy (LIBS) and random forest regression (RFR). *J. Anal. At. Spectrom.*, **29** (12), 2323–2329.

76 Pontes, M.J.C., Cortez, J., Galvão, R.K.H., Pasquini, C., Araújo, M.C.U., Coelho, R.M., Chiba, M.K., deAbreu, M.F., and Madari, B.E. (2009), Classification of Brazilian soils by using LIBS and variable selection in the wavelet domain. *Anal. Chim. Acta*, **642** (1–2), 12–18.

77 Duan, F., Fu, X., Jiang, J., Huang, T., Ma, L., and Zhang, C. (2018), Automatic variable selection method and a comparison for quantitative analysis in laser-induced breakdown spectroscopy. *Spectrochim. Acta Part B At. Spectrosc.*, **143**, 12–17.

78 Guodong, W. A. N. G., Lanxiang, S. U. N., Wei, W. A. N. G., et al. A feature selection method combined with ridge regression and recursive feature elimination in quantitative analysis of laser induced breakdown spectroscopy. *Plasma Science and Technology*, 2020, vol. **22**, no 7, p. 074002.

79 Lu, S., Shen, S., Huang, J., Dong, M., Lu, J., and Li, W. (2018), Feature selection of laser-induced breakdown spectroscopy data for steel aging estimation. *Spectrochim. Acta Part B At. Spectrosc.*, **150** (April), 49–58.

80 Yao, S., Qin, H., Wang, Q., Lu, Z., Yao, X., Yu, Z., Chen, X., Zhang, L., and Lu, J. (2020), Optimizing analysis of coal property using laser-induced breakdown and near-infrared reflectance spectroscopies. *Spectrochim. Acta Part A Mol. Biomol. Spectrosc.*, **239**, 118492.

81 Yan, C., Liang, J., Zhao, M., Zhang, X., Zhang, T., and Li, H. (2019), Analytica Chimica Acta A novel hybrid feature selection strategy in quantitative analysis of laser-induced breakdown spectroscopy. *Anal. Chim. Acta*, **1080**, 35–42.

82 Pořízka, P., Klus, J., Képeš, E., Prochazka, D., Hahn, W., and Kaiser, J. (2018), On the utilization of principal component analysis in laser-induced breakdown spectroscopy data analysis, a review. *Spectrochim. Acta Part B At. Spectrosc.*, **148**, 65–82.

83 Anderson, R.B., Bell, J.F., Wiens, R.C., Morris, R. V., and Clegg, S.M. (2012), Clustering and training set selection methods for improving the accuracy of quantitative laser induced breakdown spectroscopy. *Spectrochim. Acta Part B At. Spectrosc.*, **70**, 24–32.

84 Moncayo, S., Manzoor, S., Navarro-Villoslada, F., and Caceres, J.O. (2015), Evaluation of supervised chemometric methods for sample classification by laser induced breakdown spectroscopy. *Chemom. Intell. Lab. Syst.*, **146**, 354.

85 Wang, Q., Cui, X., Teng, G., Zhao, Y., and Wei, K. (2020), Evaluation and improvement of model robustness for plastics samples classification by laser-induced breakdown spectroscopy. *Opt. Laser Technol.*, **125** (December 2019), 106035.

86 Kim, G., Kwak, J., Kim, K.R., Lee, H., Kim, K.W., Yang, H., and Park, K. (2013), Rapid detection of soils contaminated with heavy metals and oils by laser induced breakdown spectroscopy (LIBS). *J. Hazard. Mater.*, **263**, 754–760.

87 Lee, Y., Ham, K.S., Han, S.H., Yoo, J., and Jeong, S. (2014), Revealing discriminating power of the elements in edible sea salts: line-intensity correlation analysis from laser-induced plasma emission spectra. *Spectrochim. Acta Part B At. Spectrosc.*, **101**, 57–67.

88 Sirven, J.B., Bousquet, B., Canioni, L., Sarger, L., Tellier, S., Potin-Gautier, M., and Le Hecho, I. (2006), Qualitative and quantitative investigation of chromium-polluted soils by laser-induced breakdown spectroscopy combined with neural networks analysis. *Anal. Bioanal. Chem.*, **385** (2), 256–262.

89 Bohling, C., Hohmann, K., Scheel, D., Bauer, C., Schippers, W., Burgmeier, J., Willer, U., Holl, G., and Schade, W. (2007), All-fiber-coupled laser-induced breakdown spectroscopy sensor for hazardous materials analysis. *Spectrochim. Acta Part B At. Spectrosc.*, **62** (12), 1519–1527.

90 Sahoo, T.K., Negi, A., and Gundawar, M.K. (2015), Study of preprocessing sensitivity on laser induced breakdown spectroscopy (LIBS) spectral classification. *2015 International Conference on Advances in Computing, Communications and Informatics, ICACCI 2015*, Kochi, India, 10–13 August (2015), pp. 137–143.

91 Yang, Y., Hao, X., Zhang, L., and Ren, L. (2020), Application of scikit and keras libraries for the classification of iron ore data acquired by laser-induced breakdown spectroscopy (LIBS). *Sensors*, **20** (5), 1393.

92 Anabitarte, F., Mirapeix, J., Portilla, O.M.C., Lopez-Higuera, J.M., and Cobo, A. (2012), Sensor for the detection of protective coating traces on boron steel with aluminium-silicon covering by means of laser-induced breakdown spectroscopy and support vector machines. *IEEE Sens. J.*, **12** (1), 64–70.

93 Vítková, G., Novotný, K., Prokeš, L., Hrdlička, A., Kaiser, J., Novotný, J., Malina, R., and Prochazka, D. (2012), Fast identification of biominerals by means of stand-off laser-induced breakdown spectroscopy using linear discriminant analysis and artificial neural networks. *Spectrochim. Acta Part B At. Spectrosc.*, **73**, 1–6.

94 Huang, J., Dong, M., Lu, S., Li, W., Lu, J., Liu, C., and Yoo, J.H. (2018), Estimation of the mechanical properties of steel via LIBS combined with canonical correlation analysis (CCA) and support vector regression (SVR). *J. Anal. At. Spectrom.*, **33** (5), 720–729.

95 Shin, S., Moon, Y., Lee, J., Jang, H., Hwang, E., and Jeong, S. (2019), Signal processing for real-time identification of similar metals by laser-induced breakdown spectroscopy. *Plasma Sci. Technol.*, **21** (3), 034011.

96 Vítková, G., Prokeš, L., Novotný, K., Pořízka, P., Novotný, J., Všianský, D., Čelko, L., and Kaiser, J. (2014), Comparative study on fast classification of brick samples by combination of principal component analysis and linear discriminant analysis using stand-off and table-top laser-induced breakdown spectroscopy. *Spectrochim. Acta Part B At. Spectrosc.*, **101**, 191–199.

97 Yueh, F.Y., Zheng, H., Singh, J.P., and Burgess, S. (2009), Preliminary evaluation of laser-induced breakdown spectroscopy for tissue classification. *Spectrochim. Acta Part B At. Spectrosc.*, **64** (10), 1059–1067.

98 Merdes, D.W., Suhan, J.M., Keay, J.M., Hadka, D.M., and Bradley, W.R. (2007), The investigation of laser-induced breakdown spectroscopy for detection of biological contaminants on surfaces. *Spectrosc.*, **22** (4), 28–38.

99 Li, X., Yang, S., Fan, R., Yu, X., and Chen, D. (2018), Discrimination of soft tissues using laser-induced breakdown spectroscopy in combination with k nearest neighbors (kNN) and support vector machine (SVM) classifiers. *Opt. Laser Technol.*, **102**, 233–239.

100 Wei, Jiao, Dong, Juan, Zhang, Tianlong, et al. Quantitative analysis of the major components of coal ash using laser induced breakdown spectroscopy coupled with a wavelet neural network (WNN). *Anal. Methods*, 2016, vol. **8**, no 7, p. 1674–1680.

101 Holland, J.H. (1992), *Adaptation in Natural and Artificial Systems: An Introductory Analysis with Applications to Biology, Control, and Artificial Intelligence*. MIT Press, Cambridge, MA.

102 Chiang, L.H., and Pell, R.J. (2004), Genetic algorithms combined with discriminant analysis for key variable identification. *J. Process Control*, **14** (2), 143–155.

103 Pontes, M.J.C., Galvão, R.K.H., Araújo, M.C.U., Moreira, P.N.T., Neto, O.D.P., José, G.E., and Saldanha, T.C.B. (2005), The successive projections algorithm for spectral variable selection in classification problems. *Chemom. Intell. Lab. Syst.*, **78** (1), 11–18.

104 Fink, H., Panne, U., and Niessner, R. (2002), Process analysis of recycled thermoplasts from consumer electronics by laser-induced plasma spectroscopy. *Anal. Chem.*, **74** (17), 4334–4342.

105 Zhu, Z., Li, J., Guo, Y., Cheng, X., Tang, Y., Guo, L., Li, X., Lu, Y., and Zeng, X. (2018), Accuracy improvement of boron by molecular emission with a genetic algorithm and partial least squares regression model in laser-induced breakdown spectroscopy. *J. Anal. At. Spectrom.*, **33** (2), 205–209.

106 Nespeca, M.G., Vieira, A.L., Júnior, D.S., Neto, J.A.G., and Ferreira, E.C. (2020), Detection and quantification of adulterants in honey by LIBS. *Food Chem.*, **311**, 125886.

107 Li, W., Dong, M., Lu, S., Li, S., Wei, L., Huang, J., and Lu, J. (2019), Improved measurement of the calorific value of pulverized coal particle flow by laser-induced breakdown spectroscopy (LIBS). *Anal. Methods*, **11** (35), 4471–4480.

108 Kumar Myakalwar, A., Spegazzini, N., Zhang, C., Kumar Anubham, S., Dasari, R.R., Barman, I., and Kumar Gundawar, M. (2015), Less is more: Avoiding the LIBS dimensionality curse through judicious feature selection for explosive detection. *Sci. Rep.*, **5** (July), 1–10.

109 Nunes, L.C., daSilva, G.A., Trevizan, L.C., Santos Júnior, D., Poppi, R.J., and Krug, F.J. (2009), Simultaneous optimization by neuro-genetic approach for analysis of plant materials by laser induced breakdown spectroscopy. *Spectrochim. Acta Part B At. Spectrosc.*, **64** (6), 565–572.

110 Nunes, L.C., Batista Braga, J.W., Trevizan, L.C., Florêncio De Souza, P., Arantes De Carvalho, G.G., Júnior, D.S., Poppi, R.J., and Krug, F.J. (2010), Optimization and validation of a LIBS method for the determination of macro and micronutrients in sugar cane leaves. *J. Anal. At. Spectrom.*, **25** (9), 1453–1460.

111 Shen, Q., Zhou, W., and Li, K. (2010), Quantitative analysis of Ni, Zr and Ba in soil by combing neuro-genetic approach and laser induced breakdown spectroscopy. *Infrared, Millim. Wave, Terahertz Technol.*, **7854**, 78543Q.

112 Liu, L., Liu, J., Zhao, N., Wang, Y., Shi, H., Wang, C., Zhang, Y., and Liu, W. (2011), Optimization of laser induced breakdown spectroscopy system by neurogenetic method for multi-elemental analysis of heavy metals in soil. *2011 International Conference on Optical Instruments and Technology: Optical Systems and Modern Optoelectronic Instruments* (OIT2011), Beijing, China (6–9 November 2011), **8197**, 81970Y.

113 Li, K., Guo, L., Li, J., Yang, X., Yi, R., Li, X., Lu, Y., and Zeng, X. (2017), Quantitative analysis of steel samples using laser-induced breakdown spectroscopy with an artificial neural network incorporating a genetic algorithm. *Appl. Opt.*, **56** (4), 935.

114 Lu, Z., Mo, J., Yao, S., Zhao, J., and Lu, J. (2017), Rapid determination of the gross calorific value of coal using laser-induced breakdown spectroscopy coupled with artificial neural networks and genetic algorithm. *Energy Fuels*, **31** (4), 3849–3855.

115 Yao, S., Mo, J., Zhao, J., Li, Y., Zhang, X., Lu, W., and Lu, Z. (2018), Development of a rapid coal analyzer using laser-induced breakdown spectroscopy (LIBS). *Appl. Spectrosc.*, **72** (8), 1225–1233.

116 He, J., Pan, C., Liu, Y., and Du, X. (2019), Quantitative analysis of carbon with laser-induced breakdown spectroscopy (LIBS) using genetic algorithm and back propagation neural network models. *Appl. Spectrosc.*, **73** (6), 678–686.

117 Pinkham, D.W., Bonick, J.R., and Woodka, M.D. (2012), Feature optimization in chemometric algorithms for explosives detection. *Proc. SPIE*, **8357**, 1–8.

118 Zhang, W., Zhuo, Z., Lu, P., Tang, J., Tang, H., Lu, J., Xing, T., and Wang, Y. (2020), LIBS analysis of the ash content, volatile matter, and calorific value in coal by partial least squares regression based on ash classification. *J. Anal. At. Spectrom.*, **35** (8), 1621–1631.

119 Senesi, G.S., Manzari, P., Consiglio, A., and DePascale, O. (2018), Identification and classification of meteorites using a handheld LIBS instrument coupled with a fuzzy logic-based method. *J. Anal. At. Spectrom.*, **33** (10), 1664–1675.

120 Mei, Y., Cheng, S., Hao, Z., Guo, L., Li, X., Zeng, X., and Ge, J. (2019), Quantitative analysis of steel and iron by laser-induced breakdown spectroscopy using GA-KELM. *Plasma Sci. Technol.*, **21** (3).

121 Trygg, J., and Wold, S. (1998), PLS regression on wavelet compressed NIR spectra. *Chemom. Intell. Lab. Syst.*, **42** (1–2), 209–220.

122 Shao, X., Wang, F., Chen, D., and Su, Q. (2004), A method for near-infrared spectral calibration of complex plant samples with wavelet transform and elimination of uninformative variables. *Anal. Bioanal. Chem.*, **378** (5), 1382–1387.

123 Yan, C., Zhang, T., Sun, Y., Tang, H., and Li, H. (2019), A hybrid variable selection method based on wavelet transform and mean impact value for calorific value determination of coal using laser-induced breakdown spectroscopy and kernel extreme learning machine. *Spectrochim. Acta Part B*, **154** (February), 75–81.

124 Cisewski, J., Snyder, E., Hannig, J., and Oudejans, L. (2012), Support vector machine classification of suspect powders using laser-induced breakdown spectroscopy (LIBS) spectral data. *J. Chemom.*, **26** (5), 143–149.

125 Yang, P., Zhu, Y., Tang, S., Hao, Z., Guo, L., Li, X., Lu, Y., and Zeng, X. (2018), Analytical-performance improvement of laser-induced breakdown spectroscopy for the processing degree of wheat flour using a continuous wavelet transform. *Appl. Opt.*, **57** (14), 3730.

126 Liu, K., Tian, D., Deng, X., Wang, H., and Yang, G. (2019), Rapid classification of plastic bottles by laser-induced breakdown spectroscopy (LIBS) coupled with partial least squares discrimination analysis based on spectral windows (SW-PLS-DA). *J. Anal. At. Spectrom.*, **34** (8), 1665–1671.

127 Ding, Y., Zhang, W., Zhao, X., Zhang, L., and Yan, F. (2020), A hybrid random forest method fusing wavelet transform and variable importance for the quantitative analysis of K in potassic salt ore using laser-induced breakdown spectroscopy. *J. Anal. At. Spectrom.*, **35** (6), 1131–1138.

5

Principal Component Analysis

Mohamed Abdel-Harith and Zienab Abdel-Salam

National Institute of Laser Enhanced Science, Cairo University, Cairo, Egypt

5.1 Introduction

5.1.1 Laser-Induced Breakdown Spectroscopy (LIBS)

Laser-induced breakdown spectroscopy (LIBS) is now a worldwide well-known spectrochemical analytical technique. Some tens of nano-, pico-, or femtosecond laser pulses are focused onto a solid target or on a liquid or even gas in the LIBS's most straightforward configuration. The tightly focused, moderate energy laser pulse induces the so-called plasma plume, consisting of a collection of positive ions and swirling electrons at highly elevated temperatures, namely some thousands of Kelvin. The plasma temperature depends on the laser pulse parameters and the physical properties of the material of the target and the surrounding atmospheric conditions [1–4]. As the plasma plume cools down, recombination and de-excitation of the ions and electrons take place. During this process, the plasma plume gets rid of the previously absorbed laser energy by the emission of light photons. The spectroscopic analysis of such emitted light provides the required LIBS spectrum, which includes the characteristic spectral lines of the elements existing in the plasma plume and, consequently, in the target material, in the case of stoichiometric ablation. Qualitatively, the spectrum provides the elemental constituents of the analyzed target (elements fingerprint spectrum). Quantitatively, there is a direct relationship between the spectral lines intensities and the corresponding element concentration, taking into consideration self-absorption and the matrix effect [5–7].

Compared to other spectrochemical analytical techniques, such as Laser Ablation-Inductively Coupled Plasma (LA-ICP), Inductively Coupled Plasma-Mass Spectrometry (ICP-MS), and X-Ray Fluorescence (XRF), LIBS is fast, simple, needs no or minimal target preparation, provides low and high Z multi-elemental spectra, besides the relatively low cost for the LIBS systems. Moreover, nowadays, available commercial portable and mobile LIBS systems facilitate in situ and real time for in-field measurements [8, 9].

LIBS has been exploited successfully in numerous applications in various scientific and applied fields. Besides its essential use in analytical chemistry, it has been used in industrial [10, 11], geochemical [12, 13], archaeological [14, 15], biological [16, 17], medical [18, 19], and agricultural applications [20, 21]. A few years ago, LIBS was used for extraterrestrial researches, too [22].

Though the LIBS spectrum analysis looks simple and straightforward, the information in a single spectrum is enormous. Therefore, such a tremendous immense amount of data need to be simplified through special statistical processing. LIBS fundamentals, instrumentations, and applications are detailed in the first three chapters of this book. The present chapter will focus on simplifying LIBS data via one of the chemometric techniques, namely the principal component analysis (PCA).

5.2 The Principal Component Analysis (PCA)

Historically, PCA was introduced in 1901 [23] by the British mathematician and biostatistician Karl Pearson. The American mathematical statistician Harold Hotelling presented further improvement and elaboration of PCA in 1933 [24]. This PCA statistical technique suggested by both scientists could be exploited in one of the recently known algorithms that can convert multivariate data to be envisaged on a low-dimensional scale.

Dealing with small data sets facilitates, quickens, and simplifies data analysis via various algorithms without additional data processing procedures compared to large data sets. In PCA, the main idea is transforming a broad set of variables to a smaller set without losing the significant important information characterizing the original one. Of course, the data set simplification by reducing its content of variables is on account of the accuracy. However, the dimensionality reduction strategy is to achieve simplicity through the acceptance of somewhat lower accuracy.

PCA loadings include information indicating how the variables are related to each other. Typically, the first two to three principal component loadings elucidate most of the covariance among different samples. The PCs plot indicates the variables (spectral lines; consequently, the relevant elements) effectively impact the sample clustering.

Since its introduction to the scientific community in the early years of the last century, PCA has been widely applied successfully in different disciplines. Concerning the spectrochemical analysis researches, PCA attracted many applications. During the last 15 years, numerous papers have been published adopting PCA in LIBS data's statistical treatment. This was mainly because of modern, state-of-the-art spectroscopic techniques used in the experimental LIBS set-ups, which provide spectra no more as simple as in the early times of the LIBS researches. One experiment can provide thousands of complex LIBS spectra, especially with high repetition rate lasers and the Echelle spectrometers.

Besides adopting PCA to obtain a fast data visualization on a lower dimensional scale and to examine prominent data variables, PCA has been utilized in processing the mapping of the developed elements via LIBS. Variances between the patterns and their structures are the main interest of the PCA's data display, as an unsupervised method, i.e. not related to the investigated sample class or its composition neither qualitatively nor quantitatively. In unsupervised pattern recognition, the distances between different nonsimilar objects are large in multidimensional space. On the other hand, the distance between similar objects is small, facilitating classification between unknown objects.

5.3 PCA in Some LIBS Applications

PCA has been applied in the statistical treatment of LIBS data in several disciplines. This includes industrial, archaeological, geochemical, forensic applications, among others. Elaboration of some such applications will be presented in the following.

5.3.1 Geochemical Applications

It is advantageous to perform spectrochemical remote analysis of various geological samples such as rocks [25, 26] and mineral sediments [27]. LIBS is quite a suitable technique for such a task. The adoption of LIBS in geochemical and environmental samples analysis is well-known known as GEOLIBS [28]. El-Saied et al. [25] reported on the discrimination between some igneous and sedimentary rocks. They applied PCA (the three first principal components) in the statistical analysis of the IR-LIBS spectra. The results revealed good discrimination between different individual types within each species of rocks, such as sedimentary rocks, as shown in Figure 5.1, with a total variance of 92.4%. However, discrimination was not available between the major species, namely igneous and sedimentary [25]. The same group demonstrated that LIBS represents a rapid method for classifying the studied igneous and sedimentary rocks once the proper statistical tools are used, which was PCA in their work. It is worth mentioning that this group has used both conventional LIBS and

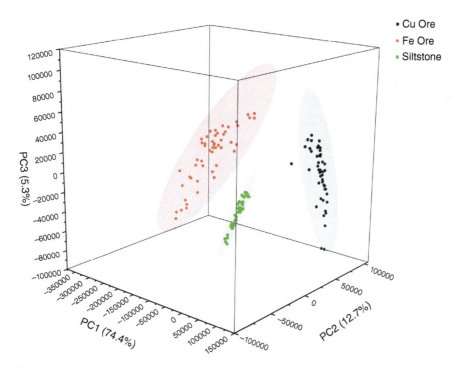

Figure 5.1 Principal components score plot of PC1, PC2, and PC3 for NELIBS of three types of sedimentary rocks. *Source:* From reference [25] / With permission of Elsevier.

nanoparticle-enhanced LIBS (NELIBS), and the PCA discrimination results were nearly the same in both cases.

Earlier, in 2007, Sirven et al. [28] exploited LIBS to analyze six rock types in the lab and used three chemometric methods, including PCA, for the spectroscopic data analysis. This experiment was in preparation for the Mars Science Laboratory (MSL) rover, launched in 2009, equipped with the ChemCam instrument that uses a LIBS system to investigate geochemical contents of Mars surface. The lab experiment's chosen rock types simulate the expected rocks on the planet (Basalt, Limestone, Gabbro, Obsidian, Trachyte, and Trachy-andesite). Figure 5.2 shows the first two PCs' PCA score plots to analyze the whole data set that accounted for 92.41% of the total variance. Rocks very much different in their composition and physical properties from the other rocks like limestone and obsidian were distinguishable from the other four rock types.

Moreover, the six rocks' obtained PCA analysis was sensitive to the composition and physical properties and the matrix where it was possible to separate trachyte and trachy-andesite. It separated basalt and gabbro partially, which have the same composition but not the matrix. The depicted loading plot in this work demonstrated the elements influencing the discrimination between different rock types, as shown in Figure 5.3. The authors concluded that PCA is a fast and accurate statistical method for itemizing and classifying rocks and determining the elements influencing their discrimination. However, they indicated the need for complementary techniques or further steps to classify rocks via the PCA automatically.

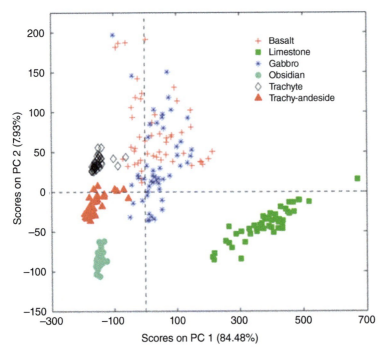

Figure 5.2 PCA score plot of spectra from six rock types. *Source:* From reference [28] / With permission of Royal Society of Chemistry.

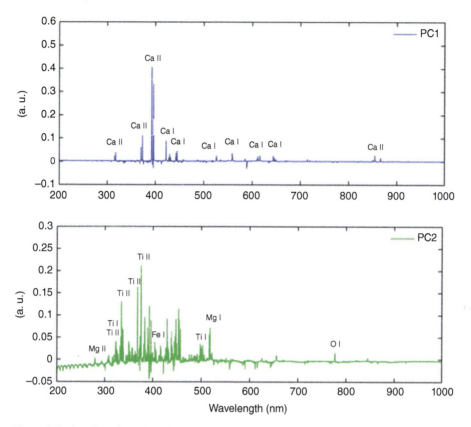

Figure 5.3 Loading plots of the first two principal components of the score plot shown in Figure 5.5. *Source:* From reference [28] / With permission of Royal Society of Chemistry.

Rai et al. [29] utilized LIBS and PCA to analyze and discriminate between sedimentary, metamorphic, and igneous rock samples. The presented score plot of the first two principal components, PC1, and PC2, explains 94% of the data set's total variance. The authors successfully interpreted the discrimination between different samples through the loading plots that indicate the major spectral lines influencing sample differentiation. This work demonstrated LIBS's potential, coupled with PCA in the characterization and classification of rock samples.

5.3.2 Food and Feed Applications

Many LIBS papers in food and feed have utilized PCA in the statistical analysis of the spectroscopic data. One of the interesting published works studied butter adulteration experimentally using LIBS and chemometrics [30]. High-dimensional spectral data of LIBS were analyzed statistically via PCA. The authors' goal was to discriminate between butter and margarine in addition to mixed samples, namely butter adulterated by margarine. Broad LIBS spectra extending from 186 to 900 nm have been analyzed using PCA. A well-separated clustering for each of the butter and margarine samples has been obtained in the obtained PCA score plot. The two first PCs explain the 97.92% variance among the data set. The authors presented loading plots for the wavelength versus PC1 and PC2, which depicted the elements spectral lines influencing the butter and margarine samples' PCA discrimination, especially Na, Ca, K, and Mg. The intensity differences of such spectral lines were decisive in the clustering of different sample types.

In this work, PCA's use to analyze the obtained complex LIBS spectra of both types of samples, and their mixtures were beneficial in their discrimination. It can be successfully applied to detect butter adulteration.

A work that demonstrated the possibility of utilizing LIBS with PCA for routinely in situ quality control of meat was presented by Bilge et al. [31]. In this work, the authors analyzed samples of different meat types: beef, pork, chicken and pork-beef, and chicken-beef mixtures. PCA has been used to discriminate between these species. Figure 5.4 depicts the principal component scores, where the two first principal components explain 96.3% variance among the data set. As shown from the figure, PC1 provides the maximum variance in the data set (83.37%), while PC2 gives the data set's rest variance (12.93%).

The obtained overall discrimination between the three meat types is generally good; however, beef and pork are better discriminated. The results achieved in this interesting work could be applied for field quality control inspections of meat, i.e. in slaughterhouses, supermarkets, and meat distribution centers. This can be performed using a portable LIBS system provided with suitable software for PCA analysis of the obtained spectra to have decisive discrimination and/or classification between different types of meat in a relatively short time compared to using other conventional techniques.

Abdel-Salam et al. [32] have exploited LIBS combined with PCA to evaluate sheep colostrum proteins compared to their mature milk. The goal of such a study was to support

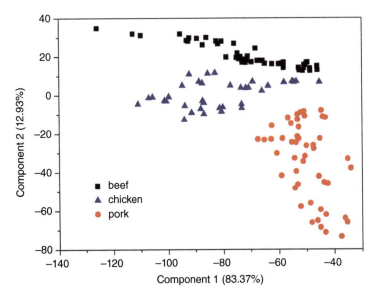

Figure 5.4 PCA score plot for the discrimination of pure beef, chicken, and pork meat species. *Source:* From reference [31] / With permission of Elsevier.

strategies of nutritional planning for newly born lambs. The authors considered the CN bands and the C spectral line at 247.8 nm in the LIBS spectra to indicate the colostrum and milk samples' proteins. Colostrum samples were obtained from 33 ewes, milked three times every 12 hours after birth. On the seventh day postpartum, the ewes were milked for mature milk representing the fourth sample. PCA has been utilized for the statistical analysis of the obtained LIBS spectra. Score plots of the first and second principal components, PC1 and PC2, were used to differentiate between colostrum and milk from their LIBS spectra. The variables were reduced by analyzing three wavelength ranges that only include the primary spectral lines influencing the discrimination process to improve the discrimination procedure. Such three wavelength ranges extended from 200 to 250 nm, 385 to 390 nm, and 392 up to 431 nm, including the carbon atomic line, the first CN bands, and many significant calcium lines. The first two PCs' score plots for the whole spectral range (200–750 nm) provided good discrimination between colostrum and milk, explaining 80.8% of the total variance. On the other hand, using the merged three limited spectral range provided a score plot with excellent discrimination representing 95.8% of the total variance.

The results obtained from this work demonstrate LIBS's potential combined with PCA in the discrimination between colostrum and milk. This could be very beneficial for feeding strategies in animal production farms, considering the possibility of performing such measurements in situ using the nowadays commercially available portable or mobile LIBS systems.

In [17], the authors utilized nanoparticle-enhanced laser-induced breakdown spectroscopy (NELIBS) and PCA to study the feeding effects on recent and ancient bovine bones. The work aimed to relate the elements content of both bovine bone types with the elements existing in the fodder that has been fed to the cattle during their life. Consequently, information about the environmental conditions, lifestyle, and agriculture could be disclosed at such ancient eras. Noble metals nanoparticles (Au, Ag, and Pt) have been used since about six years ago to enhance the LIBS technique's sensitivity [34–36]. In this work of Abdel-Salam et al., biosynthesized silver nanoparticles have been sprinkled onto the bone surface before performing the LIBS measurements. A pronounced improvement in the signal to noise ratio in the bone's NELIBS spectra has been achieved compared to the same samples' conventional LIBS spectra, as shown in Figure 5.5. The NELIBS spectra of the 4600 years ancient Egyptian cow were distinguishable from the recent cow bones' LIBS spectra. However, the authors used PCA as a multivariate statistical analysis technique to validate such discrimination between archaeological and contemporary bones. Besides, PCA has been used to analyze the spectra of different fodders.

In the PCA, 50 spectra have been analyzed for each bone type and each spectrum extended over the whole measured wavelength range (200–750 nm). Only the first principal components, PC1 and PC2, were sufficient to obtain the score plot shown in Figure 5.6. Distinct discrimination between ancient and contemporary bones has been obtained. The PC1 and PC2 represented 90.3% of the total variance of the data set. Of course, this was a satisfactory qualitative spectroscopic divergence between the two bone types, given their ages.

NELIBS spectra of clover, barley, and feed samples were also used to plot the two PCs. The obtained score plots for each type of fodders and the two bone types depict the bone samples' accumulation near each other, separated from the fodder's data points in each score plot. This means that PCA could not correlate fodders to the bones and even discriminate

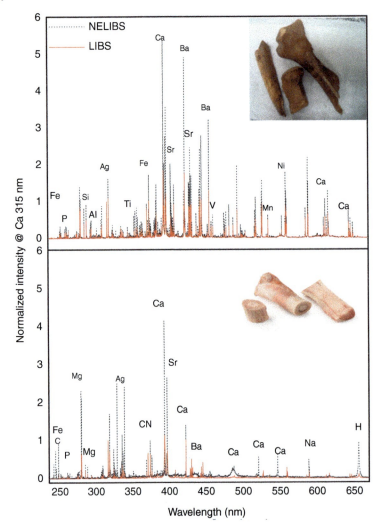

Figure 5.5 Typical LIBS and NELIBS spectra of the ancient (a) and recent (b) bovine bone. *Source:* From reference [17]/ CC BY-NC-ND 4.0 / ELSEVIER.

different fodders. The reason is that fresh fodders provided to cattle in the ancient Egyptian era contain many components such as clover and barley, utterly different from the composition of the artificial feed provided in the nowadays animal production farms. Hence, it was more reliable to rely on the NELIBS results for comparing the contents of the element of fodders and their correlation to each of the bone types.

5.3.3 Microbiological Applications

In 2003, Hybl et al. [37] exploited LIBS and PCA to successfully discriminate *Bacillus globigii* (BG) from fungus/mold spores, growth media, and ovalbumin. The authors used the first

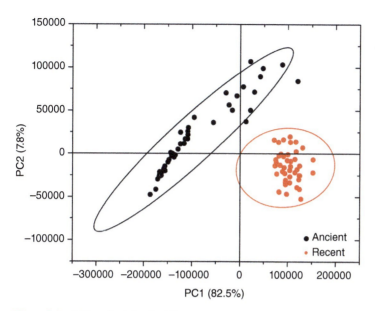

Figure 5.6 PCA analysis for the LIBS spectra of ancient and recent bovine bone (for the whole spectral range 200–700 nm). *Source:* From reference [17]/ With Permission of Elsevier.

three principal components to analyze the LIBS spectra. Each single-shot LIBS spectrum was restricted to the most thirty strong spectral lines to improve the discrimination procedure.

The score plot of the first three PCs, shown in Figure 5.7, demonstrates excellent discrimination between BG and the other biotypes. The authors justified the pollen data set's significant spread as being mainly due to the low signal to noise ratio of the pollen spectra. Hence, given the authors, this low S/N is relevant to the larger size of the pollen particles (20–60 μm), which results in complicated full aerosolizing and ablation. Clusters of the other samples are also spreading, but less than the pollen; the reason refers to the spectra reproducibility and internal variations in each class.

Prochazka et al. [38] presented a detailed study on the multivariate classification of bacteria from separate and merged LIBS and Raman spectra. The investigated samples were five *Staphylococcus* bacterial strains and one strain of *Escherichia coli*. This chapter focusses on the results of using LIBS and PCA only. The LIBS spectra obtained from bacteria cultured on Agar media were pretreated by normalization and filtering to reduce the fluctuations resulting from the matrix effect and the inhomogeneity of the samples surface and surface ripples affecting the distance between the lens and sample. Applying the PCA to the LIBS spectra, only the first two principal components PC1 and PC2 have been considered because of the higher PCs' insignificant contribution to the data discrimination. Full LIBS spectral range (200–900 nm) has been subjected to PC analysis. The scores plot shown in Figure 5.8 accounts for 18% of the data set's total variance, with PC1 = 16% and PC2 = 2%. Three bacterial strains were very well separated, while the other three were overlapped and clustered together.

5 Principal Component Analysis

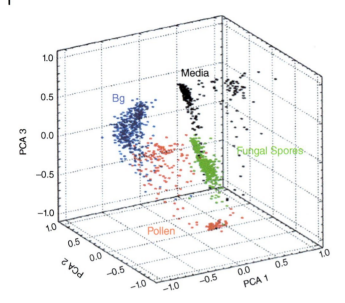

Figure 5.7 The score plot of the first three principal components of the studied samples' LIBS spectra. *Source:* From reference [37]/ With permission of Sage publicaions.

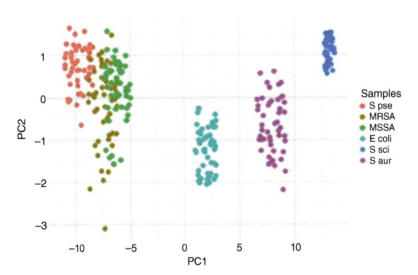

Figure 5.8 Score plot of PC1 vs. PC2 of LIBS data. The variance of PC1 is 16%, PC2 is 2%. *Source:* From reference [38]/ With permission of Elsevier.

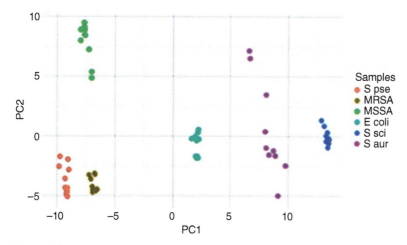

Figure 5.9 Score plot of PC1 vs. PC2 for merged LIBS and Raman spectra. The variance of PC1 is 25%, and of PC2 is 12.5%. *Source:* From reference [38] / With permission of Elsevier.

PC1 and PC2 loadings plot is depicted in Figure 5.9. The authors interpreted the shift from the zero level of the background in the PC1 loading to be due to the matrix effect. On the other hand, prominent calcium and sodium spectral lines influence the discrimination more than the trace elements, both in PC1 and PC2 loading plots.

This work demonstrated the utility of using chemometrics, namely PCA, combined with LIBS for discrimination and identification of bacterial strains.

5.3.4 Forensic Applications

In the field of nuclear forensic, Bhatt et al. [39] showed the prospects of using PCA in the analysis of LIBS spectra of mineral ores bearing trace levels of uranium that have been collected from different regions in Kenya. The first two principal components score plot depicted clusters of samples from three regions (South Ruri, Lake Magadi, and Coast). The authors have chosen certain spectral parts that include the prominent uranium spectral lines to facilitate and improve the discrimination between different samples, given their contents of elements. The score plot accounted for 95% of the data set variance, with PC1 65% and PC2 30%. The loading plots shown in Figure 5.10 point out that some elements, namely calcium, strontium, sodium, and titanium, in addition to other rare-earth elements, specially neodymium, scandium, samarium, promethium, terbium, and praseodymium, have the primary influence in attributing the measured uranium-bearing ores to their mineral mines regions in Kenya.

The authors mentioned that such LIBS-PCA results could help attribute unknown samples to their origin, using an established fingerprint library of uranium-bearing ores from various Kenya regions.

Imam et al. [40] discriminated between old and new Malaysian coins using LIBS and PCA. The samples were 10, 20, and 50 cents coins from each category. The spectra of the

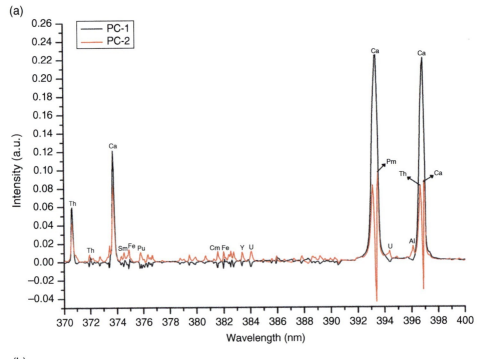

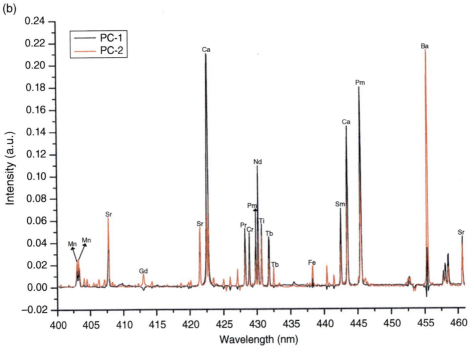

Figure 5.10 LIBS spectra of uranium trioxide in cellulose and pure cellulose (a) 340–370 nm and (b) 370–400 nm. *Source:* From reference [39]/ With permission of Royal Society of Chemistry.

old coins were exactly similar since they have almost the same elemental composition. Hence, discrimination between old coins of different values was not available via PCA. The first three principal components score plot depicted clustering of the data points for the 10, 20, and 50 cents old coins in nearly one group, without any possibility to discriminate between them. However, the authors obtained a perfect classification between old and new coins in the 3-D principal components score plot, as shown in Figure 5.11a. Of course, this was because of the difference between the alloying elements used to manufacture each of the two categories of coins.

Contrary to the old coins, it was possible to attain relatively good discrimination between new coins Figure 5.11b. The 20 and 50 cents (made of almost the same elements)

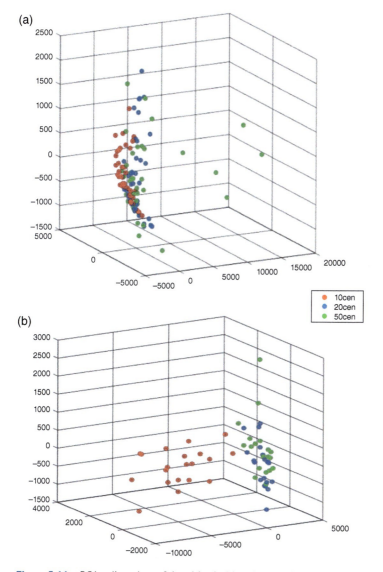

Figure 5.11 PC loading plots of the old coin (a) and new coins (b). *Source:* From reference [40] / Public Domain / CC BY 3.0.

accumulated in one cluster. In contrast, the 10 cent coin data points have clustered in a well-separated group.

Although the authors did not mention this, their work can be utilized in forensic applications to detect fake coins, whether old or new, rapidly.

5.4 Conclusion

Multivariate data analysis methodologies are indispensable and decisive for tackling statistically complex spectra such as those obtained in LIBS measurements. The vast amount of data included in one LIBS spectrum should be simplified with a significant reduction of the variables to ease the analytical procedure. PCA as an unsupervised multivariate technique has been proven to be one of the most beneficial and widely applied for the analysis of LIBS spectra. In this chapter, it has been demonstrated the successful application of PCA in numerous LIBS applications. The PCA facilitates discrimination, classification, and identification of different samples from their LIBS spectra.

References

1. D.A. Cremers, L.J. Radziemski, *Handbook of Laser-Induced Breakdown Spectroscopy*, 2nd edn., Wiley, Chichester, (2006).
2. S. Musazzi, U. Perini, *Laser-Induced Breakdown Spectroscopy, Theory and Applications*, Springer Series in Optical Sciences, **182**, Springer, Berlin, Heidelberg, (2014).
3. J.P. Singh, S.N. Thakur (Eds.), *Laser-Induced Breakdown Spectroscopy*, Elsevier, Amsterdam, (2007).
4. R. Noll, *Laser Induced Breakdown Spectroscopy Fundamentals and Applications*, Springer-Verlag, Berlin, Heidelberg, (2012).
5. A.W. Miziolek, V. Palleschi, I. Schecter, *Laser-Induced Breakdown Spectroscopy (LIBS) Fundamentals and Applications*, Cambridge University Press, New York, (2006).
6. M.A. Kasem, M.A. Harith, Laser-induced breakdown spectroscopy in Africa, *J. Chem.*, **648385** (2015), 1–10.
7. V.K. Singh, A.K. Rai, P.K. Rai, P.K. Jindal, Prospects for laser-induced breakdown spectroscopy for biomedical applications: a review, *Lasers Med. Sci.*, **24** (2009), 749–759.
8. A. Bertolini, G. Carelli, F. Francesconi, M. Francesconi, L. Marchesini, P. Marsili, F. Sorrentino, G. Cristoforetti, S. Legnaioli, V. Palleschi, L. Pardini, A. Salvetti, Modì: a new mobile instrument for in situ double-pulse LIBS analysis, *Anal. Bioanal. Chem.*, **385** (2) (2006), 240–247.
9. J. El Haddad, L. Canioni, B. Bousquet, Good practices in LIBS analysis: review and advices, *Spectrochim. Acta B*, **101** (2014), 171–182.
10. N. K. Rai, A.K. Rai, LIBS – an efficient approach for the determination of Cr in industrial wastewater, *J. Hazard. Mater.*, **150** (2008), 835–838.
11. S. Legnaioli, B. Campanella, F. Poggialini, S. Pagnotta, M.A. Harith, Z.A. Abdel-Salam, V. Palleschi, Industrial applications of laser-induced breakdown spectroscopy: a review, *Anal. Methods*, **12** (2020), 1014–1029.

12 R.S. Harmon, R.E. Russo, and R.R. Hark, Applications of laser induced breakdown spectroscopy for geochemical and environmental analysis: a comprehensive review, *Spectrochim. Acta B*, **87** (2013), 11–26.

13 G.S. Senesi, Portable hand held Laser-Induced Breakdown Spectroscopy (LIBS) instrumentation for in field elemental analysis of geological samples, *Int. J. Earth Environ. Sci.*, **2** (2017), 146–148.

14 O. Abdel-Kareem, M.A. Harith, Evaluating the use of laser radiation in cleaning of copper embroidery threads on archaeological Egyptian textiles, *Appl. Surf. Sci.*, **254** (18) (2008), 5854–5860.

15 A. Tonazzini, E. Salerno, Z.A. Abdel Salam, M.A. Harith, L. Marras, A. Botto, B. Campanella, S. Legnaioli, S. Pagnotta, F. Poggialini, V. Palleschi, Analytical and mathematical methods for revealing hidden details in ancient manuscripts and paintings: a review, *J. Adv. Res.*, **17** (2019), 31–42.

16 Z. Abdel-Salam, S.A.M. Abdel-Salam, M.A. Harith, Application of laser spectrochemical analytical techniques to follow up spoilage of white meat in chicken, *Food Anal. Methods*, **10** (2017), 102365–102372.

17 Z.A. Abdel-Salam, V. Palleschi, M.A. Harith, Study of the feeding effect on recent and ancient bovine bones by nanoparticle-enhanced laser-induced breakdown spectroscopy and chemometrics, *J. Adv. Res.*, **17** (2019), 65–72.

18 R. Gaudiuso, N. Melikechi, Z.A. Abdel-Salam, M.A. Harith, V. Palleschi, V. Motto-Ros, B. Busser, Laser-induced breakdown spectroscopy for human and animal health: a review, *Spectrochim. Acta B*, **152** (2019), 123–142.

19 Z.A. Abdel-Salam, S.A. Attala, M.A. Harith, Characterization of milk from mastitis-infected cows using laser-induced breakdown spectrometry as a molecular analytical technique, *Food Anal. Methods*, **10** (2017), 2422–2428.

20 G. Nicolodelli, J. Cabral, C.R. Menegatti, B. Marangoni, G.S. Senesi, Recent advances and future trends in LIBS applications to agricultural materials and their food derivatives: an overview of developments in the last decade (2010–2019). Part I. Soils and fertilizers, *Trends Anal. Chem.*, **105** (2019), 70–82.

21 G. Nicolodelli, J. Cabral, C.R. Menegatti, B. Marangoni, G.S. Senesi, Recent advances and future trends in LIBS applications to agricultural materials and their food derivatives: an overview of developments in the last decade (2010–2019). Part II. Crop plants and their food derivatives, *Trends Anal. Chem.*, **118** (2019), 453–469.

22 S. Giorgio, S. Senesi, Laser-induced breakdown spectroscopy (LIBS) applied to terrestrial and extraterrestrial analogue geomaterials with emphasis to minerals and rocks, *Earth Sci. Rev.*, **139** (2014), 231–267.

23 K. Pearson, On lines and planes of closest fit to systems of points in space, *Philos. Mag.*, **2** (11) (1901), 559–572.

24 H. Hotelling, Analysis of a complex of statistical variables into principal components, *J. Educ. Psychol.*, **24** (6) (1933), 417–441.

25 R.H. El-Saied., Z.A. Abdel-Salam, S. Pagnotta, V. Palleschi, M.A. Harith, Classification of sedimentary and igneous rocks by laser induced breakdown spectroscopy and nanoparticle-enhanced laser induced breakdown spectroscopy combined with principal component analysis and graph theory, *Spectrochim. Acta B*, **158**, (2019), 105622.

26 R.H. El-Saied, M. Abdelhamid, Z.A. Abdel Salam, M.A. Harith, Exploiting LIBS to analyze selected rocks and to determine their surface hardness based on the diagnostics of laser-induced plasma, *Appl. Phys. B Lasers Opt.*, **126** (2020), 10.

27 R.S. Harmon, J. Remus, N.J. McMillan, C. McManus, L. Collins, J.L. Gottfried, F.C. DeLucia, and A.W. Miziolek, LIBS analysis of geomaterials: geochemical fingerprinting for the rapid analysis and discrimination of minerals, *J. Appl. Geochem.*, **24** (2009), 1125.

28 J.-B. Sirven, B. Sallé, P. Mauchien, J.-L. Lacour, S. Maurice, G. Manhès, Feasibility study of rock identification at the surface of Mars by remote laser-induced breakdown spectroscopy and three chemometric methods, *J. Anal. At. Spectrom.*, **22** (12) (2007), 1471.

29 Ab.Kr. Rai, G.S. Maurya, R. Kumar, A.K. Pathak, J.K. Pati, Aw.K. Rai, Analysis and discrimination of sedimentary, metamorphic, and igneous rocks using laser induced breakdown spectroscopy, *J. Appl. Spectrosc.*, **83** (2017), 1089–1095 (Russian Original Vol. 83, No. 6, November–December 2016)

30 H.T. Temiza, B. Sezera, A. Berkkanb, U. Tamerb, H. Boyacia, Assessment of laser induced breakdown spectroscopy as a tool for analysis of butter adulteration, *J. Food Compos. Anal.*, **67** (2018), 48–54.

31 G. Bilge, H.M. Velioglu, B. Sezer, K.E. Eseller, I.H. Boyaci, Identification of meat species by using laser-induced breakdown spectroscopy, *Meat Sci.*, **119** (2016), 118–122.

32 Z.A. Abdel-Salam, S.A.M. Abdel-Salam, I.I. Abdel-Mageed, M.A. Harith. Evaluation of proteins in sheep colostrum via laser-induced breakdown spectroscopy and multivariate analysis, *J. Adv. Res.*, **15** (2019), 19–25.

33 A. De Giacomo, R. Gaudiuso, C. Koral, M. Dell'Aglio, G. Valenza, Perspective on the use of nanoparticles to improve LIBS analytical performance: nanoparticle enhanced laser induced breakdown spectroscopy (NELIBS), *J. Anal. At. Spectrom.*, **31** (2016), 1566–1573.

34 C. Koral, A. De Giacomo, Xi. Mao, V. Zorba, R.E. Russo, Nanoparticle enhanced laser induced breakdown spectroscopy for improving the detection of molecular bands, *Spectrochim. Acta B*, **125** (2016), 11–17.

35 F. Poggialini, B. Campanella, S. Giannarelli, E. Grifoni, S. Legnaioli, G. Lorenzetti, S. Pagnotta, A. Safi, V. Palleschi, Green-synthetized silver nanoparticles for nanoparticle-enhanced laser induced breakdown spectroscopy (NELIBS) using a mobile instrument, *Spectrochim. Acta*, **141** (2018), 53–58.

36 Z.A. Abdel-Salam, Sh.M.I. Alexeree, M.A. Harith, Utilizing biosynthesized nano-enhanced laser-induced breakdown spectroscopy for proteins estimation in canned tuna, *Spectrochim. Acta B*, **149** (2018), 112–117.

37 J.D. Hybl, G.A. Lithgow, S.G. Buckley, Laser-induced breakdown spectroscopy detection and classification of biological aerosols, *Appl. Spectrosc.*, **57** (2003), 1207–1215.

38 D. Prochazka, M. Mazura, O. Samek, K. Rebrošová, P. Pořízka, J. Klus, P. Prochazková, J. Novotný, K. Novotný, J. Kaiser, Combination of laser-induced breakdown spectroscopy and Raman spectroscopy for multivariate classification of bacteria, *Spectrochim. Acta B*, **139** (2018), 6–12.

39 B. Bhatt, K.H. Angeyo, A. Dehayem-Kamadjeu, LIBS development methodology for forensic nuclear materials analysis, *Anal. Methods*, **10** (2018), 791–798.

40 A.M. Imam, M.S. Aziz, K. Chaudhary, Z. Rizvi, J. Ali, LIBS-PCA based discrimination of Malaysian coins, *International Laser Technology and Optics Symposium (ILATOS 2017)*, Johor Bahru, Malaysia (26–28 September 2017), IOP Conference Series: *Journal of Physics*: Conference Series, 1027 (2018), 012012.

6

Time-Dependent Spectral Analysis

Fausto Bredice[1], Ivan Urbina[1], and Vincenzo Palleschi[2]

[1] Atomic Spectroscopy Laboratory, Centro de Investigaciones Ópticas (CIOp), La Plata, Argentina
[2] Applied and Laser Spectroscopy Laboratory, Institute of Chemistry of Organometallic Compounds, Research Area of National Research Council, Pisa, Italy

6.1 Introduction

The laser-induced breakdown spectroscopy (LIBS) technique is acquiring a great importance in science and technology. Since its introduction, exactly 40 years ago at the date of preparation of this book [1, 2], the researchers working in LIBS have struggled for improving its analytical performances. The LIBS technique has thus evolved both through the improvement of the experimental set-up and, mostly, by the development of new methods for the interpretation of the laser-generated spectra.

One aspect that has been somewhat overlooked, up to a few years ago, is the important information that is carried by the temporal evolution of the LIBS spectrum.

In laser-induced plasmas, the temporal evolution of the intensities of the spectral lines and the physical parameters such as temperature and electron density depend, among other factors, on the energy of the pulse, the wavelength of the laser, the region of the plasma observed, and the confinement conditions of the target. The determination of the atomic parameters used in plasma spectroscopy greatly simplifies if the following two conditions are assumed: thin plasma, this means that radiation reabsorption is negligible [3], and local thermodynamic equilibrium (LTE), which means that the electron/excitation/ionization temperatures and the electron number density, can be described through the Saha–Eggert relation and the Maxwell–Boltzmann distributions [4]. For these approximations to be valid, it is necessary to know how the actual sources may deviate from ideal behavior or when in time these approaches are valid. It should be considered that the transient nature of laser-induced plasmas and the presence of spatial gradients introduce nonlocal and time-dependent effects on populations of different energy levels; therefore, the thin plasma and LTE conditions must be verified in order to determine if it is possible to discard the self-absorption in the spectral lines and express their intensities through the Saha–Eggert relation and the Maxwell–Boltzmann distributions. Therefore, an adequate characterization of plasmas is essential to understand the mechanisms of the processes involved in their generation, evolution, and extinction. An extensive bibliography on this topic has been published [2–8].

The objective of this work is to show alternative methods for the analysis of transient plasmas such as those produced by laser pulses, and in which the temporal variation of the intensities of the spectral lines is considered as a means of analysis.

To this purpose, the spectra must be recorded at different times from the start of the plasma and with time windows narrow enough so that the plasma can be considered as stationary during that time interval. At the same time, the plasma must be also spatially homogeneous, so that the temperature variations within the different points of the observation region can be assumed to be negligible.

6.2 Time-Dependent LIBS Spectral Analysis

6.2.1 Independent Component Analysis

The independent component analysis (ICA) method belongs to the class of multiway data analysis algorithms, which can be applied when the data are represented as multidimensional arrays. ICA can be considered an evolution of principal component analysis (PCA) and, similarly to PCA, can be used for simplifying the analysis of LIBS spectra.

The highly dynamic nature of the laser-induced plasmas reflects in a strong temporal dependence of the corresponding optical emission. The spectral features of a LIBS spectrum evolve in time depending on the temporal evolution of the plasma parameters (size, temperature, atomic and electron number density), and different spectral components (continuum, ionic emission, neutral emission, and molecular emission) appear in the spectrum at different delays after the laser pulse and then evolve with different characteristic times.

The ICA approach separates the multidimensional signal under study (the LIBS spectrum acquired at different delays after the laser pulse) in statistically independent components, i.e. portions of the spectrum that temporally evolves in a similar way and differently with respect to the other components.

In ICA analysis, the temporal evolution of a LIBS spectrum can thus be represented as the linear combination of several independent components:

$$S(\lambda, t) = \sum_{i=1}^{n} S_i(\lambda) f_i(t) \tag{6.1}$$

The ICA algorithm is designed to decompose the complex temporal dependence of the LIBS spectrum $S(\lambda, t)$ in a series of n independent components $S_i(\lambda)$, whose temporal evolution is defined by the corresponding $f_i(t)$ functions. One of the most used numerical algorithms for decomposing the spectra in independent components is the fast ICA algorithm [5].

To obtain n independent components, the algorithm must be provided with at least n LIBS spectra taken at different delay times. The first to use ICA for modeling the temporal evolution of LIBS plasmas were El Rakwe et al. [6] in 2017, analyzing 21 spectra from a pure aluminum sample in the interval between 200 ns and 15 μs after the laser pulse. The authors determined three main independent components, which were associated to ionic aluminum (Al II), neutral aluminum (Al I) and aluminum oxide (AlO) emissions, with the first component (Al II lines) having a maximum at the delay of 200 ns, then decaying within 1 μs after the laser pulse, the second component (Al I lines) having its maximum after about 500 ns and disappearing after about 5 μs and the third (continuum + molecular emission) showing two maxima, one at short delays (200 ns), which is due to the continuum emission and the other after about 1 μs, which is due to the appearance of AlO emission lines in the LIBS spectrum at that delay.

Because of the different time scales of the optical emission processes involved, the application of ICA algorithms simplifies the analysis and interpretation of the LIBS spectra, by removing, for example, the contribution of the continuum emission, eliminating possible spectral interferences, and reducing the spectral noise by fitting the temporal evolution functions $f_i(t)$ with smooth functions (exponentials or combination of exponentials).

As an example of the application of the ICA in LIBS, we can consider a set of spectra acquired on pure silver at different delays (some of the spectra are shown in Figure 6.1).

From the results shown in Figures 6.2 and 6.3, we can observe that the neutral silver lines, mainly represented in IC 1 and IC 2 components, are present in the LIBS spectra for a longer time compared to IC 3 and IC 4 components, which are associated to Ag II ionic emission.

From Figure 6.3, we can also estimate the decaying time of these two groups of lines, which is of about 3.5 μs for neutral lines and 800 ns for ionic lines. It is interesting to note that the neutral and ionic contribution to the spectra are associated to two independent components, instead of one, because of the presence at early times of strong self-absorption effects [7] that disappear at longer delays.

The light emitted from the hot internal parts of the plasma travels to the cold exterior regions. The light can be absorbed by the same type of emitting atoms and molecules. This phenomenon is called self-absorption. In plasmas, self-absorption is one of the main causes of error in the measurements of intensities and line widths. Lines emitted by neutral or ionized species of the same element may be subject to this effect, depending on the value of their lower energy level and abundance in the plasma during the observation window. In addition to reducing the line intensity, self-absorption also generates a distortion of the line profile, which leads to a widening of the spectral line that can affect measurements based on linear relationships, such as the estimation of the electron number density from the Stark broadening of a given line [8]. In some spectra, if the plasma is not homogeneous, self-absorption can also result in the formation of an absorption dip at the top of the line profile. This effect is known as *self-reversal*.

The ICA algorithm recognizes this behavior and associates different components to the peak and to the wings of the self-reversed lines (Figure 6.4).

A statistical method similar to the ICA algorithm is the parallel factor analysis (PARAFAC), which has been recently proposed by Castro et al. [9] for the analysis of LIBS spectra of electronic waste taken at different depth under the surface. The same algorithm can be easily adapted to the study of the temporal evolution of LIBS spectra.

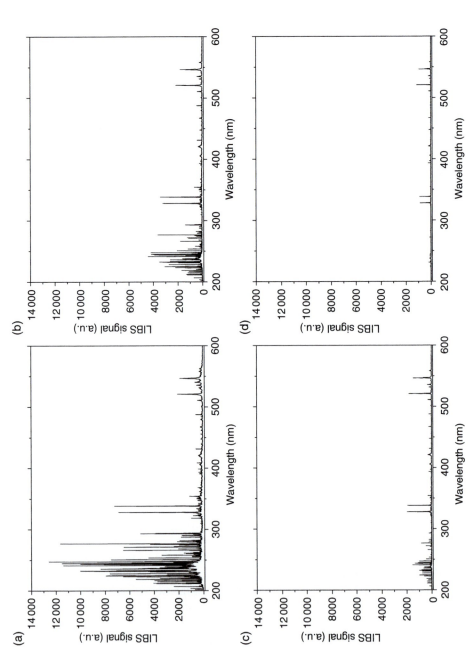

Figure 6.1 LIBS spectra on pure silver, acquired at different delay times after the laser pulse (a): 400 ns, (b): 1400 ns, (c): 2400 ns, (d): 4400 ns. The gate time was 2 μs.

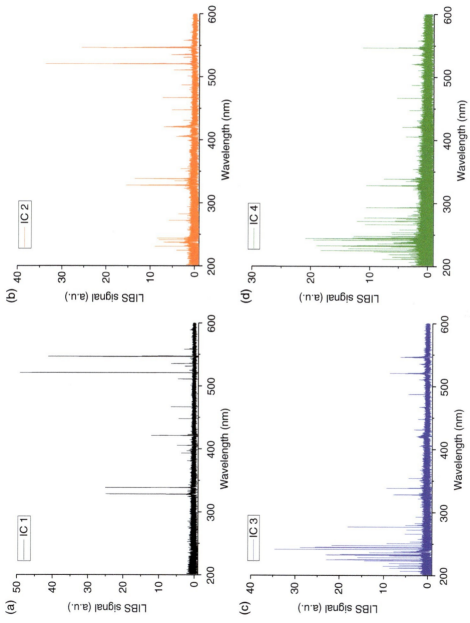

Figure 6.2 (a–d) The first four independent components of the LIBS spectrum of silver.

6 Time-Dependent Spectral Analysis

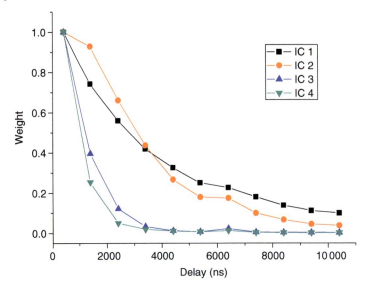

Figure 6.3 Time evolution of the first four independent components. For an easier comparison, the curves are normalized at their maximum (400 ns).

6.2.2 3D Boltzmann Plot

The ICA method described earlier has been developed for generic applications of multiway data analysis, and then adapted to LIBS research. A few years ago, however, Bredice et al. [10] proposed a method that takes full account of the physical aspects underlying the optical emission of laser-induced plasmas.

This method is known as 3D Boltzmann plot because the time axis is added to the conventional Boltzmann diagram in order to analyze the temporal evolution of the plasma spectrum.

During the time interval in which plasma can be considered under LTE conditions [4], its temperature is defined and can be determined, for example, through graphical methods called Boltzmann [11] or Saha-Boltzmann [12] plot.

In a Boltzmann plot, the plasma electron temperature is obtained from the slope of the best fit line representing the dependence of $\ln(I_j/g_j A_{jl})$ (with I_j = integral intensity of the line of a given species and g_j and A_{jl} degeneracy and transition probability of the transition) on the energy of the upper level of the corresponding transition.

For transient plasmas, as the LIBS plasmas are the slope changes over time depending on the experimental conditions at which the plasma was generated.

The variation in time of the Boltzmann diagram describes a surface in the space defined by the coordinates $X = Energy$, $Y = Time$, and $Z = \ln(I_j/g_j A_{jl})$. Hereafter, we will refer to this surface as "Boltzmann surface," as schematically shown in Figure 6.5.

This surface is determined by the temporal evolution of $\ln(I_n/g_n A_{nl})$ for each line; we will see in the following that it can be fitted using a reduced number of constants B_i and δ_i. In this way, for example, it is possible to describe the temporal evolution of the plasma temperature as a function the constants B_i and δ_i [13]. Another way to define the Boltzmann plane is to

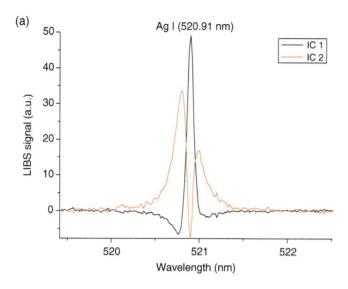

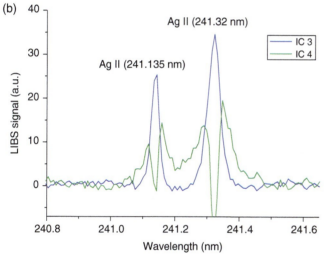

Figure 6.4 Strongly self-reversed Ag I lines at 520.91 nm (a) and Al II at 241.135 nm and 241.32 nm (b). The ICA algorithm recognizes the different behavior in time of the peak and the wings of these lines, associating them to different independent components.

define the Z axis as $\ln(I/I_0)$, where I_0 represents the integrated intensity of the line at a time $t = t_0$. In this way, the relative temporal evolution of the intensity of a line can be determined without knowing the transition probability and the degeneracy of the upper level of the transition.

6.2.2.1 Principles of the Method

The 3D Boltzmann plot method is based on the assumption that the plasma is optically thin and in LTE in the time interval considered.

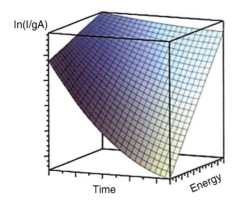

Figure 6.5 Schematic representation of a Boltzmann surface.

Under these conditions the integrated intensity of a spectral line, at each time t can be expressed as

$$I_{\lambda_n}(t) = F(\lambda_n) g_n A_{nl} N(t) \frac{\exp(-E_n/kT(t))}{U(T(t))} \qquad (6.2)$$

where λ_n is the central wavelength of the transition, A_{nl} is the transition rate from level n to level l, E_n and g_n are the energy and the statistical weight of level n, N is the total density number of emitting atoms, $U(T)$ is the atomic species partition function at temperature T, k is the Boltzmann constant, and $F(\lambda)$ is a factor that takes into account the spectral efficiency of the system.

If a laser-induced plasma (LIP) is in LTE, the temperature can be determined for example by the Boltzmann plot. That is, in the space defined by the coordinates $y = \ln(I/gA)$ and E (energy), Eq. (6.2) represents a straight line, and its slope is proportional to $-1/kT$

$$y = aE + b$$

$$y = \ln\left(\frac{I_{\lambda_n}(t)}{F(\lambda_n g_n A_{nl})}\right) \quad a = \frac{-1}{kT(t)}; \quad b = \ln\left(\frac{N(T(t))}{U(T(t))}\right)$$

The integrated intensity of a spectral line can be written as

$$I_n(t) = I_{0n} \exp\left(\sum_{i=1}^{S} b_i^n (t-t_0)^i\right) \qquad (6.3)$$

where

$$\sum_{i=1}^{S} b_i^n (t-t_0)^i$$

is a polynomial of degree S and I_{0n} is the intensity of the line at a time $t = t_0$. Without loss of generality, we can take $t_0 = 0$, then

$$\sum_{i=1}^{S} b_i^n t^i = \ln\left(\frac{I_n(t)}{I_{0n}}\right) \qquad (6.4)$$

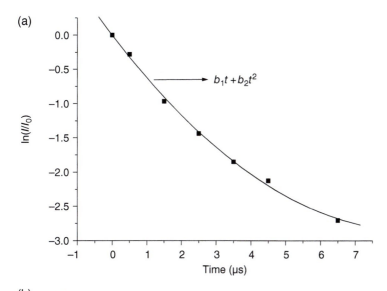
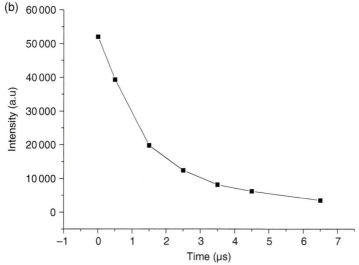

Figure 6.6 Temporal dependence of the integrated intensity of the Cu I line at 515.32 nm (a) and $\ln(I(t)/I_0)$ (b) for the same line.

The number of constants b_i^n is equal to the degree of the polynomial used to fit Eq. (6.3). As an example, Figure.6.6 shows the temporal evolution of the Cu I line at 515.32 nm, together with the temporal evolution of $\ln(I(t)/I_{on})$ of the same line.

In this case, two constants, b_1^n and b_2^n are necessary to fit the temporal dependence of $\ln(I(t)/I_{on})$

On the other hand, we have from Eqs. (6.2) and (6.4) that

$$\sum_{i=1}^{S} b_i^n t^i = \ln\left(\frac{I_n(t)}{I_n(0)}\right) = E_n\left(\frac{1}{kT(0)} - \frac{1}{kT(t)}\right) + \ln\left(\frac{N(t)U(T(0))}{N(0)U(T(t))}\right) \qquad (6.5)$$

Rewriting Eq. (6.5) for another spectral line and subtracting the two quantities, we have

$$\sum_{i=1}^{S} b_i^n t^i - \sum_{i=1}^{S} b_i^k t^i = (E_n - E_k)\left(\frac{1}{kT(0)} - \frac{1}{kT(t)}\right) \tag{6.6}$$

Eq. (6.6) can be put in the form:

$$\frac{\sum_{i=1}^{S}(b_i^n - b_i^k)t^i}{E_n - E_k} = \left(\frac{1}{kT(0)} - \frac{1}{kT(t)}\right)$$

$$\therefore \frac{\sum_{i=1}^{S}(b_i^n - b_i^k)t^i}{E_n - E_k} = \frac{\sum_{i=1}^{S}(b_i^j - b_i^l)t^i}{E_j - E_l} \tag{6.7}$$

The coefficients corresponding to any pair of spectral lines of the same species must fulfil Eq. (6.7). The unique solution that fulfils Eq. (6.7) for any set of lines k, n, j, l is

$$\frac{b_1^n - b_1^k}{E_n - E_k} = \frac{b_1^j - b_1^l}{E_j - E_l} = B_1 \quad \frac{b_2^n - b_2^k}{E_n - E_k} = \frac{b_2^j - b_2^l}{E_j - E_l} = B_2 \quad \ldots\ldots \quad \frac{b_S^n - b_S^k}{E_n - E_k} = \frac{b_S^j - b_S^l}{E_j - E_l} = B_S$$

$$\therefore \sum_{i=1}^{S} B_i t^i = \left(\frac{1}{kT(0)} - \frac{1}{kT(t)}\right)$$

Thus, for any generic line x, the constants B_i are given by

$$B_i = \frac{b_i^x - b_i^R}{E_x - E_R} \tag{6.8}$$

Then,

$$b_i^x = B_i(E_x - E_R) + b_i^R$$

$$\therefore \sum_{i=1}^{S} b_i^x \cdot t^i = \ln\left(\frac{I_x(t)}{I_{0x}}\right) = \sum_{i=1}^{S}\left(B_i(E_x - E_R) + b_i^R\right) \cdot t^i \tag{6.9}$$

The superscript R represents any arbitrary line of the same species that is taken as a reference. It is convenient to choose as reference line one that is free from self-absorption and does not interfere with other spectral lines.

If the coefficients b_i^x are plotted in function of $E_x - E_R$ according to Eq. (6.8), they must be aligned in a straight line with slope B_i.

In many cases, only the b_1 coefficients must be determined because a linear relation is often sufficient to fit well the graph of $\ln(I(t)/I(t_0))$ at all the times considered. A typical graph of b_1 vs. energy is represented in Figure 6.7.

In the few cases where the adjustment of $\ln(I(t)/I(t_0))$ requires the use of the constants b_2, according to Eq. (6.8), they are also arranged on a straight line, as a function of the upper energy of the transitions. A typical graph of b_2 vs. energy is represented in Figure 6.8.

Taking into account Eq. (6.5), if E_x tends to zero, then

$$E_x \to 0 \therefore \sum_i b_i^x t^i \to \ln\left(\frac{N(t)U(T(0))}{N(0)U(T(t))}\right)$$

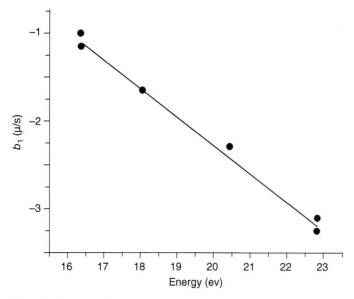

Figure 6.7 Typical graph of b_1 vs. energy.

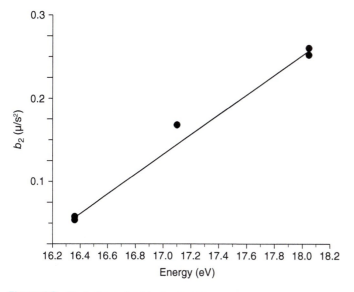

Figure 6.8 Typical graph of b_2 vs. energy.

We can define

$$\ln\left(\frac{N(t)U(T(0))}{N(0)U(T(t))}\right) = \sum_{i=1}^{S}\delta_i t^i \qquad (6.10)$$

and from Eqs (6.9) and (6.10), we have

$$\sum_{i=1}^{S}\delta_i t^i = \sum_{i=1}^{S}(b_i^R - B_i E_R)t^i = \ln\left(\frac{N(t)U(T(0))}{N(0)U(T(t))}\right)$$

and then

$$\sum_{i=1}^{S} b_i^x t^i = \sum_{i=1}^{S}(B_i E_x + \delta_i)t^i$$

According to Eq. (6.8), for each set of coefficients b_i corresponding to the same species, a graph can be constructed (b_i vs. $E-E_R$) in which, as previously stated, they must be aligned on a straight line. The δ_i coefficients can be obtained from the intersection of these lines with the y-axis at $E = 0$

$$\therefore I_n(t) = I_{0n}\exp\left(\sum_{i=1}^{S}(B_i E_n + \delta_i)t^i\right) \qquad (6.11)$$

Therefore, the Boltzmann surface in the space defined by the coordinates $X =$ Energy, $Y =$ Time, and $Z = \ln(I_n/I_{0n})$ is defined by the equation

$$\ln\left(\frac{I_n(t)}{I_{0n}}\right) = \sum_{i=1}^{S}(B_i E_n + \delta_i)t^i \qquad (6.12)$$

$B_1, ..., B_S$ and $\delta_1, ..., \delta_S$ represent the characteristic constants of the plasma.

From the knowledge of these constants, it is possible to determine the values of b_i of any line if its upper energy level is known, and therefore to obtain all the information about its temporal evolution.

It must be noted that the errors in the measurements of the spectral line intensities at different times due to the uncertainties on the spectral response of the detector or other factors that are independent of time can be considered as a multiplicative constant of these intensities and therefore do not affect the values of the coefficients b_i^n and, consequently, of the constants B_i.

The 3D Boltzmann plot approach can be especially useful in those cases, where the spectral line information is not fully available for analytical purposes.

This procedure has a direct application for the determination of the LTE conditions in multielemental plasmas and in the determination of the time interval during which this situation occurs. Indeed, if the multielemental plasma is in LTE conditions, it means that the temporal evolutions of the $\ln(I/I_0)$ of each line belonging to the different elements present in the sample should stay on the Boltzmann surface.

At the same time, after the determination of the constants B_i and δ_i that define the Boltzmann surface, it is possible to calculate the degree of departure of a spectral line with respect to this ideal plane.

6.3 Applications

6.3.1 3D Boltzmann Plot Coupled with Independent Component Analysis

In ref. [13], Bredice et al. applied the main ideas of the 3D Boltzmann plot and ICA for the determination of the electron temperature of a 60% Cu–40% Zn plasma without calculating the intensities of the lines under study and without using any information about the transition probabilities and degeneracies of the corresponding transitions.

The procedure for determining the plasma temperature involved the acquisition of LIBS spectra at 1, 3, and 5 µs after the laser pulse (gate time of the spectrometer 250 ns). After that, the ICA algorithm was used for determining the first three independent components; the third was not contributing too much to the spectra, while the first two were associated to ionic and neutral emission, respectively.

At this point, the authors realized that the time evolution of $\ln(I(t)/I(t_0))$ for the two independent components IC1 and IC2 was well described by a single exponential decay in the range of delays of interest (1–5 µs), a fact which allowed a quick calculation of the B factor from Eq. (6.12). In this case, only a single factor is needed when $\ln(I(t)/I(t_0))$ decreases linearly with the delay time.

The authors of results of the [13] reported that the results obtained for the plasma electron temperature obtained using the ICA + 3D Boltzmann plot procedure were in an excellent agreement with the ones derived by the application of a conventional Boltzmann plot. Most of all, the ICA + 3D Boltzmann plot method did not require the determination of the intensities of the spectral lines used neither the knowledge of the transition probabilities of the transitions.

6.3.2 Analysis of a Carbon Plasma by 3D Boltzmann Plot Method

In ref. [10], with the aim of analyzing the existence of LTE conditions in a carbon plasma, the LIBS spectra were recorded at low pressure (10^{-5} Torr) and at different distances ($\Delta x = 1$, 5, and 10 mm) from the target.

Figure 6.9 shows the time evolution of the integrated intensity of the 723.13 nm line belonging to the spectrum of C II, collected at $\Delta x = 1$ mm from the target together with the logarithm of the normalized intensities. In this case, the line intensity decreases monotonously with time, whereas for larger distances, see Figure 6.10 and Figure 6.11, the intensity shows a maximum that corresponds to the time at which most of the emitting species reach the observation point.

At each position, the graphs of $\ln(I(t)/I_0)$ vs. $(t - t_0)$ for each analyzed spectral line can be fitted, according to Eq. (6.4), by a polynomial of the form

$$y = \sum_{i=1}^{S} b_i^n t^i \tag{6.13}$$

For $\Delta x = 5$ mm and $\Delta x = 10$ mm, an excellent fit of $\ln(I(t)/I_0)$ was obtained with a polynomial of degree 3; therefore, a set of coefficients b_1^n, b_2^n, and b_3^n was obtained for each spectral line and for each distance analyzed. This means that three graphs corresponding to b_1, b_2, and b_3 can be constructed. If a number r of spectral lines is used in the analysis, this

6 Time-Dependent Spectral Analysis

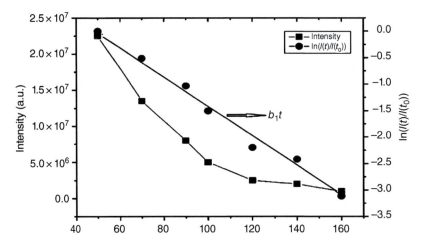

Figure 6.9 Time evolution of the integrated intensity of 723.13 nm line collected at $\Delta x = 1$. The logarithm of the normalized intensities is also shown.

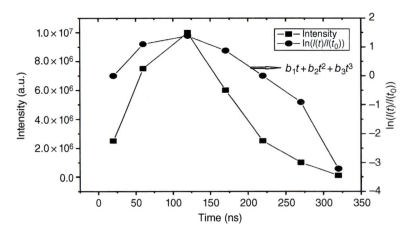

Figure 6.10 Time evolution of the integrated intensity of 723.13 nm line collected at $\Delta x = 5$. The logarithm of the normalized intensities is also shown.

means that each graph will contain r points. In Ref. [10], six lines of C II lines were used whose highest energy levels are in the range of approximately 16–22 eV. These lines were chosen because they are isolated and free from interactions with other lines, so they are good candidates for applying this procedure. In the case of $\Delta x = 1$ mm, only the coefficient b_1 was determined because a linear polynomial was sufficient to fit the dependence of $\ln(I(t)/I(t_0))$ vs. energy.

It is important to note that for the different distances to the target, the graphs of b_i as a function of energy always fit a straight line quite well and therefore the corresponding B_i could also be determined. This fact means that a Boltzmann plot can be constructed at any

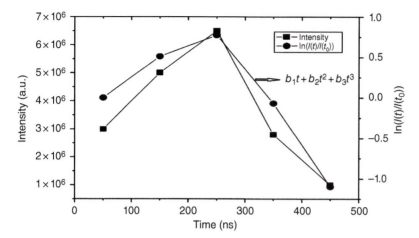

Figure 6.11 Time evolution of the integrated intensity of 723.13 nm line collected at $\Delta x = 10$. The logarithm of the normalized intensities is also shown.

instant in the time interval of the analysis and, therefore, the plasma can be considered to be in LTE conditions.

According to Eqs (6.7) and (6.8)

$$\frac{\sum_{i=1}^{S}\left(b_i^n - b_i^k\right)t^i}{E_n - E_k} = \left(\frac{1}{kT(0)} - \frac{1}{kT(t)}\right) = \sum_i B_i t^i \qquad (6.14)$$

Then,

$$kT(t) = \frac{kT(0)}{1 - kT(0)\sum_{i=1}^{S} B_i t^i}$$

This means that the plasma temperature can be determined at each time knowing the initial temperature $T(0)$ and the constants B_i. As it was previously pointed out, the uncertainty in the determination of the intensities of the spectral lines due to the spectral response of spectrometer or other wavelength-dependent, but otherwise time-independent, factors do not affect the value of B_i constant. Once the B_i constants are known, the only requirement for the correct evaluation of Eq. (12) is an accurate determination of kT_0.

On the other hand, if the spectral efficiency of the spectrometer is known, the 3D Boltzmann plot method offers a way to determine $kT(0)$, too.

In fact, the ratio of the intensities of two spectral lines is, according to Eq. (6.2):

$$\frac{I_{kj}(t)}{I_{lm}(t)} = \frac{g_k A_{kj}}{g_l A_{lm}} \exp\left[-(E_k - E_l)\frac{1}{kT(t)}\right]$$

According to Eq. (6.14) and assuming that the recording system is calibrated in intensity (the dependence on the wavelength of the $F(\lambda_n)$ factor is known and can included in the evaluation of the line intensity), we have that

$$\frac{I_{kj}(t)}{I_{lm}(t)} = \frac{g_k A_{kj}}{g_l A_{lm}} \exp\left[-(E_k - E_l)\left(\frac{1}{kT(0)} - \sum_{i=1}^{S} B_i t^i\right)\right]$$

Then,

$$\frac{1}{kT(0)} = \frac{\ln\left[\frac{g_l A_{lm} I_{kj}(t)}{g_k A_{kj} I_{lm}(t)}\right]}{(E_l - E_k)} + \sum_{i=1}^{S} B_i t^i \tag{6.15}$$

According to Eq. (6.15), if the constants B_i are known, it is possible to determine the plasma temperature at $t = 0$ from the line intensities obtained for $t > 0$. The product of the number of delays at which the LIBS spectrum was recorded multiplied by the number of pairs of lines that can be included within Eq. (6.15) is large, and this means that, in principle, a good precision can be achieved in the determination of $kT(0)$.

Figure 6.12a–c represent the temporal variations of the temperature of the carbon plasma at the distance of $\Delta x = 1$ mm, $\Delta x = 5$ mm, and $\Delta x = 10$ mm from the target, as obtained from the 3D Boltzmann plot method.

In all these cases, the C II lines at 658.288, 657.805, 723.132, and 723.642 nm were used in the Boltzmann plot procedure.

It is important to note that in this procedure, the existence of B_2 greater than zero for the spectra that are recorded on positions from which the intensity of their spectral lines begins to decline once the laser pulse has ended, for example, on the target surface, limits the validity time of the LTE condition. Since B_1 is always less than zero, the predicted plasma temperature should reach a minimum value and then it would grow again, which is obviously impossible. The time in which the derivative of temperature with respect to time becomes zero can be considered as the maximum time that this method can be applied. It can also be considered as the time from which the LTE conditions are not met. According to Eq. (6.14), the derivative of temperature with respect to time becomes zero for $t = -B_1/2B_2$.

It is interesting to note that according to Eq. (6.14), if two plasmas are generated in the same target (same B_i) but at different temperatures T_1 and T_2, with $T_1 > T_2$, when the hottest plasma reaches the temperature T_2 from that point, its temporal evolution it will be exactly the same as the coldest plasma. This fact is valid only in the case in which the evolution of the plasma is described by a single constant $B(B_1)$. The presence of the constants b_2 in the adjustment of the curves $\ln(I/I_0)$ and therefore $B_2 \neq 0$ indicates that the speed with which the intensities of the lines decrease in time decreases as time passes. The constants b_2 are always positive and at long delay times their weight within the adjustment polynomial begins to be important. It can be assumed that this slowdown in the speed of extinction of the spectral lines is due to the repopulation of the energy levels with electrons from ion recombination. According to Saha–Eggert's equation, the number of ions generated in the plasma strongly depends on the temperature reached and the ionization limit of the target. Possibly for these reasons, the presence of the constants b_2 depends on each experiment.

The relative error in the plasma temperature due the error in the determination of B_i constant is giving by (Figure 6.13)

$$\frac{\Delta T(t)}{T(0)} = \frac{kT(0)\sum_{i=1}^{S}|\Delta B_i t^i|}{\left(1 - kT(0)\sum_{i=1}^{S} B_i t^i\right)^2} \tag{6.15}$$

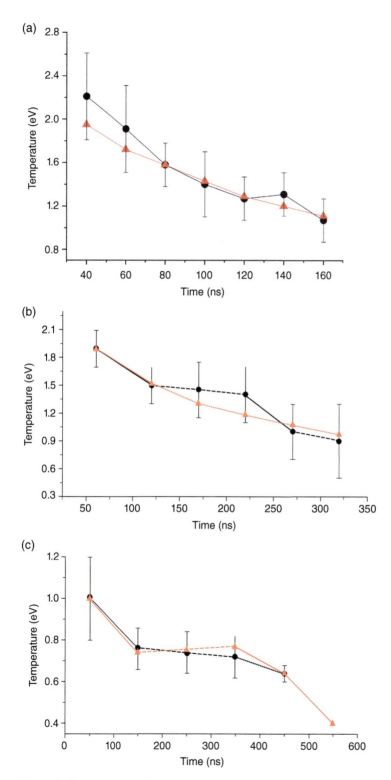

Figure 6.12 Variation of the carbon plasma temperature at (a) $\Delta x = 1$ mm, (b) $\Delta x = 5$ mm and (c) $\Delta x = 10$ mm. The black circles are the values obtained with the traditional Boltzmann plot.

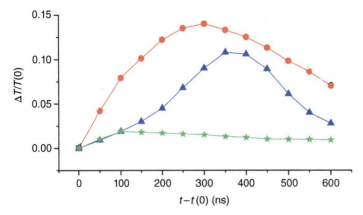

Figure 6.13 Relative errors in the plasma temperature according to Eq. (6.15) corresponding to the spectra of CII recorded at 1 mm (green stars), 5 mm (blue triangles), and 10 mm (red circles) from the target.

It is important to note that both in the traditional Boltzmann plot as in this procedure, the errors in the calculation of the temperatures are due to uncertainty on the linear fit of the graphs of the $\ln(I/gA)$ or the b_i^x vs. energy. In both cases, the error in the fit can be reduced by increasing the number of lines considered, and in particular those with the largest difference between their upper energy levels.

6.3.3 Assessment of the LTE Condition Through the 3D Boltzmann Plot Method

Another example on the use of the 3D Boltzmann plot method for determining the existence of LTE in the analysis of a four-element alloy (an amalgam used for dental treatments), whose composition was Hg (50%), Ag (23%), Cu (12%), and Sn (15%) (in weight).

The LIBS spectra were recorded at delay times from 1 to 5 μs, with increments of 0.5 μs. It should be noted that for Hg lines, the effects of self-absorption can be considered negligible (excluding the resonant line at 253.65 nm), but the more intense lines belonging to the other elements of the amalgam are affected by self-absorption. In this case, the lines intensities were corrected for self-absorption using the procedure developed by the Pisa group in 2005 [14].

Figures 6.14 and 6.15 show the values of the constants b_1 and b_2 corresponding to emission lines belonging to the different materials in the amalgam.

The values of the b_1 and b_2 constants obtained from the best fit of the variation of relative intensity of the lines in the interval between 1 and 5 μs, lies on a straight line. Therefore, it is possible to construct a Boltzmann plot, defining the plasma temperature at each instant in that time interval, which indicates that the plasma is in LTE conditions.

6.3.4 Evaluation of Self-Absorption

The self-absorption coefficient SA for an homogenous plasma rod of length l is defined as the quotient between of the intensity of the emission line at its maximum $I_p(\lambda_0)$ (in counts

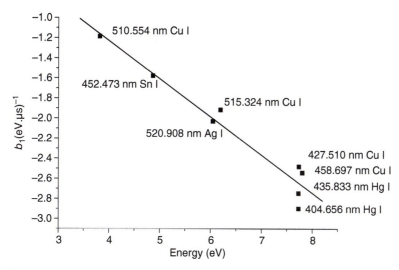

Figure 6.14 b_1 constants vs. energy for the different elements in the plasma.

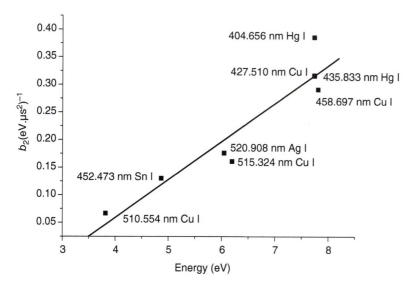

Figure 6.15 b_2 constants vs. energy for the different elements in the plasma.

per second), over the value $I_{p0}(\lambda_0)$ (where $I_{p0}(\lambda_0)$ is obtained, for example, extrapolating the corresponding curve of growth to the optically thin limit) [15].

$$SA = \frac{I_p(\lambda_0)}{I_{p0}(\lambda_0)} = \frac{(1 - \exp(-k(\lambda_0)l))}{k(\lambda_0)l} \tag{6.16}$$

where $k(\lambda_0)$ is the absorption coefficient defined as

$$k(\lambda_0) = \frac{(\lambda_0)^4}{8\pi c} A_{nk} g_n N_0 \frac{\exp(-E_k/KT)}{U(T)} \frac{1}{\Delta \lambda_0} \qquad (6.17)$$

where N_0 is the number density of the emitting species, λ_0 is the central wavelength of the line, $\Delta\lambda_0$ is the full width at half maximum (FWHM) of the emission line, and E_k is the lower energy level of the considered transition.

As demonstrated by the Pisa group in 2015 [14], the relationship between the integrated intensity of a self-absorbed line $I(\lambda)$ with its optically thin limit $I^*(\lambda)$, scale as

$$\frac{I(\lambda)}{I^*(\lambda)} = \frac{\int I_p(\lambda) d\lambda}{\int I_{p0}(\lambda) d\lambda} = (SA)^\beta \qquad (6.18)$$

with $\beta = 0.46$.

In the above-mentioned work, it is also shown that the FWHM of the Lorentzian profile lines, such as the typical lines generated in the laser produced plasmas, scales as

$$\Delta\lambda = \Delta\lambda_0 (SA)^\alpha$$

with $\alpha = -0.54$.

Assuming, as is often appropriate in laser-plasma applications, that the main source of broadening of the spectral lines is given by the Stark effect, we can assume that $\Delta\lambda_0 \approx 2 w_s n_e$, where n_e is the electron number density, and w_s is the Stark coefficient of the transition [8].

Several years ago, two of the authors of this work (FB and VP) presented a new method to determine the self-absorption coefficients of emission lines in laser-induced breakdown spectroscopy experiments [16]. This method does not need to knowledge of the parameters of the plasma such as temperature and electron density and of the spectral coefficients of the emission line such as transition probability, lower level energy, and degeneracy of the upper and lower transition levels. The author proposed, instead, to consider the logarithm of the relationship between the integral intensities of the same emission line at two different temperatures T_a and T_b. Considering the plasma in local thermal equilibrium, in the absence of self-absorption, there would be a linear dependence between this logarithm and the energy E_k of the upper level of the transition (Figure 6.16). In other words, different emission lines of the same species would be aligned along a straight line with a slope:

$$\begin{aligned} Tg\alpha &= \frac{\ln((I_{0_2}(T_a))/I_{0_2}(T_b)) - \ln((I_{0_1}(T_a))/(I_{0_1}(T_b)))}{(E_2 - E_1)} \\ &= \frac{E_2((1/T_b)-(1/T_a)) - E_1((1/T_b)-(1/T_b))}{k(E_2 - E_1)} = \frac{1}{k}\left(\frac{1}{T_b} - \frac{1}{T_a}\right) \end{aligned} \qquad (6.19)$$

Putting $E_1 = 0$, we obtain

$$Tg\alpha = \frac{\ln((I_0(T_a))/(I_0(T_b)))}{E} = \frac{1}{K}\left(\frac{1}{T_b} - \frac{1}{T_a}\right) \qquad (6.20)$$

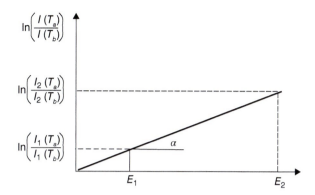

Figure 6.16 Linear plot of the logarithm of the intensity ratios vs. E_k.

If an emission line is self-absorbed, the logarithm of the ratio of the measured intensities must be corrected taking into account the value of the self-absorption parameter at the temperatures considered:

$$\frac{\ln\left(I(T_a)/I(T_b)(((SA)(T_b))/((SA)(T_a)))^\beta\right)}{E} = \frac{1}{K}\left(\frac{1}{T_b} - \frac{1}{T_a}\right) \qquad (6.21)$$

If at the temperature T_a the term $kl \ll 1$, the plasma becomes optically thin for this transition and $SA(T_a) \approx 1$
Then,

$$\frac{\ln\left((I(T_a)/I(T_b))((SA)(T_b))^\beta\right)}{E} \approx \frac{1}{K}\left(\frac{1}{T_b} - \frac{1}{T_a}\right) \qquad (6.22)$$

From which the calculation of SA is straightforward.

However, if a nonself-absorbed line of the same species is available, Eq. (6.21) can be simplified using the relation

$$\frac{\ln\left(I_R(T_a)/I_R(T_b)\right)}{E_R} = \frac{1}{K}\left(\frac{1}{T_b} - \frac{1}{T_a}\right) \qquad (6.23)$$

where $I_R(T_a)$ and $I_R(T_b)$ are the integrated intensities of this "reference" line at the temperatures T_a, and T_b and E_R are the energy of the upper level of the corresponding transition.
Substituting Eq. (6.23) in Eq. (6.21) gives:

$$\frac{\ln\left((I(T_a)/I(T_b))((SA)(T_b))^\beta\right)}{E} \approx \frac{\ln\left(I_R(T_a)/I_R(T_b)\right)}{E_R} \qquad (6.24)$$

Therefore,

$$SA(T_b) = \left[\frac{I(T_b)}{I(T_a)}\right]^{1/\beta} \left[\frac{I_R(T_a)}{I_R(T_b)}\right]^{E/\beta E_R} \qquad (6.25)$$

In ref. [16], the time evolution of $\ln[(SA(\lambda, t)/SA(\lambda, t_0)]^\beta$ was considered for four Ag I lines (328.068, 338.289, 447.604, and 546.550 nm) calculated using Eq. (6.25). It can be seen that for the 328.068 and 338.289 nm resonant lines, the self-absorption coefficient quickly drops to zero (self-absorption increases) as time progresses.

The current approach does not require the knowledge of parameters such as the instrumental broadening or the spectral response of the spectrometer because their effect on the spectral lines do not depend on the temperature of the plasma and are cancelled in the quotient of the intensities. It can also be seen that according to the definition of Eq. (6.20), which is equivalent to Eq. (6.7):

$$tg\alpha = \sum_{i=1}^{S} B_i t^i = \left(\frac{1}{kT(0)} - \frac{1}{kT(t)}\right)$$

On the other hand, and according to Eq. (6.18)

$$\frac{I(\lambda, t)}{I^*(\lambda, t)} = SA(\lambda, t)^\beta; \quad \frac{I(\lambda, t_0)}{I^*(\lambda, t_0)} = SA(\lambda, t_0)^\beta$$

where $I(\lambda, t)$ is the intensity of the absorbed line of wavelength λ at time t, and $I^*(\lambda, t)$ is the intensity of the same line without absorption.

Applying logarithms and subtracting

$$\ln\left(\frac{I(\lambda, t)}{I^*(\lambda, t)}\right) - \ln\left(\frac{I(\lambda, t_0)}{I^*(\lambda, t_0)}\right) = \ln\left(\frac{I(\lambda, t)}{I(\lambda, t_0)}\right) - \ln\left(\frac{I^*(\lambda, t)}{I^*(\lambda, t_0)}\right)$$

$$= \sum_{i=1}^{S}(b_i^\lambda - b_i^{\lambda*})t^i = \ln\left[\frac{SA(\lambda, t)}{SA(\lambda, t_0)}\right]^\beta \quad (6.26)$$

This expression could be used to measure the degree of self-absorption of any line.

In ref. [17], the time evolution of the $\ln[(SA(\lambda, t)/SA(\lambda, t_0)]^\beta$ was determined, for seven lines of W I.

It is important to note that, according to the definition of the self-absorption parameter SA, it might depend strongly on the plasma temperature; consequently, in transient plasmas its value would depend on time.

This means that the effect of self-absorption on the intensity of a spectral line will be greater as time passes and therefore the b_i coefficients that describe its temporal evolution will be modified. However, when the value of SA for some line is small (strong self-absorption) but constant in time, the coefficients b_i that describe its temporal evolution are not affected because they are defined by the evolution of the logarithm of the quotient of I/I_0 and not by absolute intensity.

Therefore, the b_i coefficients are only affected by the derivative of SA with respect to time.

6.3.5 Determination of Transition Probabilities

Laser-induced plasmas have proven to be a useful source for the measurement of line emission transition probabilities, which are relevant not only for the investigation of atomic structures but also for applications in plasma and laser physics, astrophysics, and for other analytical purposes. A review on the application of the LIBS technique to the determination of plasma spectral parameters of the emission lines of several elements was recently published by Aberkane et al. [18].

The 3D Boltzmann plot method can be used for the precise determination of the transition probabilities of the elements that can be studied by LIBS, as demonstrated in Ref. [17] in the determination of the transition probabilities of some tungsten neutral lines.

Assuming that the plasma is optically thin and in LTE condition and according to Eqs (6.2) and (6.14), the ratio of the integrated intensity of two spectral lines at time t can be written as

$$\frac{I_{kj}(t)}{I_{lm}(t)} = \frac{g_k A_{kj}}{g_l A_{lm}} \exp\left[-(E_k - E_l)\frac{1}{kT(t)}\right]$$

then

$$\frac{I_{kj}(t)}{I_{lm}(t)} = \frac{g_k A_{kj}}{g_l A_{lm}} \exp\left[-(E_k - E_l)\left(\frac{1}{kT(0)} - \sum_{i=1}^{S} B_i t^i\right)\right]$$

The following expression is thus obtained for A_{kj}:

$$A_{kj} = \frac{g_l A_{lm} I_{kj}(t)}{g_k I_{lm}(t)} \exp\left[(E_k - E_l)\left(\frac{1}{kT(0)} - \sum_{i=1}^{S} B_i t^i\right)\right] \quad (6.27)$$

From Eq. (6.27), it is possible to determine the relative transition probabilities of the lines, assuming that the parameters A_{lm}, E_k, and E_l are known and the temperature $T(0)$, the constants B_i, and the intensities of the lines I_{kj} and I_{lm} were measured. We can also see from Eq. (6.27) that there are as many equations for each line quotient as the values that the time t can take.

If M represents the number of reference lines, N is the number of times in which the spectrum was registered, and a set of $N \times M$ equations is generated for the determination of each transition probability.

The application of this equation to spectral lines with known transition probabilities allows to adjust, iteratively, the values of $T(0)$ and B_i, and the knowledge of these two parameters can be exploited for the determination of unknown transition probabilities.

From Eq. (6.27), it can be observed that the intensity and the probability of transition of the line I_{lm} is taken as the "reference" line to calculate the probability of transition corresponding to the line I_{kj}.

The correction for self-absorption of the quotient of intensities that appears in Eq. (6.27) was carried out according to the equation

$$\ln\left(\frac{I^*(\lambda, t)}{I^*(\lambda_{REF}, t)}\right) = \ln\left(\frac{I(\lambda, t)}{I(\lambda_{REF}, t)}\right)\left[\frac{SA(\lambda_{REF}, t)}{SA(\lambda, t)}\right]^\beta$$

where the W I line at 429.46 nm was taken as reference. This equation results from the combination of Eqs (6.18) and (6.26).

According to Figures 6.17 and 6.18, it is interesting to see that the linear fit of the points in the 3D Boltzmann graph reveals that the slopes of the curves, that is the constants B_1 and B_2, are the same for the two laser energies considered.

6 Time-Dependent Spectral Analysis

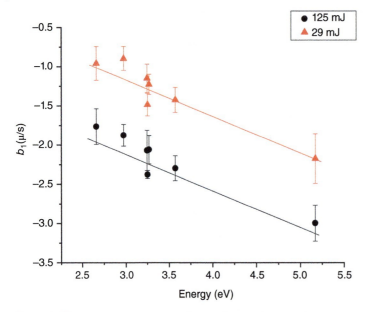

Figure 6.17 b_1 constant vs. energy for two laser pulse energies.

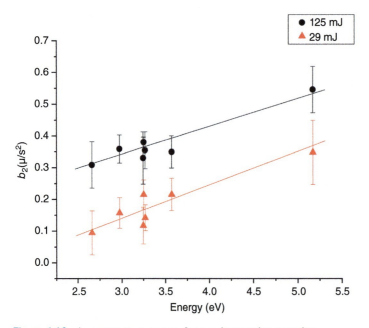

Figure 6.18 b_2 constant vs. energy for two laser pulse energies.

This constitutes an interesting result since it shows that the constants B_1 and B_2 do not depend on the energy of the laser pulse, and this allows predicting the temporal evolution of the temperature and line intensities within a certain range of laser pulse energies.

As previously mentioned, the presence of a positive B_2 coefficient limits the temporal interval in which the plasma can be considered in LTE conditions. Since B_1 is always less than zero, the predicted plasma temperature would reach a minimum value and then it would grow again, which is obviously impossible. Then, the time when the derivative of temperature with respect to time becomes zero can be taken as the time when the LTE conditions are no longer met. According to Eq. (6.14), the derivative of temperature with respect to time becomes zero for $t = -B_1/2B_2$. In our case, it means that the LTE approximation is valid up to about 2 µs from our zero time, that is, up to 3 µs after the laser pulse.

6.3.6 3D Boltzmann Plot and Calibration-free Laser-induced Breakdown Spectroscopy

The calibration-free laser-induced breakdown spectroscopy (CF-LIBS) technique is an analytical method, proposed in 1999 by the Pisa group [19] for the analysis of unknown materials without standards.

The main idea at the bases of the CF-LIBS method is the determination of the plasma parameters (electron temperature and number density) directly from the LIBS spectrum, under suitable assumptions, and using this information for linking the LIBS spectroscopic signal to the elemental composition of the plasma.

The plasma temperature is usually determined using the conventional Boltzmann plot method, from the slope of the best linear fit of the dependence of $\ln(I/g_k A_{ki})$ vs. the energy of the upper level of the corresponding transition. The intercept $q_a(T)$ of this line is proportional to the logarithm of the numerical abundance of the corresponding species in the plasma

$$\ln\left(\frac{I}{g_k A_{kj}}\right) = -\frac{E_k}{kT} + \ln\left(\frac{F n_a}{U_a(T)}\right)$$

$$q_a(T) = \ln\left(\frac{F n_a}{U_a(T)}\right) \rightarrow q_a(T) = \ln\left(\frac{I}{g_k A_{kj}}\right) + \frac{E_k}{kT} \tag{6.28}$$

$q_a(T)$ is, thus, a function of the number density (n_a) of the individual species (a) in the plasma, $U_a(T)$ is the value of the partition function of the species at the temperature T, and F is an unknown parameter, which accounts for the instrumental response of the system.

The number density of each species can be thus derived from Eq. (6.28) from the knowledge of the plasma electron temperature and the spectral parameters of the transition.

For a two-component plasma (let us call generically these two components "a" and "b"), we have that

$$n_a = \frac{e^{q_a(T)} U_a(T)}{F}$$

$$C_a = \frac{n_a}{n_a + n_b} = \frac{e^{q_a(T)} U_a(T)}{e^{q_a(T)} U_a(T) + e^{q_b(T)} U_b(T)}$$

Then,

$$\therefore C_a = \frac{1}{1 + (e^{q_b(T) - q_a(T)})(U_b(T)/U_a(T))}$$

Starting from Eq. (6.28), and assuming a stationary plasma, we have that the intercept of the best fit line on the Boltzmann plot changes with the plasma temperature as

$$q_a(T(t_0)) - q_a(T(t)) = \ln\left(\frac{n_a(t_0)U_a(T(t))}{n_a(t)U_a(T(t_0))}\right)$$

where T_0 is the temperature corresponding to the time $t_0 = 0$ (the instant from which the plasma begins to be analyzed)

$$q_a(T(t_0)) = \ln\left(\frac{I(t)}{g_k A_{kj}}\right) + \frac{E_k}{kT(t)} = \ln\left(\frac{n_a(0)U_a(T(t))}{n_a(t)U_a(T(0))}\right) \quad (6.29)$$

It is important to note that the number density of the species n_a (Eq (6.28)) may vary in time because of the changes in plasma temperature and electron number density.

Assuming that only neutral species and once ionized are present in the plasma, the elemental number density can be written as

$$N_a = n_a^I(T_0) + n_a^{II}(T_0)$$

where N_a is the total number density of the element, $n_a^I(T_0)$ is the number density of the neutral species, and $n_a^{II}(T_0)$ is the number density of the ionized species, both at temperature T_0.

The ratio between the number densities of successive ionization stages of the same element is given by the Saha-Boltzmann equation [20]

$$\frac{n_a^{II}(T)}{n_a^I(T)} = \left(\frac{2\pi mkT}{h^2}\right)^{3/2} \frac{2U_a^{II}(T)}{U_a^I(T)} \frac{e^{-(E^I/KT)}}{n_e} \quad (6.30)$$

where n_e is the electron number density, and E^I is the ionization energy of the atom. The procedure for calculating the ratio in Eq. (6.30) may be cumbersome. Fortunately, the variation of the number densities and the partition functions can be calculated using the 3D Boltzmann plot formalism as follows. According to Eq. (6.5)

$$E_n \to 0 \therefore \sum_i b_i^n t^i \to \ln\left(\frac{n(t)U(T(0))}{n(0)U(T(t))}\right) \quad (6.31)$$

Considering Eqs (6.8) and (6.9)

$$\sum_{i=1}^{S}(b_i^R - B_i E_R)t^i = \ln\left(\frac{n(t)U(T(0))}{n(0)U(T(t))}\right)$$

and thus

$$\therefore q_a(T(0)) = \ln\left(\frac{I_k(t)}{g_k A_{kj}}\right) + \frac{E_k\left(1 - kT(0)\sum_{i=1}^{S} B_i t^i\right)}{kT(0)} - \sum_{i=1}^{S}(b_i^R - B_i E_R)t^i \quad (6.32)$$

As previously discussed, for the determination of $q_a(T_0)$, we can use $N \times M$ equations, where N is the number of lines used, and M is the number of spectra. The temperature is expressed according to Eq. (6.14).

The reliability of the application of the 3D Boltzmann plot method together with CF-LIBS has been demonstrated in Ref. [21] studying various binary Pb–Sn alloys.

The LIBS spectra were acquired at various delay times, from 1 to 4 µs at 0.5 µs steps after the breakdown. It turned out that was not necessary to include the b_2 coefficients to fit correctly the temporal decay of the intensity of the spectral lines. Furthermore, the dependence of the b_1 values vs. the energy of the upper level is well represented by a straight line, which is an indication that plasma can be considered under LTE conditions in the time interval between 1 and 4 µs.

The quantitative results obtained with the 3D-CF-LIBS method on the composition of the binary samples under study were in an excellent agreement with the nominal concentrations. Moreover, the trueness of the 3D-CF-LIBS results, derived from the analysis of the spectral data set as a whole, was substantially better that the one obtained by conventional CF-LIBS analysis, in which the spectra acquired at different delay times were treated separately, averaging the concentrations obtained at the end of the calculation.

6.4 Conclusion

The LIBS spectra are typically rich of spectral information that, however, sometimes is difficult to manage properly. In many occasions, the study of the temporal variation of these spectra may simplify the interpretation of the results.

This can be done either grouping the emission lines in independent components, each one characterized by a definite time behavior, as in the ICA or exploiting the peculiarity of the laser-induced plasma emission for determining the plasma parameters (specifically, the plasma electron temperature), as in the 3D Boltzmann plot approach developed by Bredice et al.

The 3D Boltzmann approach is one of the few chemometric tools explicitly derived for LIBS applications; it can be used for estimating the temporal evolution of the plasma temperature and/or provide information about the degree of self-absorption in the plasma and derive the transition probabilities of the emission lines. Coupled with calibration-free LIBS, this method can be used for simplifying the spectral analysis and obtaining quantitative analytical results that are more accurate with respect to the application of conventional CF-LIBS methods.

References

1 Radziemski, L.J., Loree, T.R. Laser-induced breakdown spectroscopy: Time-resolved spectrochemical applications. *Plasma Chem. Plasma Process.* (1981), **1**, 281–293, doi:10.1007/BF00568836.
2 Loree, T.R., Radziemski, L.J. Laser-induced breakdown spectroscopy: Time-integrated applications. *Plasma Chem. Plasma Process.* (1981), **1**, 271–279, doi:10.1007/BF00568835.
3 Rezaei, F., Cristoforetti, G., Tognoni, E., Legnaioli, S., Palleschi, V., Safi, A. A review of the current analytical approaches for evaluating, compensating and exploiting self-absorption in

laser induced breakdown spectroscopy. *Spectrochim. Acta Part B At. Spectrosc.* (2020), **169**, 105878, doi:10.1016/j.sab.2020.105878.

4 Cristoforetti, G., De Giacomo, A., Dell'Aglio, M., Legnaioli, S., Tognoni, E., Palleschi, V., Omenetto, N. Local thermodynamic equilibrium in laser-induced breakdown spectroscopy: Beyond the McWhirter criterion. *Spectrochim. Acta Part B At. Spectrosc.* (2010), **65**, 86–95, doi:10.1016/j.sab.2009.11.005.

5 Hyvärinen, A., Oja, E. Independent component analysis: algorithms and applications. *Neural Netw.* (2000), **13**, 411–430, doi:10.1016/S0893-6080(00)00026-5.

6 El Rakwe, M., Rutledge, D.N., Moutiers, G., Sirven, J.-B. Analysis of time-resolved laser-induced breakdown spectra by mean field-independent components analysis (MFICA) and multivariate curve resolution-alternating least squares (MCR-ALS). *J. Chemom.* (2017), **31**, e2869, doi:10.1002/cem.2869.

7 Bredice, F., Borges, F.O., Sobral, H., Villagrán-Muniz, M., Di Rocco, H.O., Cristoforetti, G., Legnaioli, S., Palleschi, V., Pardini, L., Salvetti, A., et al. Evaluation of self-absorption of manganese emission lines in laser induced breakdown spectroscopy measurements. *Spectrochim. Acta Part B At. Spectrosc.* (2006), **61**, 1294–1303, doi:10.1016/j.sab.2006.10.015.

8 Griem, H. *Spectral Line Broadening by Plasmas.*; Elsevier Science, Saint Louis, MI, (1974), ISBN 9780323150941.

9 Castro, J.P., Rodrigues Pereira-Filho, E., Bro, R. Laser-induced breakdown spectroscopy (LIBS) spectra interpretation and characterization using parallel factor analysis (PARAFAC): A new procedure for data and spectral interference processing fostering the waste electrical and electronic equipment (WEEE) recycling process†. *J. Anal.At. Spectrom.* (2020), **35**, 1115–1124, doi:10.1039/d0ja00026d.

10 Bredice, F., Pacheco Martinez, P., Sánchez-Aké, C., Villagrán-Muniz, M. Temporal evolution of the spectral lines emission and temperatures in laser induced plasmas through characteristic parameters. *Spectrochim. Acta Part B At. Spectrosc.* (2015), **107**, 25–31, doi:10.1016/J.SAB.2015.02.012.

11 Tognoni, E., Palleschi, V., Corsi, M., Cristoforetti, G., Omenetto, N., Gornushkin, I., Smith, B. W., Winefordner, J.D. From sample to signal in laser-induced breakdown spectroscopy: A complex route to quantitative analysis, in: *Laser-Induced Breakdown Spectroscopy*, A.W. Miziolek, V. Palleschi, I. Schechter, Eds., Cambridge University Press, Cambridge, (2006), Vol. **9780521852**, ISBN 9780511541261.

12 Yalçin, Ş., Crosley, D.R.R., Smith, G.P.P., Faris, G.W.W. Influence of ambient conditions on the laser air spark. *Appl. Phys. B Lasers Opt.* (1999), **68**, 121–130, doi:10.1007/s003400050596.

13 Bredice, F., Pacheco Martinez, P., Sarmiento Mercado, R., Sánchez-Aké, C., Villagrán-Muniz, M., Sirven, J.B., El Rakwe, M., Grifoni, E., Legnaioli, S., Lorenzetti, G., et al. Determination of electron temperature temporal evolution in laser-induced plasmas through independent component analysis and 3D Boltzmann plot. *Spectrochim. Acta Part B At. Spectrosc.* (2017), **135**, 48–53, doi:10.1016/j.sab.2017.07.004.

14 El Sherbini, A.M., El Sherbini, T.M., Hegazy, H., Cristoforetti, G., Legnaioli, S., Palleschi, V., Pardini, L., Salvetti, A., Tognoni, E. Evaluation of self-absorption coefficients of aluminum emission lines in laser-induced breakdown spectroscopy measurements. *Spectrochim. Acta Part B At. Spectrosc.* (2005), **60**, 1573–1579, doi:10.1016/j.sab.2005.10.011.

15 Gornushkin, I.B., Anzano, J.M., King, L.A., Smith, B.W., Omenetto, N., Winefordner, J. D. Curve of growth methodology applied to laser-induced plasma emission spectroscopy.

Spectrochim. Acta Part B At. Spectrosc. (1999), **54**, 491–503, doi:10.1016/S0584-8547(99)00004-X.

16 Bredice, F.O., Di Rocco, H.O., Sobral, H.M., Villagrán-Muniz, M., Palleschi, V. A new method for determination of self-absorption coefficients of emission lines in laser-induced breakdown spectroscopy experiments. *Appl. Spectrosc.* (2010), **64**, 320–323, doi:10.1366/000370210790918454.

17 Urbina, I., Carneiro, D., Rocha, S., Farias, E., Bredice, F., Palleschi, V. Measurement of atomic transition probabilities with laser-induced breakdown spectroscopy using the 3D Boltzmann plot method. *Spectrochim. Acta Part B At. Spectrosc.* (2019), **154**, 91–96, doi:10.1016/j.sab.2019.02.008.

18 Messaoud Aberkane, S., Safi, A., Botto, A., Campanella, B., Legnaioli, S., Poggialini, F., Raneri, S., Rezaei, F., Palleschi, V. Laser-induced breakdown spectroscopy for determination of spectral fundamental parameters. *Appl. Sci.* (2020), **10**, 4973, doi:10.3390/app10144973.

19 Ciucci, A., Corsi, M., Palleschi, V., Rastelli, S., Salvetti, A., Tognoni, E. New procedure for quantitative elemental analysis by laser-induced plasma spectroscopy. *Appl. Spectrosc.* (1999), **53**, 960–964, doi:10.1366/0003702991947612.

20 Miziolek, A.W., Palleschi, V., Schechter, I. *Laser-Induced Breakdown Spectroscopy (LIBS): Fundamentals and Applications*, Cambridge University Press, Cambridge, (2006), ISBN 9780511541261.

21 Urbina, I., Carneiro, D., Rocha, S., Farias, E., Bredice, F., Palleschi, V. Study of binary lead-tin alloys using a new procedure based on calibration-free laser induced breakdown spectroscopy. *Spectrochim. Acta Part B At. Spectrosc.* (2020), **170**, 105902, doi:10.1016/j.sab.2020.105902.

Part III

Classification by LIBS

7

Distance-based Method

Hua Li and Tianlong Zhang

Key Laboratory of Synthetic and Natural Functional Molecule of the Ministry of Education, College of Chemistry and Materials Science, Northwest University, Xi'an, China

Analytical chemistry [1] is a science that uses various methods and instruments to obtain information about the composition, content, and structure of substances, which plays an indispensable role in life science, environmental science, military field, industrial agriculture, etc. Modern analytical chemistry has two remarkable characteristics: (i) The continuous development of various analytical instruments. On the one hand, the combination of instruments and the continuous update and development of high-order instruments make the analytical samples more comprehensive and detailed. At the same time, it also produces complex multidimensional analysis data with rich information. How to make full use of these data to achieve the purpose of research has become a new challenge for scientists. (ii). More and more complex analysis objects. The interdisciplinary research and multidisciplinary development of environmental monitoring, biopharmaceutical, food engineering, chemical, and other technical fields make the research system and research objects more complicated and have higher requirements for qualitative and quantitative analysis [2, 3]. Therefore, it has become an important research hotspot to extract useful information from the huge and complex chemical measurement data as much as possible. The emergence of chemometrics is just to solve this problem, which is a comprehensive interdisciplinary method that integrates chemistry, mathematics, computer science, and other disciplines to solve the problems of data processing of modern analytical instruments.

Chemometrics [4–6] is first proposed by Swedish scientist Svante Wold in his fund in 1971. International Chemometrics Society (ICS) was founded by Svante Wold and Bruce R. Kowalski in the United States in 1974. By ICS, chemometrics is defined as the application of data and statistical methods, the design and improvement of chemical measurement process, and the optimal method to extract the useful information of related targets from the chemical measurement data as far as possible. In 1984, the MATLAB software of Mathworks company promoted the development of this subject to a great extent. In 1987, the two professional journals establishment of *Journal of Chemometrics* and *Chemometrics and*

Intelligent Laboratory Systems marked the maturity of the field. Nowadays, chemometrics has become an indispensable part of analytical chemistry. The research content of chemometrics almost includes the whole process of chemical measurement, including sampling, scheme design, signal processing, qualitative and quantitative analysis, pattern recognition, process simulation, database retrieval, and so on [7–12].

There are many chemometrics methods used to solve the classification or discrimination problem have been proposed and applied in many fields such as principal component analysis (PCA), independent component analysis (ICA), partial least squares (PLS), support vector machine (SVM), random forest (RF), and artificial neural network (ANN) [13–18]. Based on the principle of algorithm, chemometrics algorithm used to solve the classification or discrimination problem can be divided into the following categories, distance-based approach, classification method of decision tree, Bayesian classification method, and rule induction algorithm.

1) Classification methods based on distance [19, 20]

    Set up a database $D = \{t_1, t_2, t_3, ..., t_n\}$ and group class $C = \{C_1, C_2, C_3, ..., C_m\}$. Suppose that each cell group contains some numeric attribute values: $t_i = \{t_{i1}, t_{i2}, t_{i3}, ..., t_{ik}\}$, and each class also contains some numeric attribute values: $C_j = \{C_{j1}, C_{j2}, C_{j3}, ..., C_{jk}\}$. Then the classification problem is to assign each t_i to a class C_j satisfying the following conditions: $\text{sim}(t_i, C_j) \geq (t_i, C_1), \forall C_1 \in C, C_1 \neq C_j$. In the above formula, $\text{sim}(t_i, C_j)$ is defined as similarity. In general, the distance is often used as the evaluation index in practical application. The closer the distance, the higher the similarity, and the farther the distance, the lower the similarity. There are many classification methods based on distance proposed in this years that has been applied to solve the relative problem in many research field, especially in spectroscopy data analysis. The following in this chapter will be a detailed introduction of several classic methods involved in the classification methods based on distance.

2) Classification methods based on decision tree [21, 22]

    Each internal node of the decision tree represents the prediction on the attribute value, each branch represents a test output, each leaf node represents a class or class distribution, and the top node of the tree is the root node. A top-down recursive method is used in the classification methods based on decision tree. The attribute comparison is carried out in the inner node of the decision tree, and the downward branch from the node is judged according to the attribute value of the node. The conclusion is obtained at the leaf node of the decision tree. Therefore, a path from the root node to the leaf node of the decision tree corresponds to a coalescence rule, and the whole decision tree corresponds to a disjunctive expression rule.

    One of the biggest advantages of the classification method based on decision tree is that it does not need users to know a lot of relevant knowledge that also is the biggest problems of the algorithm. As long as the training examples can be expressed by attribute conclusion, the algorithm can be used to solve this problems. The construction procedure of a classification model based on decision tree can be divided into two main procedures, the generation and pruning of decision tree. There are many decision tree methods that have been developed. Meanwhile, the application of decision tree methods has been involved many fields, such as biological medicine, chemical industry, energy, and environment.

3) Bayesian classification methods [23, 24]

X is a kind of data sample with unknown label. Suppose H is a hypothesis. If the unknown sample X belongs to a class C. For the classification problem, we want to determine P(H/X), which is the probability of hypothesis H given the sample observation data.

$$P(H/X) = \frac{P(X/H)P(H)}{P(X)} \tag{7.1}$$

In formula 1, $P(X)$ is the prior probability, also known as the prior probability of H. $P(X/H)$ represents the probability that X is observed under the assumption H. $P(H/X)$ is the posterior probability, or becomes the posterior probability of H under condition X. Bayesian algorithm has good performance for the following two categories of classification problems: (i) completely independent data and (ii) function-dependent data.

4) Rule induction algorithm [25, 26]

The common construction methods of rule induction algorithm include the following: (i) generating rules directly based on rule induction algorithm, (ii) generating decision tree first, and then transforming decision tree into rules, (iii) generating rules based on rough set, and (iv) generating rules based on the classifier technology in genetic algorithm. The methods of rule induction algorithm are not limited to the above four ideas. The four strategies (subtraction, addition, addition before subtraction, and subtraction after addition) are contained in rule induction algorithm.

For subtraction, starting from the concrete examples, it is generalized or generalized, generalize and subtract the condition (attribute value), or subtract the conjunctive term, which is the extended example without covering any counter examples. For addition, the initial assumption rule is empty, if the rule overlaps the counterexample, it will continue to add conditions or conjunctions to the rule until the rule does not cover the counterexample. For addition before subtraction, due to the correlation between attributes, it is possible that the addition of a certain condition will make the later condition invalid. Therefore, the former condition needs to be subtracted. For subtraction after addition, similar to the previous method, it is also used to solve the problem of correlation between attribute values. Some classic algorithms of the four algorithms mentioned earlier are listed in Table 7.1.

Table 7.1 The representative algorithm of the classification method.

Classification methods	Methods
Classification methods based on distance	Cluster analysis (CA), independent components analysis (ICA), K-nearest neighbor (KNN), principal components analysis (PCA), and so on.
Classification methods based on decision tree	ID3 algorithm, random forest (RF), Adaboosting algorithm, gradient boosting machine (GBM), and so on.
Bayesian classification methods	Bayesian belief network (BBN), Naive Bayesian algorithm, and so on.
Rule induction algorithm	AQ, CN2, FOIL, and so on.

LIBS [27–29], a new type of atomic spectroscopy, has many advantages, such as online analysis, real-time analysis, nondestructive analysis, multielement simultaneous analysis, and so on. Nowadays, it has been widely used in metallurgical analysis, environmental analysis, and many other fields. This technology has potential for development and is recognized as a potential technology in extreme environmental analysis. Combined with chemometrics, it provides a new way for quantitative and qualitative solutions in many fields. The following part summarizes the application of LIBS combined with distance-based classification methods in various fields in recent years. First, the development of each method is briefly introduced. Second, the principle of each algorithm is introduced. Finally, the method in LIBS is introduced.

7.1 Cluster Analysis

7.1.1 Introduction

CA is an unsupervised classification method by that the samples were grouped naturally according to the characteristics of the samples. Different from the training data with labeled samples, the clustering model can be built on the unlabeled data. Clustering analysis is based on the principle that the distance between samples pairs in measurement space that is inversely proportional to their similarity. There are many kinds of clustering analysis methods (partition clustering, hierarchical clustering, mutually exclusive clustering, fuzzy clustering, complete clustering, etc.), but the most popular methods are K-means clustering and hierarchical clustering [30]. These two clustering analysis methods will be introduced in the following section.

K-means clustering, also known as fast clustering method, was first proposed by MacQueen in 1967. On the basis of minimizing error function, the data are divided into predetermined number K. The principle of this algorithm is simple and easy to process a large number of data. However, this method needs to determine the number of clusters K in advance, and the selection of the initial clustering center point, which greatly affects the classification results. For K-means clustering, people continue to improve. For example, American Bureau of standards proposed an improved algorithm based on K-means iterative self-organizing ISODATA algorithm, in which the global optimization methods (such as genetic algorithm and particle swarm optimization algorithm) were used to improve the K-means algorithm to obtain the optimal cluster number and cluster center. K-means is also widely used in practical classification, such as LIBS combined with K-means algorithm to classify objects. Guo et al. [31] proposed to combine LIBS with iterative K-means clustering to classify polymers. Lukas Brunnbauer et al. [32] proposed to use PCA and K-means clustering combined with LIBS to classify the spatial distribution of five different synthetic polymers (ABS, PLA, PE, Pak, PVC). Peng et al. [33] proposed a hybrid classification scheme of coal, municipal sludge, and biomass using LIBS combined with SVM and K-means clustering algorithm. Li et al. [34] used LIBS combined with an algorithm for choosing the seeds of the K-means algorithm (K-means^{++}) to shorten the clustering time to classify breast cancer tissues and compared the classification performance of K-means clustering based on two distance functions (Euclidean distance and angle cosine distance).

Hierarchical clustering is a hierarchical clustering, and its clustering results are represented by clustering tree graph. Hierarchical clustering can be divided into two ways: one is agglomerative (first, each point of all samples is regarded as a cluster, and each step merges a pair of nearest clusters, and repeats to the expected cluster); the other is top-down splitting method (divisive: all samples are regarded as a whole cluster, and each step is split into clusters and repeated until the last single cluster is left cluster of points) [35]. Hierarchical clustering can effectively divide samples in different levels without prior knowledge of samples. Combined with a variety of analytical techniques, it can be widely used in chemical, biological, environmental analysis, and other fields. In practical classification, agglomerative hierarchical clustering has a wider application. He et al. [36] proposed LIBS combined with hierarchical clustering to classify industrial waste polymers and achieved good classification results. Wang et al. [37] proposed to use LIBS combined with hierarchical clustering to classify three kinds of explosives and three kinds of plastics. The hierarchical clustering methods used in these works are all agglomerative hierarchical clustering.

7.1.2 Theory

Clustering analysis divides the data without class label into several classes according to the relationship in the data, which makes the similarity of samples in the same class very large, and the similarity of samples in different classes very small. In clustering analysis, it is very important to define the similarity (distance) between samples. There are many ways to define distance, three types of that were introduced the following:

Euclidean distance:

$$\mathrm{dis}(x,y) = \sqrt{\sum_{i=1}^{n}(x_i - y_i)^2} \tag{7.2}$$

Manhattan distance:

$$\mathrm{dis}(x,y) = \sum_{i=1}^{n}|x_i - y_i| \tag{7.3}$$

Minkowski distance:

$$\mathrm{dis}(x,y) = \left(\sum_{i=1}^{n}|x_i - y_i|^p\right)^{1/p} \tag{7.4}$$

In formula (7.4), the commonly used value of p is 1 or 2, where $p = 1$ is Manhattan distance and $p = 2$ is Euclidean distance [38].

7.1.2.1 K-means Clustering

K-means clustering is a clustering method based on distance iteration, which belongs to one of the classic clustering analysis algorithms in partition method. In K-means clustering analysis, the value of parameter K should be specified first, and k points should be selected randomly as the centroid of K clusters. According to the distance function, the distances between all the objects and the centroid are calculated, and all the objects are allocated to K clusters according to the principle of the nearest distance, so that the objects in the cluster should be as close to each other as possible and as far away from the objects in other clusters [39].

The specific process of K-means clustering algorithm is as follows:

1) First, the initial centers of K sample clusters are selected from n samples $\{x_1, x_2, ..., x_n\}$;
2) For each sample point, the distance between them and K centers is calculated, and it is classified into the cluster where the center with the smallest distance is located;
3) The mean value of each sample point in each cluster is calculated as the new center point of the cluster;
4) The distances between each sample point and these new centers are calculated, and the clusters are re-divided according to the minimum distance principle;
5) Repeat the above process until the cluster of sample points is no longer changed.

Attention should be paid to K-means clustering:

1) K value selection. Clustering algorithm is divided into several classes, the selection of the initial center.
2) The choice of distance function and criterion function. How to measure the distance from sample point to centroid [40]. As shown in Table 7.2, when Euclidean distance is used, sum of the squared error (SSE) can be selected as the criterion function.

$$\text{SSE} = \sum_{i=1}^{K} \sum dis(c_i, x)^2 \tag{7.5}$$

7.1.2.2 Hierarchical Clustering

The basic idea of hierarchical clustering method is to calculate the similarity between nodes through a certain similarity measure, rank them from high to low according to the similarity, and gradually re connect the nodes. Hierarchical clustering has two main forms of clustering, namely the bottom-up aggregation clustering and the top-down splitting clustering [35]. Here, we will introduce the principle of these two kinds of clustering. The bottom-up agglomerative method: first, each point of all samples is regarded as a cluster, and each step merges a pair of nearest clusters and repeats to the expected cluster. The main operation steps of agglomerative clustering are as follows:

Table 7.2 Distance function and criterion function of K-means algorithm.

Distance function	Centroid	Criterion function
Euclidean distance	Mean value	Minimizes the sum of squares of distances from an object to its centroid
Cosine distance	Mean value	Minimizing the cosine similarity between an object and its centroid
Manhattan distance	Median	Minimize the sum of distances from an object to its centroid

1) Each sample is classified into one class, and the distance between each two classes is calculated, that is, the similarity between samples;
2) Find the two nearest classes between classes and group and put them into one class (so that the total number of classes is reduced by one);
3) The similarity between the new class and each old class is recalculated;
4) Repeat 2) and 3) until all sample points are grouped into one group.

Divisive method (split from top to bottom): all samples are regarded as a whole cluster, each step splits a cluster, and repeats to the last cluster with single point. The main steps are as follows:

1) Each individual sample is classified into one class, and the distance between each sample in the class is calculated;
2) Find the farthest distance between the samples and split them;
3) Repeat the calculation and split into smaller and smaller clusters until each object becomes a cluster.

In order to find the nearest sample point pair in agglomerative hierarchical clustering and the farthest sample point pair in split hierarchical clustering, the similarity between sample points should be calculated in both kinds of clustering. In hierarchical clustering analysis, there are three different similarity measurement methods: single link, complete link, and average link [41].

The single link method calculates the similarity between the sample points and the classes by measuring the distance between the sample points and the nearest points in the cluster. Complete linkage method is used to measure the distance between sample points and clusters and calculate the similarity between sample points and classes.

7.1.3 Application

Guo et al. [31] combining LIBS technology and K-means clustering algorithm proposed a new polymer classification method. On the basis of the initial clustering analysis, the method is implemented by calculating the ARSD value of each cluster. Iterative K-means clustering has an iterative process to calculate the ARSD value of clustering. T-test ($P = 0.01$) is used as the critical value to identify abnormal clustering. If the ARSD value of a cluster is greater than the critical value, the cluster will be judged as abnormal. For the identified abnormal clustering, further clustering analysis is carried out, iterating until there is no outlier, and then CA is carried out. The operation of iterative K-means clustering is as follows:

1) The initial cluster number k is determined by calculating DB index.
2) K spectral vectors are randomly selected as the initial centroid from the spectral data set to cluster the spectral data.
3) Calculate the ARSD value of each cluster obtained by step (2).
4) Based on t-test, the clustering with abnormally large ARSD value was determined. Repeat steps (1)–(3) for the abnormal clusters until no abnormal clusters are found.

As shown in Figure 7.1, LIBS combined with iterative K-means clustering, for 1000 spectra of 20 polymers, only 4 spectra were misclassified. The classification accuracy was 99.6%. This method greatly improves the accuracy of classification results. Using the proposed

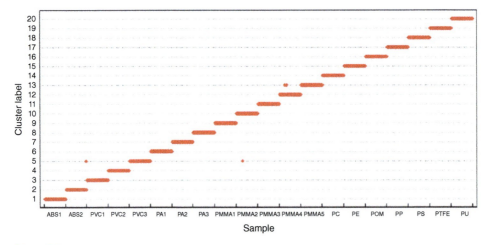

Figure 7.1 Analysis results of iterative clustering.

method, 20 kinds of polymers can be classified under atmospheric conditions without any prior knowledge.

In Brunnbauera's work, LIBS was used to analyze two structured synthetic polymer samples (2D structured and multilayer systems) [32]. The spatial distribution of five synthetic polymers (acrylonitrile butadiene styrene [ABS], polylactic acid [PLA], polyethylene [PE], polyacrylate [PAK], and polyvinyl chloride [PVC]) were classified by multivariate statistical methods (PCA and K-means clustering).

In this research, PCA is used to classify the polymer multilayer system of a double-sided adhesive (PAK, PVC, and PE), the results for PVC and PAK were poor. Then, the K-means method was used to classify the data sets and produce the clustering results of each layer of the multilayer polymer. The combination of PCA and K mean value makes the analysis of 2D structure polymers get better classification results.

In Peng's paper, LIBS combined with SVM and K-means clustering algorithm is used to classify coal, municipal sludge, and biomass [33]. Therefore, SVM combined with K-means is used to classify the biomass fuels with impurities. Three coal samples, one municipal sludge sample and six biomass samples were prepared. Spectral data (200 spectra per sample) were obtained by LIBS. Nonmetallic elements (carbon, hydrogen, nitrogen, and oxygen) and main metal elements (calcium, silicon, magnesium, aluminum, iron, sodium, and potassium) were selected as characteristic elements. Eleven analytical lines were selected according to the principle of analysis line selection. First, K-means clustering is performed. Every 40 spectra of each sample are averaged as a point, that is, each sample is divided into five points: 10 samples, 50 spots in total. Then, K value is selected by comparing the average value of contour value. The K value is set to 3, and 10 samples are clustered.

As shown in Figure 7.2, 50 points (10 samples) using K-means are well divided into three categories: coal, municipal sludge, and biomass, with a classification accuracy of 100%. For fuels with large characteristic differences (such as coal, biomass, and municipal sludge),

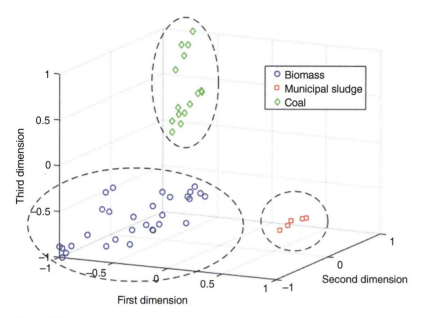

Figure 7.2 Results of clustering with K-means.

K-means is used to distinguish fuels with large characteristic differences, and SVM is used to distinguish fuels with small characteristic differences (such as several biomass). The results show that the comprehensive accuracy of the mixed classification model is over 98%. Compared with single SVM classification model, the hybrid classification model can save 58.92% of the operation time while ensuring the accuracy. The results show that the hybrid classification model can classify coal, municipal sludge, and biomass efficiently, quickly, and accurately and has a high accuracy in the detection of various biomass fuels.

LIBS combined with self-organizing map (SOM) and K-means of unsupervised learning algorithm was used to distinguish industrial polymers outdoors [42]. The experimental samples were 20 polymers (ABS, PVC, PC, PS, PE, POM, PP, PTFE, PU, PMMA, covering plastics of different colors). Each polymer had 100 spectra, with a total of 2000 spectral data. Two molecular bands (C—N (0, 0) 388.3 nm and C2 (0, 0) 516.5 nm) and four atomic emission lines (C I 247.9 nm, H—I 656.3 nm, N I 746.9 nm, and O I 777.3 nm) were selected as analytical lines. First, the self-organizing neural network (ASW) was used to separate 20 kinds of polymers. The results showed that 18 kinds of polymers were separated except polycarbonate and polystyrene. Then, K-means clustering algorithm is used to separate PC and PS. Spectral data not distinguished by SOM are used as input of K-means, and the number of initial centroids is set to 2. Figure 7.3 shows the results obtained by K-means clustering. 50 spectra of each sample are selected for K-means clustering. Each diamond represents the spectral data of a sample. The classification accuracy of K-means clustering for PC and PS is 100%. By combining SOM and K-means, the final accuracy of 20 kinds of industrial polymers is 99.2%, which proves the feasibility of clustering industrial polymers by LIBS.

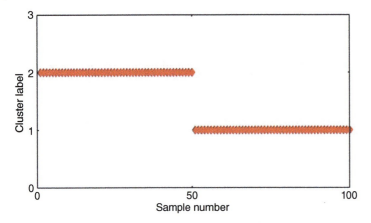

Figure 7.3 K-means clustering PC (label 1) and PS (label 2).

7.2 Independent Components Analysis

7.2.1 Introduction

The idea and method of ICA originated from the 1980s, known as H-J algorithm [43]. In the early 1990s, some scholars extended the problem of blind source separation in some fields [44, 45]. Among them, Cichochi and Unbehauen [46–48] jointly proposed the popular ICA algorithm; in 1994, after Comon's summary [49], a stricter and clearer mathematical framework was given. Since then, ICA has become a formal term in the literature. American scholars Bell and Sejnowski published their literary theories in the 1990s, which became a milestone in the development of ICA and attracted wide attention of the academic circles at that time [50–53]. So far, ICA has become a research hotspot in signal processing, neural computing, data statistics, and other disciplines.

ICA is considered as an extension of PCA. PCA is a common preprocessing step, which decomposes data into orthogonal principal components, and independence is a higher order statistics much stronger than orthogonality. Therefore, ICA can provide more chemical explanations than PCA. ICA is a widely used unsupervised blind source separation technology. ICA linearly separates n-dimensional data sets into n components. It assumes that the observed data are generated by a linear mixture of independent signal sources and non-Gaussian distribution and mixes these signals into the original signal source. It does this by calculating a linear transformation, which decomposes the mixed signal sources. If the mixed transformation is regarded as a matrix, ICA calculates a matrix equivalent to its inverse matrix. ICA has three basic models: generalized model, noise model, and noiseless model. Noise-free model is the most widely used model in ICA, which is usually called classic model.

7.2.2 Theory

Fast ICA algorithm, also known as fixed point algorithm, was proposed by Hyvarinen et.al, University of Helsinki, Finland [54]. It is a fast optimization iterative algorithm. Different from the common neural network algorithm, this algorithm adopts batch processing, that is, a large number of sample data participate in the operation in each iteration step. But from the point of view of distributed parallel processing, this algorithm can still be called a neural network algorithm. Fast ICA algorithm has the form of kurtosis, likelihood maximum, and negative entropy maximum. This paper mainly introduces the fast ICA algorithm based on the maximum negative entropy [52, 55]. It takes the maximum negative entropy as a search direction and can extract independent sources in sequence, which fully embodies the traditional linear transformation idea of projection pursuit. In addition, the algorithm uses the fixed-point iterative optimization algorithm, which makes the convergence faster and more robust.

Since the fast ICA algorithm takes the maximum negative entropy as a search direction, first review the negative entropy criterion: according to the information theory, among all the random variables with equal variance, the entropy of Gaussian variable is the largest. Therefore, entropy is used to measure non-Gaussian property, and the modified form of entropy is commonly used, namely negative entropy. According to the central limit theorem, if the random variable x is composed of many independent random variables S_i ($i = 1, 2, 3, \ldots, n$), as long as S_i has finite mean and variance, no matter what kind of distribution, the random variable x is closer to the Gaussian distribution than S_i. In other words, S_i is more non-Gaussian than x. In the process of separation, the non-Gaussian measure can be used to express the mutual independence of the separation results. When the non-Gaussian measure reaches the maximum, it indicates that the independent components have completed the separation.

The derivation of fast ICA algorithm: first, the maximum approximate value of negative entropy of WX can be obtained by optimizing $E\{g(WX)\}$ (where g is a nonlinear function). After a series of derivation, the approximate Newton iterative formula can be obtained. After simplification, the iterative formula of fast ICA algorithm can be obtained.

$$W^* = E\{Xg(W^TX)\} - E\{g'(W^TX)\}W \\ W = W^*/\|W^*\| \tag{7.6}$$

In the formula (7.6), W^* is the new value of W. The basic steps of monadic fast ICA algorithm are as follows:

1) Centralize the observation data X so that its mean value is 0.
2) Whiten the observed data X, $X \to Z$.
3) Select an initial weight vector (random) W.
4) $W^* = E\{Xg(W^TX)\} - E\{g'(W^TX)\}W$.
5) Let $W = W^*/\|W^*\|$.
6) If not, return to step 4.

Convergence means that the values of the kth iteration and the $k-1$ iteration of W point in the same direction, that is, their point product is 1. There is no need to converge to a point because W and $-W$ refer to the same direction.

In practice, the expectations used in fast ICA algorithm must be replaced by their estimates. Of course, the best estimate is the corresponding sample average. Under ideal conditions, all valid data should be involved in the calculation, but this will reduce the calculation speed. Therefore, the average of some samples is usually used to estimate.

The number of samples has a great influence on the accuracy of the final estimation. If the convergence is not ideal, the number of samples can be increased. When estimating multiple components, the steps of fast ICA algorithm are the same as those of unitary fast ICA algorithm.

7.2.3 Application

The classification and identification of fly ash are helpful to the recycling of metallurgical waste. Zhang et al. [56] discussed the classification and analysis method of coal ash based on LIBS technology and ICA wavelet neural network (ICA-WNN). First, 45 fly ash samples were prepared by mixing 7 different reagents. Then, according to the K-S algorithm, the Euclidean distance is used to uniformly select the training set, 36 samples are selected as the training set, and the remaining 9 samples are used as the test set to verify the performance of the model. Before testing, a series of coal ash samples need to be compressed into pellets for LIBS measurement. In addition, in order to improve the signal-to-noise ratio, the spectrum was collected at ten positions of each coal ash sample.

For the collected spectra, first, PCA is combined with Mahalanobis distance to identify and remove abnormal spectra to optimize the training set of WNN. After optimization, 365 training sets and 90 test sets are obtained. However, the selection of input variables is also very important for the performance of the model. PCA and ICA are used to select input variables. For ICA, fast ICA is often used to optimize and select input variables. Figure 7.4 shows the mean square error (MSE) corresponding to different independent component scores (ICs). The results show that: in the range of 1–30, MSE decreases with the increase of ICs; when ICs is equal to 30, MSE reaches the minimum; when ICs is greater than 30, MSE continues to increase. Therefore, this study used 30 ICs to establish WNN training model. For PCA, Figure 7.5 shows that when the number of independent

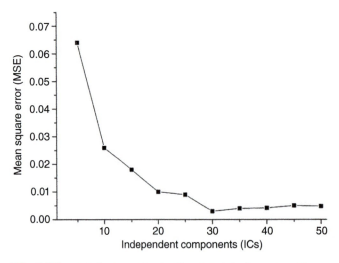

Figure 7.4 Contribution rate of different principal components.

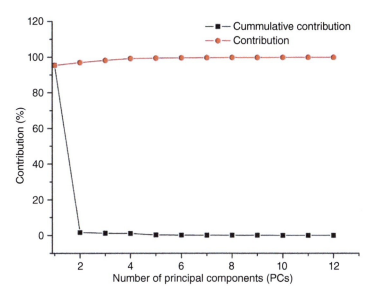

Figure 7.5 MSE of different independent components.

components (PCS) is equal to 10, the corresponding contribution rate is the largest. Then, the performance of WNN training models with different input variables obtained by PCA, ICA and full spectrum methods is compared. The results are shown in Table 7.3. The results show that the training model has the highest classification accuracy and the lowest MSE when using ICA to optimize and select input variables. So. The classification model of coal ash samples was established by ICA.

After optimizing the parameters of WNN model, the training model of WNN and ANN is verified by using fivefold cross-validation (CV) and training sample set. The prediction performance of ICA-WNN is shown in Table 7.4. The results show that the prediction performance of ICA-WNN is better than that of ANN. Finally, WNN and Ann test sets are used to further verify the performance of the classification model. The results are shown in Table 7.4. The results show that ICA-WNN model shows better classification performance than ANN model.

Forni et al. [57] used the distance measure of ICA space, namely Manhattan and Mahalanobis distance, to classify the spectrum of an unknown rock effectively. In this study, first, the spectra were pretreated. Then the spectrum after pretreatment is normalized to make its

Table 7.3 Effects of different input variables on the performance of neural network model.

Input variable	Classification accuracy (%)	Mean square error (MSE)
PC	99.16 (353/356)	0.0067
IC	100.00 (356/356)	0.0019
Full spectra	89.33 (318/356)	0.0125

Table 7.4 The classification accuracy (%) for different classes of coal ash samples by ICA-ANN and ICA-WNN methods.

Class	Fivefold CV		Test set samples	
	ICA-ANN	ICA-WNN	ICA-ANN	ICA-WNN
Class A	98.86	100.00	96.67	96.67
Class B	98.87	98.87	96.67	100.00
Class C	100.00	100.00	96.67	100.00
Average	99.44	99.73	96.67	98.89

mean and unit variance zero. As one of the disadvantages of ICA is that the optimal number of sources is unknown, it is necessary to test different number of sources. Using JADE algorithm, it has eight components. So, eight independent components were obtained. For each component, its weight or its contribution to each original spectrum can be calculated. This is just the correlation coefficient between the components and the spectra because the spectra have been standardized.

The classification procedure is very simple. Suppose that a group of rocks has known spectra, K unknown components are selected by ICA program. The purpose of classification is to match the unknown spectrum with a category. Generally, the distance between unknown samples and each known sample or group of centroids is calculated in the k-coordinate space, and the position samples are classified as the spectrum with the smallest distance. In addition, the threshold criterion should be used: if the minimum distance exceeds the threshold, then the sample is displayed as: do not assign.

In this process, two methods are used: the first is to calculate the city block or Manhattan distance between the unknown spectrum and the known spectrum and assign the unknown spectrum to the group with the smallest distance. The second method uses Mahalanobis distance, which considers the covariance between variables. With this measure, the inherent scale and correlation of Euclidean distance are no longer a problem. However, with this minimum distance criterion, any spectrum will be assigned to a class. In other words, this method cannot identify spectra from the categories that are not included in the calibration set, so it is not suitable for field applications. The two distance measurements in ICA space are very sensitive and robust. Mahalanobis distance seems to be more robust than Manhattan distance if taken alone. In addition, it is much easier to determine the cutoff value. Mahalanobis distance is farther than Manhattan distance. However, the combination of these two measures allows a slight increase in performance. The results show that the spectrum of an unknown rock can be effectively classified by using the distance measure of ICA space, i.e. Manhattan and Mahalanobis distance. Although Manhattan distance is easier to manage, the cutoff distance of Manhattan distance is not easy to determine. However, due to the complementarity of the two techniques, the combination of the two methods will improve the accuracy and performance of the analysis.

7.3 K-Nearest Neighbor

7.3.1 Introduction

K-nearest neighbor (KNN), as a generalization of the nearest neighbor method, was first proposed by Cover and Hart [58] in 1967. It is a typical learning algorithm of analogy principle. KNN stores all the samples of the training set in the computer, then selects K adjacent samples, and then classifies them into the category with the largest proportion according to the data category. This method is different from the traditional method, that is, only one nearest sample is selected for classification, which can avoid high misclassification rate. In terms of the improved algorithm, for the problem that the classification accuracy is reduced due to the unbalanced density of samples. Shang et al. [59] adopted the concept of fuzzy sets of samples and redefined the classification function. Experiments show that the improved algorithm can improve the accuracy of classification. Saetern et al. [60] proposed an improved KNN algorithm based on neuro fuzzy K-nearest neighbor algorithm and improved the classification mode, and the classification performance was also improved. Jiang et al. [61] improved KNN algorithm for classification and clustering, and the algorithm has been verified in the classification accuracy has been significantly improved. Li et al. [62] improved the traditional decision tree diagnosis algorithm for fault tree diagnosis. Compared with the traditional classification algorithm, it has better performance in fault diagnosis accuracy.

According to the current practical needs, under the premise of ensuring the accuracy of K-nearest neighbor classification algorithm, effectively improving the classification efficiency of the algorithm is the most important problem faced by the promotion of this algorithm. At present, the research status of KNN classification algorithm is mostly to improve the classification efficiency: First, the vector space model with lower dimension is established by feature selection to reduce the amount of calculation, thus reducing the time consumption; second, cutting training samples to reduce the amount of calculation; and third, establishing efficient cable or introducing fast search algorithm to speed up the speed of finding the nearest neighbor of the test sample. The related results show that many researchers have improved KNN classification algorithm in these aspects, but there are still many insufficiencies that are difficult to overcome. Some algorithms even lead to the reduction of the original accuracy rate, and some improved algorithms do not improve the classification effect [63–68].

7.3.2 Theory

KNN is a method based on classification and regression, which is mainly used for supervised pattern recognition. The input of KNN is the feature vector of the sample, corresponding to the points in the feature space; the output is the category of the test sample, which can take multiple classes. KNN assumes that a training text set is given, in which the training sample class is given in advance. In the classification decision, the new sample points to be tested are predicted by majority voting or weight according to the class of K-nearest neighbor training sample points. Therefore, KNN does not have an explicit learning process.

KNN actually uses the training data set to partition the feature vector space and uses it as its classification model. The specific steps of KNN algorithm are given as follows:
Input: training data set

$$T = \{(x_1, y_1), (x_2, y_2), \dots, (x_N, y_N)\} \tag{7.7}$$

In the formula (7.7), $X_i \in x \subseteq RP$ is the eigenvector of training samples, $Y_i \in y = \{C_1, C_2, \dots C_N\}$ is the class of samples, $I = 1, 2, \dots, N$. The characteristic vector X of the test sample;
Output: test sample x belongs to the class y.

1) According to the given distance measure, k points nearest to X are found in the training set T. The neighborhood of X covering these k points is denoted as $N_{k(x)}$;
2) In $N_{k(x)}$, decision rules (such as majority voting) determine the class y of X;

$$y = \arg\max_{cj} \sum_{y_i \in N_{k(x)}} I(y_i = c_j)$$
$$i = 1, 2, \dots, N \tag{7.8}$$
$$j = 1, 2, \dots, |C|$$

In the formula (7.8), I is the indicating function. k is the number of neighbors, $I = 1$ when $y_i = c_i$, otherwise $I = 0$. In K-nearest neighbor method, the class of the nearest point of X in the training set is regarded as the class of X.

In Figure 7.6, there are two types of training sample set: blue square and red triangle, while the test sample in the figure is green circle. In actual situation, we first think of that category the green circle should be classified into? This is a simple text classification problem; using KNN classification algorithm, when $k = 4$, among the three samples nearest to it, the red triangle accounts for 3/4, so the green circle is classified as red to blue quadrangle; when $k = 8$, among the 8 samples closest to the green circle, there are 5 blue quadrangles, so the green circle is identified into the blue quadrangle, which is the original KNN classification a visual description of reason.

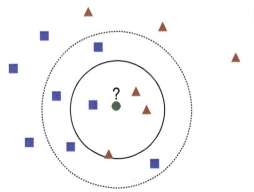

Figure 7.6 The effect of K value selection.

7.3.3 Application

Yang et al. [69] used LIBS combined with machine learning methods to classify of iron ore. KNN model, neural network (NN) model, and SVM model are used for classification in this work. The corresponding confusion matrix for KNN, NN, and SVM model is presented in Figure 7.7 The prediction accuracy (PA) of the three models is 82.96, 93.33, and 94.07%, respectively. Thus, results in Figure 7.2 suggest that LIBS technology combined with machine learning can realize the rapid and accurate classification of iron ore, which provides a new method for iron ore selection in metallurgical industry.

Alvarez et al. [70] used LIBS combined with chemometrics methods for the determination of copper-based mineral species. The samples in this work include bornite (Cu_5feS_4), chalcocite (Cu_2S), chalcopyrite ($CuFeS_2$), omphacite (Cu_3As_4), molybdenite (MoS_2), and pyrite (FeS_2). The unsupervised multivariate analysis methods, such as PCA and tree diagram analysis, and supervised pattern recognition techniques, such as soft independent class analogy modeling (SIMCA), partial least squares discriminant analysis (PLS-DA), KNN, decision tree analysis, and artificial neural network (ANN) were compared. It is shown in Table 7.5 that the average classification accuracy of KNN, SIMCA, PLS-DA, and ANN models is 96.2, 98.1, 90.6, and 100%, respectively. In the results, ANN shows better classification performance. This study shows that very similar species can also be correctly identified based on elemental composition (e.g. bornite/chalcopyrite and chalcocite/cerulene).

Michael et al. [71] used spark discharge-LIBS for the brown rice authenticity evaluation. In this work, LIBS spectral data from rice analysis were used to evaluate PDO certification of Argentine brown rice. Results of the classification metrics calculated for different evaluated models are shown in Table 7.6. The result indicates that emission lines of C, CA, Fe, Mg, and Na are selected and the best classification performance is obtained by KNN algorithm. The accuracy, sensitivity, and specificity of the method were 84, 100, and 78%, respectively. In addition, the method is simple, clean, and easy to apply to rice certification.

7.4 Linear Discriminant Analysis

7.4.1 Introduction

Linear discriminant analysis (LDA), also known as Fisher linear discriminant (FLD), is a classic algorithm of pattern recognition, which was introduced into the field of pattern recognition and artificial intelligence by Belhumeur in 1996. The basic idea is to project the high-dimensional pattern samples into the optimal discriminant vector space to extract classification information and compress the dimension of feature space. After projection, the pattern samples are guaranteed to have the maximum inter class distance and the minimum intra class distance in the new subspace (the pattern has the best separability in this space). Up to now, LIBS combining LDA has been applied in many fields. Colao et al. [72] used laser-induced breakdown spectroscopy (LIBS) and X-ray fluorescence (XRF) technology, combined with LDA to characterize the historical building materials and determined the relative content of major elements and trace elements according to the geographical source of the quarry. Gustavos. Larios et al. [73] used LIBS combined with LDA to

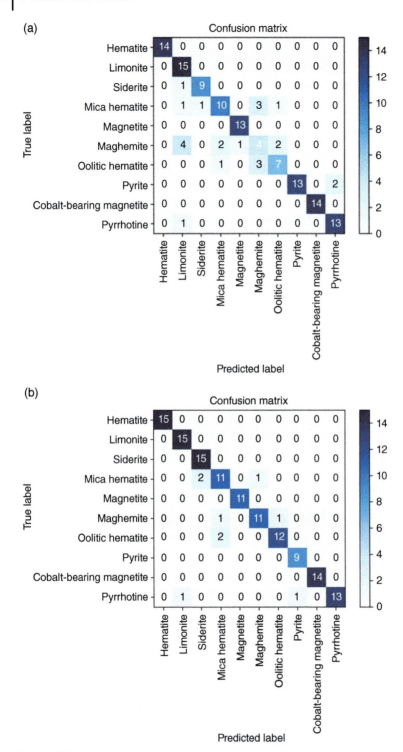

Figure 7.7 (a) Confusion matrix for the K-nearest neighbor (KNN) model. (b) Confusion matrix for neural network model. (c) Confusion matrix for the support vector machine (SVM) model.

7.4 Linear Discriminant Analysis

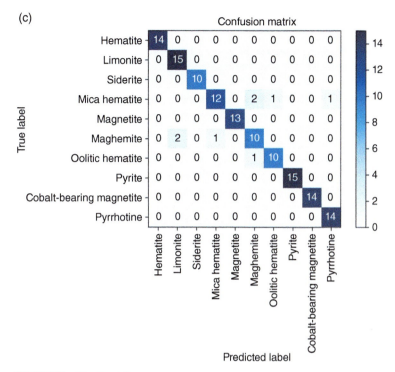

Figure 7.7 (Continued)

Table 7.5 Validation summary: accuracy, sensitivity, and precision for SIMCA, PLS-DA, KNN, and ANN.

Sample\validation	SIMCA		PLS-DA		KNN		ANN		
	Train	Test	Train	Test	Train	Test	Train	Validation	Test
Bornite	100	100	86.6	100	100	100	100	100	100
Chalcocite	92.3	75.0	61.5	75.0	69.2	75.0	100	100	100
Chalcopyrite	100	100	100	100	100	100	100	100	100
Covellite	100	100	70	80	100	100	100	100	100
Enargite	100	80.0	93.3	100	93.3	100	100	100	100
Molybdenite	100	100	100	100	100	100	100	100	100
Pyrite	100	100	91.3	100	100	100	100	100	100
Accuracy (%)	98.1		90.6		96.2		100		
Sensitivity (%)	97.6		88.1		95.1		100		
Precision (%)	98.4		98.7		95.6		100		
No Match (%)	0.63		8.18		—		—		

Table 7.6 Results of the classification metrics calculated for different evaluated models.

Algorithms	Parameters	Accuracy (%)	Sensitivity (%)	Specificity (%)
LDA	—	53	70	33
KNN	$K = 7$	84	100	78
SVM	$C = 0.8$ $\varepsilon = 0.075$	74	90	44
RF	$n_{tree} = 500$ $m_{try} = 1$	57	70	44

distinguish high and low vigor soybean seed varieties. Li. et al. [74] applied the automatic peak recognition method of LIBS combined with LDA (aspi-lda) to rock classification, and compared with the manual peak recognition method of LDA (mspi-lda). Gaudiuso et al. [75] used LIBS combined with LDA to diagnose melanoma in biomedical liquid on solid substrate. Rokhbin et al. [76] used LIBS combined with LIBS to evaluate the quality of smoke.

7.4.2 Theory

7.4.2.1 The Calculation Process of LDA (Two Categories)

In this section, the calculation of LDA for only two categories of data is mainly discussed. Now we are going to find a vector W, and we are going to project X onto w to get the new data Y. First, the absolute value of the mean difference between the two categories is used to measure to realize the distance between the two categories after projection. Second, the variance of each category is used to measure to realize the aggregation of data points within each class after projection.

Mean value of class i:

$$m_i = \frac{1}{n_i} \sum_{x \in D_i} x \tag{7.9}$$

Mean value after projection of class i (actually equal to the projection of m_i):

$$\widetilde{m} = \frac{1}{n_i} \sum_{y \in Y_i} y = \frac{1}{n_i} \sum_{x \in D_i} w^t x = \frac{1}{n_i} w^t \sum_{x \in D_i} x = w^t m_i \tag{7.10}$$

Absolute value of the mean difference after projection:

$$|\widetilde{m}_1 - \widetilde{m}_2| = |w^t(m_1 - m_2)| \tag{7.11}$$

Variance after projection (where y is the projected data of data in class i, i.e., $y = w_t \times x$:

$$\widetilde{S}i^2 = \sum_{y \in Y_i}(y - \widetilde{m}_i)^2 \tag{7.12}$$

7.4 Linear Discriminant Analysis

The objective optimization function is as follows:

$$J(w) = \frac{|\tilde{m}_1 - \tilde{m}_2|^2}{\tilde{S}1^2 + \tilde{S}2^2} \tag{7.13}$$

S_B and S_w defined by expanding m' and s':

$$Si^2 = \sum_{x \in D_i}(w^t x - w^t m_i)^2 = \sum_{x \in D_i} w^t(x - m_i)(x - m_i)^t w = w^t S_i w \tag{7.14}$$

$$\tilde{S}_1^2 + \tilde{S}_2^2 = w^t(S_1 + S_2)w = w^t S_w W \tag{7.15}$$

$$(m_1 - m_2)^2 = (w^t m_1 - w^t m_2)^2 = w^t(m_1 - m_2)(m_1 - m_2)^t w = w^t S_B w \tag{7.16}$$

The optimization objective $J(w)$ is rewritten as follows to facilitate the derivation of the method for calculating w:

$$J(w) = \frac{w^t S_B w}{w^t S_w w} \tag{7.17}$$

The results are as follows:

$$w = Sw^{-1}(m_1 - m_2) \tag{7.18}$$

Suppose that the data are n dimensional features, m data and the number of classification are 2.

So S_w is actually the sum of the covariance matrices of each category. The covariance matrices of each category are $n \times n$, so S_w is $n \times n$, and $m_1 - m_2$ is $n \times 1$. The calculated w is $n \times 1$, that is, w maps the dimension feature to 1 dimension. There is no need to tangle with the covariance matrix form of S_w. In fact, this is the splitting of w and w_t. In fact, after $w \times S_w \times w_t$, it is still a numerical value, that is, the sum of the variance after projection of the two classes.

The calculation process of LDA (multiple categories)

For S_w. That is, "the sum of covariance matrices of two categories" becomes "the sum of covariance matrices of multiple categories."

$$S_w = \sum_{i=1}^{c} S_i \tag{7.19}$$

For S_B. It used to be the absolute value of the difference between the mean values of two categories. Now that there are multiple categories, how to calculate it? Calculate the sum of the absolute values of the difference between the mean values of any two categories? In this way, for n categories, $C(n, 2)$ calculations are required, which may be a method. However, LDA uses the method to calculate the difference between the mean value of each category and the mean value of all categories and weight the data volume of each category. Where m is the mean value of all categories, m_i is the mean value of class i, n_i is the data volume of class i.

$$S_B = \sum_{i=1}^{c} n_i(m_i - m)(m_i - m)^t, m = \frac{1}{n}\sum_x x \tag{7.20}$$

For the data of n-dimensional features, c categories and m samples, n-dimensional data are mapped to $c-1$ dimension. S_w is an $n \times n$ matrix (without the sum of covariance matrices divided by the number of samples). S_B is a matrix of $c \times c$. Actually, the rank of S_B matrix is $C-1$ at most. This is because vectors are actually linearly correlated, and their sum is a constant multiple of the mean m. This leads to the later solution of w, which is actually to find a w composed of $c-1$ vectors.

Examples of three categories (Figure 7.8):

LDA algorithm can be used not only for dimensionality reduction but also for classification, but at present, it is mainly used for dimensionality reduction. LDA is a powerful tool for data analysis related to image recognition. The advantages and disadvantages of LDA Algorithm are summarized below.

The main advantages of LDA are as follows:

1) The prior knowledge experience of category can be used in dimensionality reduction, while unsupervised learning such as PCA cannot.
2) LDA is better than PCA when the sample classification information depends on mean rather than variance.

The main disadvantages of LDA are as follows:

1) LDA is not suitable for dimensionality reduction of non-Gaussian distribution samples, and PCA also has this problem.
2) LDA can be reduced to the dimension of category number $k-1$ at most. If the dimension of LDA is larger than $k-1$, LDA cannot be used. Of course, there are some LDA evolutionary algorithms that can bypass this problem.
3) LDA cannot reduce the dimension effectively when the sample classification information depends on variance rather than mean.
4) LDA may over fit the data.

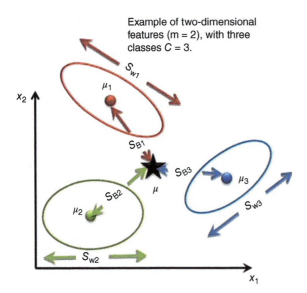

Figure 7.8 Examples of three categories.

7.4.3 Application

The LDA algorithm was used to classify 24 metamorphic rocks in this work. [77] The results show that the classification accuracy is improved from 94.38% of MLS-LDA to 98.54% of IFLAS-LDA. In addition, the time required for the whole classification process is reduced from 2768.38 seconds of MLS-LDA to 4.36 seconds of the proposed method, which greatly improves the classification efficiency. In addition, the convergence rate of IFLAS-LDA is significantly faster than that of ASPI (automatic spectra peak identification) – LDA. This study shows that LIBS-assisted image features in machine vision can promote the analysis performance of LIBS technology.

In this study, LDA algorithm is used for classification. The analysis lines selected by MLS and IFLAS are combined with LDA algorithm to establish the classification model. Take the classification model IFLAS-LDA as an example, the spectral intensities of 171 analytical lines were used as the input data of the model. In addition, the spectral number of each sample is 100, 80 of that are used to train the classification model, and the remaining 20 are used to evaluate the model, and 20 spectra were obtained by uniform sampling. This means that the first, sixth, eleventh, ... and 96th spectra of each sample are selected as test set data. Therefore, the input data size of training set and test set are 171 × 80 and 171 × 20, respectively. The classification results of the above two classification models are shown in Figure 7.9.

As shown in Figure 7.9, when MLS-LDA was used for rock classification, 27 out of 480 test spectra were misclassified, and their predicted labels were inconsistent with the actual ones. Take the red marble rock (sample number. 2) as an example, the real category tag is 2, but three of the 20 prediction tags are 4, 8, and 12 (3 asterisks in the blue ellipse, seen in Figure 7.9a). The results show that the PA of red marble is only 85%. However, when IFLAS-LDA is used, only seven spectra are misclassified, so the number of misidentified prediction tags is greatly reduced, as shown in Figure 7.9b. The PA of red marble sample is 95%, and the classification accuracy is improved by 10%. In addition, the average PA calculated by MLS-LDA and IFLAS-LDA of all metamorphic rocks is 94.38 and 98.54%, respectively. Figure 7.2 shows the results of LIBS combined with MLS-LDA and IFALS-LDA to

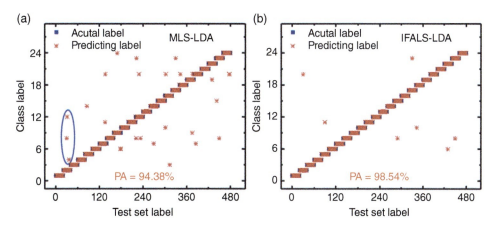

Figure 7.9 Comparisons of the classification results. (a) MLS-LDA and (b) IFALS-LDA.

classify 24 metamorphic rock samples. Compared with MLS-LDA, IFALS-LDA has better classification performance. The results show that IFLAS can effectively improve the classification accuracy of metamorphic rocks. In order to further evaluate the performance of the method, 10-fold cross-validation was carried out. The cross-validation accuracy (CVA) of MLS-LDA and IFALS-LDA were 94.01 and 98.18%, respectively. In addition, from the perspective of classification efficiency, the time required by MLS-LDA is much longer than that by IFALS-LDA. In detail, the former takes about 2768.38 seconds, while IFLAS-LDA only needs about 4.34 seconds, which shows that ifals method greatly improves the efficiency of line selection and classification.

Vítkova et al. [78] used LIBS combined with LDA and artificial neural network (ANN) to rapidly identify biominerals. LDA has correctly classified six of the eight spectra, which is a good and sufficient result. Table 7.7 shows the prediction results of a specific method for data processing and classification of spectra of 18 different samples using LIBS combined with LDA and ANN. It can be seen from Table 7.7 that LDA method has two sample prediction errors, and ANN method has only one sample prediction error, so ANN method has better prediction performance. The results show that LDA and ANN are usually used for data classification, which obviously facilitates the purpose of remote LIBS spectral classification. The new method of combining LDA and ANN with PC fractions has been proved to be a promising model.

Zhou et al. [79] proposed a pattern recognition algorithm based on PCA and LDA (PCA-LDA), which was used to rapidly detect heavy metals in granosa using LIBS data. Figures 7.10–7.12 show the results of LIBS combined with PCA-LDA, PLS-DA, and SIMCA for rapid detection of heavy metals in granosa. The optimal parameters determined by PCA-LDA and PLS-DA are 16 and 9, respectively. The optimal parameters of SIMCA and SVM are 6, 8, 3, and 0.0313, respectively. Therefore, PCA-LDA algorithm has the best classification accuracy. The results show that, compared with PCA-LDA, PLS-DA, SIMCA, and SVM, PCA-LDA has the best classification accuracy and the recognition rate is 87%. Therefore, the combination of PCA-LDA algorithm and portable LIBS spectrometer provides the

Table 7.7 The predictions for test spectra created by LDA (ML) and ANN with back propagation.

Test spectrum	Material	LDA(ML)	ANN
No.11	Soil	Soil	**Ceramics**
No.12	Brick	Brick	Brick
No.36	Bear tooth	Bear tooth	Bear tooth
No.42	Mortar	**Soil**	Mortar
No.43	Bone	Bone	Bone
No.50	Shell	Shell	Shell
No.51	Ceramics	Ceramics	Ceramics
No.93	Human tooth	**Bone**	Human tooth

In the first column, there are a number of test spectra, and in the second one there are real material characters of particular samples. The incorrect predictions are marked in bold.

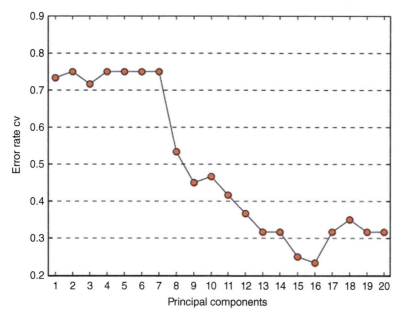

Figure 7.10 Optimal parameters of PCA-LDA.

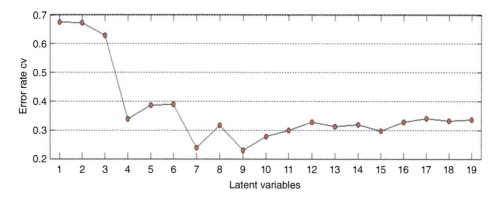

Figure 7.11 Optimal parameters of PLS-DA.

possibility for rapid onsite identification of food safety, marine environmental monitoring, and protection.

7.5 Partial Least Squares Discriminant Analysis

7.5.1 Introduction

PLS is a classical multivariate statistical data analysis method, which was first proposed by Wold and Albano in 1983. In recent decades, it has been developed rapidly in theory, method, and application. The mathematical basis of PLS is also PCA. The difference is that

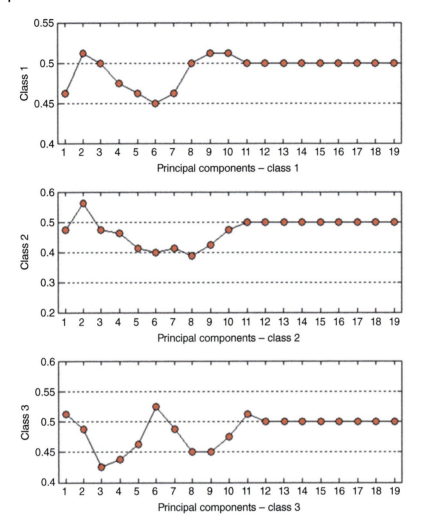

Figure 7.12 Optimal parameters of SIMCA.

PLS considers the information in matrix Y while describing the information of matrix X. In other words, it links sample data with sample label (category) in operation and realizes regression modeling (multiple linear regression), data structure simplification (PCA), and correlation analysis between variables (canonical correlation analysis). Therefore, PLS can be used as a supervised pattern recognition method to establish models and realize automatic recognition and classification of samples [80].

Since the final output of PLS is a probability matrix, further DA is needed in the modeling, which is called PLS-DA method. PLS-DA is a multivariate statistical analysis method for discriminant analysis. DA is a common statistical analysis method to judge how to classify research objects according to the observed or measured values of several variables. The principle is to train the characteristics of different processing samples (such as observation samples and control samples) to generate training sets and test the reliability of training sets.

Compared with the traditional PCA, PLS-DA method maximizes the feature deviation between different classes of samples and minimizes the feature deviation between similar individuals, so as to realize the correspondence between sample information data and sample types. Therefore, PLS-DA is a supervised discriminant analysis method, which can be used to establish the corresponding classification or quantitative model [81].

PLS-DA is widely used in various fields. Jeremiah J. Remus et al. [82] used the advanced signal of LIBS data combined with PLS-DA processing analysis to identify Obsidian sources. Castro et al. [83] used LIBS combined with PLS-DA slave hard disk LIBS (hd) to characterize the base (Al and Cu) and noble (Au and Ag) elements on printed circuit boards (PCB). De Lucia et al. [84] used LIBS combined with PLS-DA to detect explosives. Hark et al. [85] used LIBS system combined with PLS-DA to distinguish tantalum ores from different sources. The advantages of PLS-DA are that the number of sample observations is small and the influence of multicollinearity among variables can be reduced.

7.5.2 Theory

In mathematics, the columns of matrix Y participate in the calculation of matrix X factor, and the columns of matrix X also participate in the calculation of matrix Y factor. The mathematical model is $X = TP' + E$, $Y = UQ' + F$. The matrix elements of T and U are called the scores of X sum, the matrix elements of P and Q are called the loading of X and Y, and E and F are the errors introduced by PLS model to fit X and Y. Specifically, let the overall data have Q dependent variables and P independent variables. In order to study the statistical relationship between dependent variables and independent variables, n sample points are observed. Thus, the data matrix X and Y of independent variables and dependent variables are formed. PLS extracts T and U from X and Y, respectively, which requires that (i) T and U should carry as much variation information as possible in their respective data tables and (ii) the correlation between T and U can reach the maximum. After extracting the first component, PLS is used to perform the regression of X to T and Y to T, respectively. If the regression equation has reached a satisfactory precision, the algorithm will be terminated; otherwise, the residual information of X interpreted by T and Y interpreted by T will be used for the second round of component extraction until a satisfactory accuracy can be achieved. If more than one component is extracted from X, the PLS will be expressed as the equation of Y with respect to the original independent variable. Before calculating the PLS, the matrix X and Y should be standardized, namely moment. Each column of the matrix subtracts their average value and divides it by the standard deviation. Then for each principal component:

Step 1: Normalize X and Y, extract the parameters T and U, and let $U_{start} = y_i$.

For matrices X:

$$w' = \frac{u'X}{u'x} \tag{7.21}$$

$$W'_{new} = \frac{w'_{old}}{\|w'_{old}\|} \text{ (normalized)}$$

$$t = \frac{Xw}{w'w} \tag{7.22}$$

For matrices Y:

$$q' = \frac{t'Y}{t't} \tag{7.23}$$

$$Q'_{new} = \frac{q'_{old}}{\|q'_{old}\|}$$

$$u = \frac{Yq}{q'q} \tag{7.24}$$

Step 2: Convergence test.

Compare the T obtained by formula (7.22) with the previous iteration. If it is equal (including certain rounding error), proceed to the next step, otherwise, go to Eq. (7.21). If y has only one variable, then skip expressions (7.23) and (7.24) and place $q = 1$.

Step 3: Calculate the loading of X and re standardize the score and weight.

$$P' = \frac{t'X}{t't} \tag{7.25}$$

$$P'_{new} = \frac{p'_{old}}{\|p'_{old}\|} \text{ (normalized)}$$

$$t_{new} = \frac{w_{old}}{\|P_{old}\|} \tag{7.26}$$

$W'_{new} = w'_{old}/\|p'_{old}\|$ (P', Q' and w' are used for prediction; u and t are used for classification)

Step 4: Calculate the regression coefficient B for internal correlation.

$$b = \frac{u't}{t't} \tag{7.27}$$

Step 5: Calculate the residual error for the principal component H.

$$E_h = E_{h-1} - t_h p_{h'}; \quad X = E_0 \tag{7.28}$$

$$F_h = F_{h-1} - b_h t_h q_{h'}; \quad Y = F_0 \tag{7.29}$$

After that, go back to the first step to calculate the next principal component. (Note: after the first principal component operation, X and Y will be replaced by their residuals E_h and F_h in the first three steps).

The above steps are the modeling process of PLS algorithm, and the ultimate purpose of modeling is to predict the unknown samples. The sample data used for prediction also include independent variable matrix X and dependent variable matrix Y. At this time, it is assumed that there are n1 samples, and the prediction steps are as follows:

Step 1: As in the modeling process, first standardize X and Y.

$$h = 0, \quad Y = \hat{y} \text{(Normalized)} \tag{7.30}$$

Step 2:

$$h = h + 1$$

$$\hat{t}_h = X w_h$$

$$y = y + b_h \hat{t}_h q'_h \tag{7.31}$$

$$x = x - \hat{t}_h p'_h \tag{7.32}$$

If $H > A$ (main component fraction), continue with the third step, otherwise go to the second step.

Step 3: At this time, the obtained Y is the normalized matrix, so it needs to be restored to the original coordinates according to the reverse operation of the standardization step that is to get the final prediction result of the model that users want.

7.5.3 Application

Gottfried et al. [86] used LIBS to study a large number of natural carbonate, fluorite, and silicate geological materials. Single-pulse and double-pulse LIBS spectra were obtained by close contact table and confrontation ($25m$) LIBS system. PCA and PLS-DA were used to identify geological samples and classify materials. Table 7.8 shows the PLS-DA classification results of calcium carbonate samples analyzed by confrontation LIBS. Table 7.9 shows the PLS-DA classification results of all carbonate samples analyzed by single pulse, double pulse and confrontation LIBS. Table 7.10 shows the PLS-DA classification results of silicate samples obtained by laboratory single-pulse LIBS system. It can be seen from these three tables that PLS-DA has good classification results for all samples. The results show that PLS-DA has a good identification of all sample types, and several improved sample classification techniques are determined.

Ji et al. [87] used LIBS and pattern recognition method to identify artificially contaminated Tegillarca particulate samples of three toxic heavy metals, zinc (Zn), cadmium (Cd), and lead (Pb). The 30 variables are used as input variables of three classifiers: PLS-DA, SVM, and random forest (RF). This work shows that the combination of pattern recognition analysis and LIBS technology can distinguish the healthy shellfish samples of *Tegillarca granosa* from the shellfish samples polluted by toxic heavy metals, and LIBS technology only needs minimal pretreatment (Figure 7.13).

Table 7.8 PLS-DA classification results for calcium carbonate samples analysis by standoff LIBS.

#LV	#Class	Correct classification (%)	Wrong classification (%)
20	22	95.5	2.1
20	5	100	0.4

Table 7.9 PLS-DA classification results for all the carbonate samples analyzed by sigle-pulse, double-pulse, and standoff LIBS.

LIBS system	#LV	#Classes	Correct classification (%)	Wrong classification (%)
Single-pulse	15	11	100	0.07
Double	20	11	100	0.0
Standoff	19	14	100	0.1

7 Distance-based Method

Table 7.10 PLS-DA classification results silicate sample spectra acquired with the laboratory single-pulse LIBS system.

Group (subgroup)	#LV	#Classes	Correct classification (%)	Wrong classification (%)
Silicates	20	6	100.0	2.4
Quartzite	20	5	92.0	4.0
Chert	9	4	100.0	0.0
Volcanic tuff	20	5	100.0	0.0
Basaltic lava	20	6	100.0	0.7
Shales and slates	20	25	100.0%	0.7

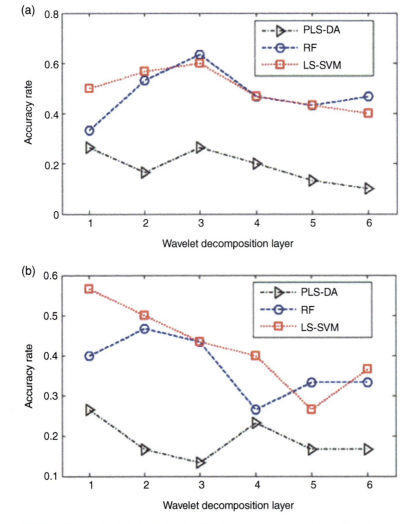

Figure 7.13 Recognition results at different decomposition layers: (a) high-frequency components; (b) low-frequency components.

Table 7.11 PLS-DA classification results for the validation set.

Validation set	Correct classification (%)	Wrong classification (%)	Unclassification (%)
Quartz Sandstone	0.533	0	0.467
Brown Mudstone	1	0	0
Green Mudstone	0.600	0.133^a	0.267
Black Mudstone	0.600	0^b	0.400
Average	0.6834	0.0333	0.2833

^{a} Two spectra of Green Mudstone are misclassified as Brown Mudstone.
^{b} One spectrum of Black Mudstone is misclassified as Quartz Sandstone.

Tian et al. [88] used LIBS and chemometrics methods to classify a series of cutting samples in geological logging field. The linear PLS-DA method is compared with the nonlinear SVM method. The optimal PLS-DA model and SVM model are established by LIBS spectrum calibration method and tested by validation spectrum. In order to further improve the classification accuracy, a new classification method is designed through the joint analysis of PLS-DA and SVM models. With this method, 95% of the verified spectra were correctly classified, and no unclassified spectra were observed. Table 7.11 shows the PLS-DA classification results of the verification set, Table 7.12 shows the SVM classification results of the verification set, and Table 7.13 shows the analysis of the classification results of the verification set through the joint analysis of PLS-DA and SVM. The results show that the performance of SVM and PLS-DA coupling is the best. This work shows that the combination of PLS-DA and SVM makes cuttings identification have excellent performance.

Porto et al. [89] used LIBS and PLS-DA to early predict the genotype resistance or susceptibility of sugarcane borer. PLS-DA was used to establish the classification model (Figure 7.14). Figure 7.15 shows the results of using LIBS combined with PLS-DA to establish the classification model, which shows that PLS-DA model can be used to classify

Table 7.12 SVM classification results for the validation set.

Validation set	Correct classification (%)	Wrong classification (%)	Unclassification (%)
Quartz Sandstone	1	0	0
Brown Mudstone	0.867	0.067^a	0.067
Green Mudstone	0.867	0.067^b	0.067
Black Mudstone	0.933	0.067^c	0
Average	0.9167	0.0500	0.0333

^{a} One spectrum of Brown Mudstone is misclassified as Green Mudstone.
^{b} One spectrum of Green Mudstone is misclassified as Brown Mudstone.
^{c} One spectrum of Black Mudstone is misclassified as Quartz Sandstone.

Table 7.13 Classification results for the validation set by the joint analysis of PLS-DA and SVM.

Validation set	Correct classification (%)	Wrong classification (%)
Quartz Sandstone	1	0
Brown Mudstone	1	0
Green Mudstone	0.867	0.133[a]
Black Mudstone	0.933	0.067[b]
Average	0.9500	0.0500

[a] Two spectra of Green Mudstone are misclassified as Brown Mudstone.
[b] One spectrum of Black Mudstone is misclassified as Quartz Sandstone.

sugarcane genotypes that are sensitive and resistant to sugarcane borer. The results showed that the established models, in addition to using LIBS made of straw particles, showed good predictive ability, which made them suitable for predicting the resistance or susceptibility of sugarcane genotypes in the early stage of plant life.

Gazmeh et al. [90] used LIBS and PLS-DA to identify healthy teeth and caries. In laser drilling, a microplasma is produced, which can be used for element analysis of ablated structures by LIBS. In this study, LIBS was used to study the possibility of differentiating healthy and carious tissues. This possibility is based on the atomic and ionic emission lines of the

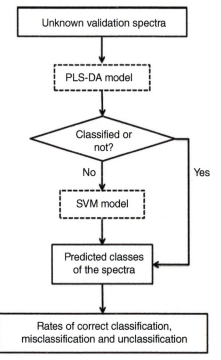

Figure 7.14 Flowchart of the joint analysis of PLS-DA and SVM models.

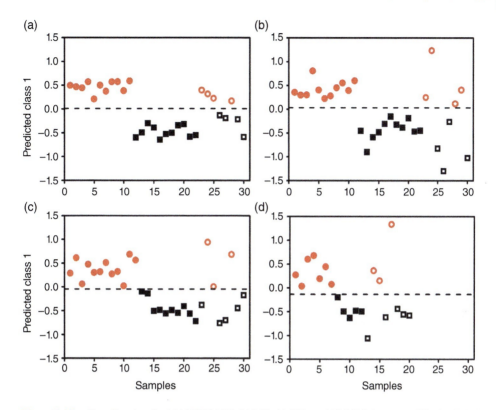

Figure 7.15 Classification for (a) UV-VIS-NIR, (b) MID, (c) NIR, and (d) LIBS data sets. Filled red circles (●) and empty red circles (○) are samples of class 1. Filled black squares (■) and empty black squares (□) are samples of class 2. Filled and empty shapes refer to calibration and prediction sets, respectively. The dashed black line (–) indicates the threshold estimated by the PLS-DA algorithm.

LIBS spectra of teeth belonging to P, Ca, Mg, Zn, K, Sr, C, Na, H, and O elements by multivariate statistical analysis, called PLS-DA. Figures 7.16 and 7.17 show the classification and prediction of different samples by PLS-DA model, which shows that PLS-DA method is a promising caries detection technology for spectral analysis of plasma emission during laser drilling. The results show that PLS-DA method is a promising method for caries detection.

7.6 Principal Component Analysis

7.6.1 Introduction

PCA is a kind of data analysis method to establish multivariate linear model for complex data. Multivariate linear PCA model is constructed by orthogonal basis vector (eigenvector), which is also called principal component. The principal components fit the statistically significant variance and random error in the data. One of the main purposes of PCA is to eliminate the random errors in principal components, so as to reduce the dimension of complex

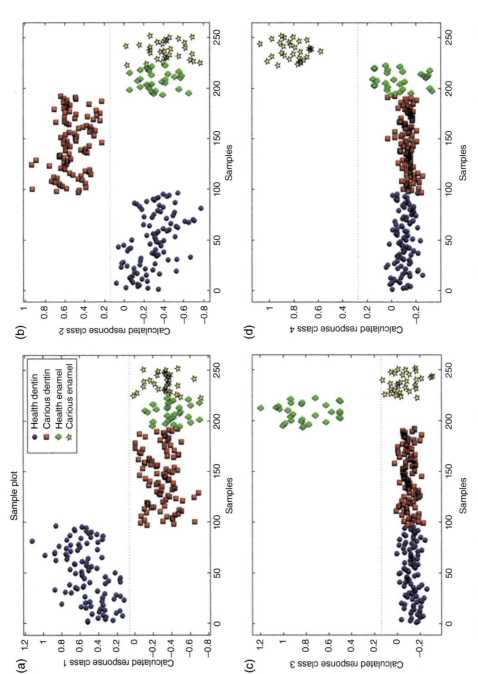

Figure 7.16 PLS-DA model: (a) 1–96 spectra of healthy enamel samples, (b) 97–192 spectra of carious enamel samples, (c) 193–222 spectra of healthy dentin samples, and (d) 223–252 spectra of carious dentin samples.

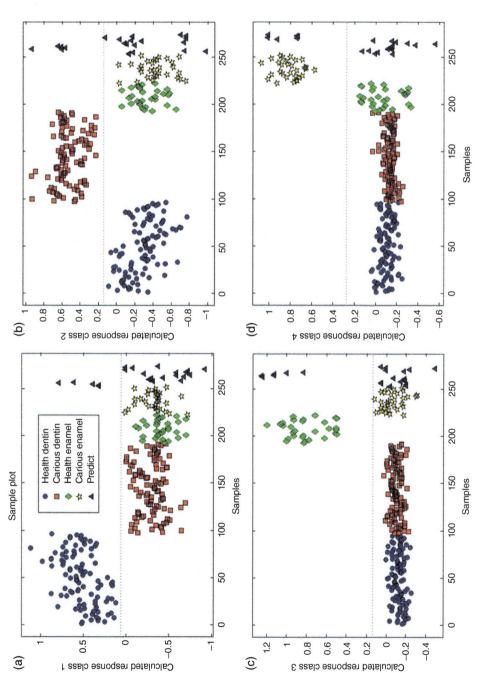

Figure 7.17 PLS-DA model: (a) 1–96 spectra of healthy enamel samples, (b) 97–192 spectra of carious enamel samples, (c) 193–222 spectra of health dentin samples, (d) 223–252 spectra of carious dentin samples, and 253–274 spectra of new prediction samples.

problems and minimize the influence of measurement errors. PCA can provide more transparent information accumulation. Factor analysis attempts to identify potential variables or explain a set of factors that observe the correlation patterns within variables. PCA is often used for data simplification to identify a small number of factors that account for most of the variance observed in a larger number of explicit variables [91].

In modern multivariate data analysis, it is inevitable to implement fast, robust, and effective algorithms. PCA algorithms are becoming popular not only in the spectral community because it meets the above quality. Therefore, PCA is often used to process the detected multiple signals (characteristic spectra). In the past decade, PCA has been adopted by the application of LIBS, and the number of scientific articles citing PCA has steadily increased. Interest in PCA is not only due to the basic need to obtain fast data visualization at lower dimensions but also to examine the most significant variables. Recently, PCA has also been applied to unconventional data analysis (deal with large-scale LIBS) [92].

7.6.2 Theory

Because the time-domain waveform information measured is not conducive to the in-depth data mining, it is necessary to denoize, standardize, and transform the collected data, and fast Fourier transform (FFT) the data to the frequency domain analysis. Due to the large amount of spectrum data obtained by FFT and the large redundancy, the spectrum data need to be further processed by PCA to improve the efficiency of feature extraction. PCA is an effective method to process, compress, and extract information based on covariance matrix, of which the aim is to reduce the number of variables, remove redundant data, and use as few variables as possible to represent most of the information in the observed data, that is to transform multiple observed variables into as few as possible unrelated comprehensive indicators. Through this transformation, the collinearity between the observed original data variables can be effectively solved, and the problems such as large amount of operation data, instability, and ill-conditioned matrix caused by this transformation can be solved [93]. PCA is a linear combination of principal components expressed as original observations.

$$Y_i = \sum_{i=1}^{n} f_{ij} X_j \tag{7.33}$$

In formula (7.33), X_j represents the matrix obtained after standardization of the original data matrix, and its variance is 1; f_{ij} represents the elements in the special transverse beam matrix corresponding to the eigenvalues of the correlation coefficient matrix, and $i = 1, 2, 3, ..., n$, n is the digits of the correlation coefficient matrix. When determining the number of principal components, the appropriate number can be selected according to the cumulative contribution rate of eigenvalues of correlation matrix. Selecting the appropriate principal component can reduce the data dimension and the main information of the original data can be retained.

PCA shows good performance in the analysis of LIBS spectral data and is widely used in metallurgical analysis, life science, environmental science, agriculture, and other fields. LIBS combined with PCA provides a powerful solution to the classification and regression problems in the above fields. The following of this section will give an example of this method in solving classification problems of the field mentioned earlier [94].

7.6 Principal Component Analysis

The calculation steps of PCA

1) Calculate the correlation coefficient matrix

$$R = \begin{pmatrix} r_{11} & r_{12} & \cdots & r_{1p} \\ r_{21} & r_{22} & \cdots & r_{2p} \\ \vdots & \vdots & \ddots & \vdots \\ r_{p1} & r_{p2} & \cdots & r_{pp} \end{pmatrix} \qquad (7.34)$$

In formula (7.35), R_{ij} $(i, j = 1, 2, ..., p)$ is the correlation coefficient of the original variable X_i and X_j, $R_{ij} = r_{ij}$, and the calculation formula is as follows:

$$r_{ij} = \frac{\sum_{k=1}^{n}(X_{ki} - \bar{x})(X_{kj} - \bar{x})}{\sqrt{\sum_{k=1}^{n}(X_{ki} - \bar{x})^2(X_{kj} - \bar{x})^2}} \qquad (7.35)$$

2) Calculating eigenvalues and eigenvectors

In order to solve the characteristic equation, Jacobian method is often used to find the eigenvalues and arrange them in order of magnitude; $\lambda_1 \geq \lambda_2 \geq \cdots \geq \lambda_p \geq 0$; The eigenvectors corresponding to the eigenvalues are obtained, respectively. $e_i(i = 1, 2, L, p)$, requirement $|e_i| = 1$, $\sum_{j=1}^{p} e_{ij}^2 = 1$, where e_{ij} is the jth component of vector e_i.

3) Calculate the contribution rate of principal component and cumulative contribution rate

The contribution rate and cumulative contribution rate are calculated according to the formulas (7.36) and (7.37):

$$\text{Contribution rate}: \frac{\lambda_i}{\sum_{k=1}^{p} \lambda_k} (i = 1, 2, L, p) \qquad (7.36)$$

$$\text{Cumulative contribution rate}: \frac{\sum_{k=1}^{i} \lambda_k}{\sum_{k=1}^{p} \lambda_k} (i = 1, 2, L, p) \qquad (7.37)$$

In formulas (7.36) and (7.37), variables with a cumulative contribution rate of 85–95% are selected. $\lambda_1, \lambda_2, \lambda_L, \lambda_p$ represent the corresponding principal component.

4) Calculation of principal component load

In formula (7.38), the principal component load is obtained from the following formula:

$$I_{ij} = p(Z_i, X_j) = \sqrt{\lambda_i} e_{ij} \quad (i, j = 1, 2, L, p) \qquad (7.38)$$

5) Scores of each principal component

In formula (7.39), the scores of each principal component can be calculated, as is shown in the following matrix.

$$Z = \begin{pmatrix} Z_{11} & Z_{12} & \cdots & Z_{1p} \\ Z_{21} & Z_{22} & \cdots & Z_{2p} \\ \vdots & \vdots & \ddots & \vdots \\ Z_{p1} & Z_{p2} & \cdots & Z_{pp} \end{pmatrix} \qquad (7.39)$$

7.6.3 Application

In Yang's work, PCA and artificial neural network are combined with LIBS to identify iron ore [95]. In PCA, the relationship between principal component fractions (PCS) has been widely used in libpca analysis. In addition, PCA also provides a method to reduce the effective dimension of data. In this work, PCA is used to reduce the dimension of LIBS spectra. The obtained PCs are used for unsupervised clustering of LIBS spectral data. The 3D diagram of PC1, PC2, and PC3 is shown in Figure 7.18. As can be seen from Figure 7.18, the samples from Australia and South Africa are divided into two groups with almost no overlap, but the samples from Brazil and South Africa overlap greatly. Due to the similarity of the matrix, it overlaps with the Australian sample. Because PCA only uses the linear combination of the original independent variables, these independent variables have the largest variation, so it can only achieve moderate classification performance. Therefore, it is necessary to classify iron ions better, and more advanced algorithms are needed according to the requirements of the country and brand of ore production.

Wang et al. [96] used LIBS to identify cervical cancer. In order to improve the accuracy of LIBS for cervical cancer recognition, chemometrics methods of PCA and SVM were combined. The recognition accuracy of PCA-SVM is 94.44 and 93.06%, respectively. PCA, as an unsupervised statistical analysis technique, is an effective information compression method, which can reduce the dimension of data sets and discover potential variables between samples. The feature decomposition method is used to replace the original variables to get the principal component, which eliminates the correlation and information

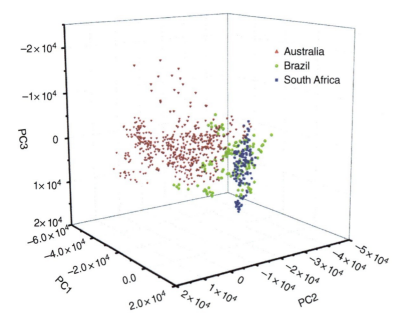

Figure 7.18 PCA plot for the iron ore samples with the first three PCs; variance and cumulative percentage of PCs; the black line in the PRESS, and the red line is the cumulative percentage of the PCs in PCA.

Figure 7.19 Scores of the first three PCs from spectral data of normal tissues and cervical cancer tissues.

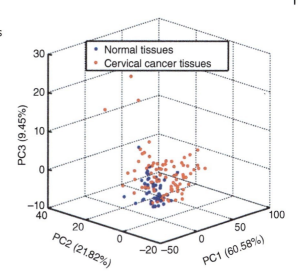

redundancy of the original data. The speed and stability of the model based on the algorithm can be improved by optimizing the number of PCs instead of the original number of variables. In addition, the data clustering graph can also be obtained on the main PC. Figure 7.19 shows the scores of the first three principal components. Although the cumulative contribution rate of the first three principal components is 91.85%, it is difficult to distinguish the spectra of normal tissues and cervical cancer tissues, as shown in Figure 7.19. Therefore, other algorithms need to be studied.

Totally, 31 new PCs were obtained by PCA. The number of PCs has a significant impact on the recognition effect of SVM training set. As shown in Figure 7.20, it can be seen that PCs with more than 10 PCs can achieve better average training recognition accuracy. Under

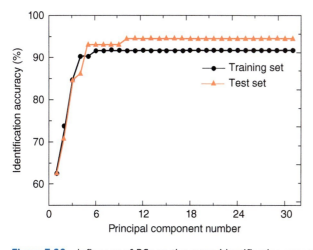

Figure 7.20 Influence of PCs on the mean identification accuracy of training set and test set.

the same training accuracy, the least number of PCs can improve the modeling efficiency of SVM algorithm. Therefore, 10 PCs are selected as input variables to construct SVM identification model. The average recognition rate of training model is 94.67%. The values of C and g were 4.00 and 9.77×10^{-4}, respectively. The results show that PCA-SVM can improve the recognition accuracy and shorten the modeling time from 21.55 to 1.97 seconds. The results showed that LIBS combined with chemometrics can identify normal tissues and cervical cancer tissues.

Pathak et al. [97] studies the LIBS of different parts of teeth. The hardness of different parts of teeth samples of different ages was studied by using the ion atomic strength ratio of Ca spectrum line. From this ratio, the authors infer that enamel is the hardest part of a tooth. PCA was used to obtain PC1 (96%) and PC2 (4%) from LIBS data of different parts of teeth, which explained the maximum variance in the data set. Therefore, this work demonstrates the applicability of LIBS and PCA in conventional in vitro/in vivo dental practice. The PCA score map shown in Figure 7.21 clearly shows that the three parts of the tooth (enamel, cementum, and dentin) are forming three different clusters. From the bar diagram shown in Figure 7.21a, it is clear that the intensity ratio (signal to background) of Ca and Mg is higher in enamel than in cementum and dentin, which is due to the matrix of the enamel. PCA score plot shown in Figure 7.21b clearly demonstrates that three parts (enamel, cementum, and dentin) of tooth are forming three different clusters. Thus, LIBS data with PCA enable us for the rapid classification and distinction of enamel, dentin, and cementum of tooth sample. Therefore, the LIBS data based on PCA enable us to quickly classify and distinguish enamel, dentin, and cementum of tooth samples.

Hanasil et al. [98] used LIBS technique assisted with PCA on the discrimination of extracted animal fats in the liquid form. In Figure 7.22, the cluster of extracted chicken fat, lamb fat, and lard were clearly distinguished in 3D compared to 2D score plot. In previous study, the discrimination of extracted animal fats has been achieved using freezing methods coupled with the ungated LIBS system and PCA approach. A good discrimination of extracted chicken fat, beef fat, lamb fat, and lard was obtained using 3D PCA score plot.

El-Saeid et al. [99] applied standard LIBS and nanoparticle reinforced LIBS for rock classification (igneous and sedimentary). PCA and graph theory were used to classify the spectra. The sedimentary rocks are shown in Figure 7.23a and b, respectively. The whole spectral range (200–700 nm) was used in the analysis since the use of specific spectral segments did not make any substantial difference in the obtained results. In the case of LIBS, the first three principal components constituted 90.5% of the total variance, where PC1, PC2, and PC3 accounted for 62.0, 27, and 5.4%, respectively. The PCA score plot for NELIBS in Figure 7.24b shows overall improved discrimination with the three PCs constituted 92.4% of the total variance, where PC1, PC2, and PC3 accounted for 74.4, 12.7, and 5.3%, respectively. It can be seen from Figure 7.24a and b that PCA provides better discrimination results in case of NELIBS for the studied sedimentary rocks. For igneous rocks, the PCA score plots are depicted in Figure 7.24a and b. The three PCs in case of LIBS (Figure 7.23a) account for a total variance of 90.9% with 66.9, 17.5, and 6.5% variance for PC1, PC2, and PC3, respectively. The plot shows some overlap between the dolerite and granite clusters. By using the NELIBS data in the PCA analysis, Figure 7.23b does not show a clear difference compared to the LIBS data. The three PCs constituted 89.4% of the total variance, where PC1, PC2, and PC3 accounted to 63.3, 15.5, and 10.6% variance,

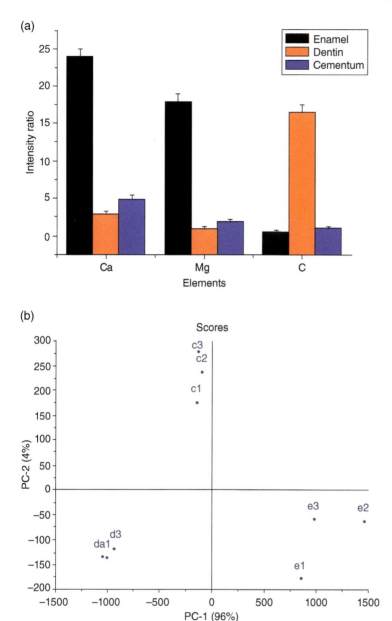

Figure 7.21 (a) A bar diagram of calcium, magnesium, and carbon in different parts of tooth samples. (b) PCA plots (scores).

respectively. The results presented in this work demonstrated that LIBS represents a rapid method for classifying the studied igneous and sedimentary rocks, once the proper statistical tools are used. Although a relatively good classification can be achieved using classical PCA, the approach based on the graph theory gave better results. Moreover, despite the improved signal to noise guaranteed by the NELIBS approach, compared to standard LIBS,

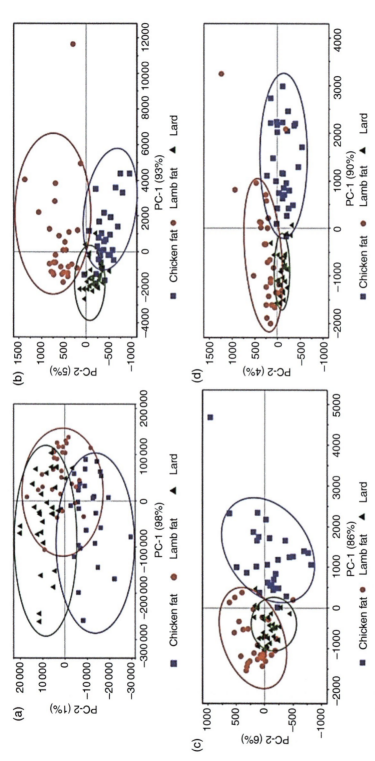

Figure 7.22 The score plots of extracted animal fats for wavelength of (a) 182–946 nm, (b) 390–400 nm, (c) 420–430 nm, (d) 510–520 nm, (e) 585–595 nm, (f) 600–659 nm and combination wavelengths of 390–400 nm, 420–430 nm, and 510–520 nm in (g) 2D and (h) 3D.

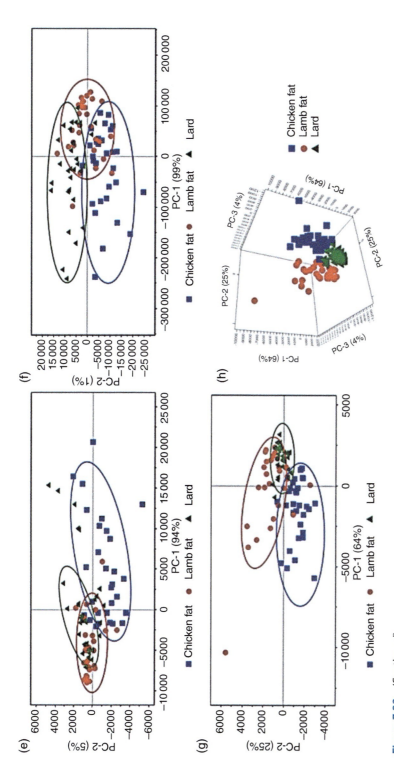

Figure 7.22 (Continued)

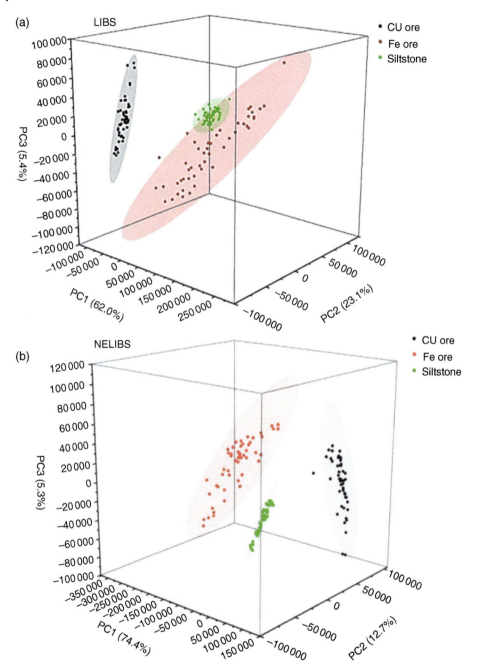

Figure 7.23 PCs score plot of the first three principal components (PC1, PC2, and PC3) for three types of sedimentary rocks for LIBS (a) and NELIBS (b).

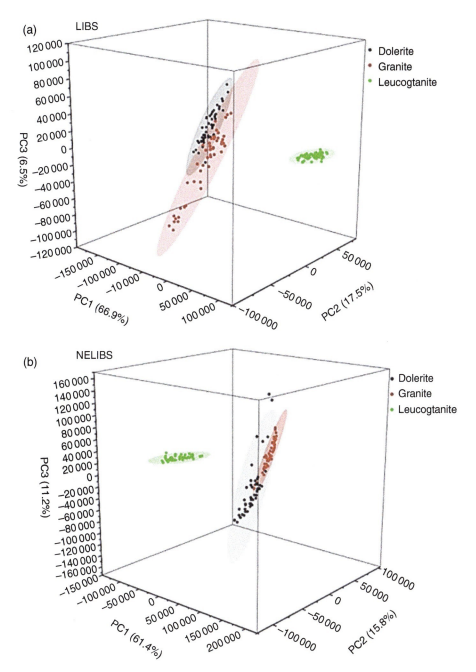

Figure 7.24 PCs score plot of the first three principal components (PC1, PC2, and PC3) for three types of igneous rocks for LIBS (a) and NELIBS (b).

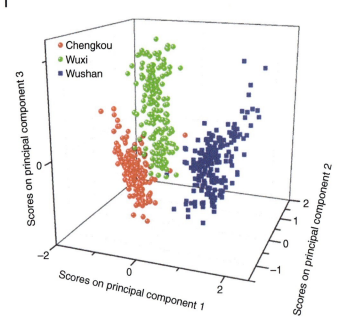

Figure 7.25 Three-dimensional score plot of the roots of Angelica pubescens. The scores for first three principal components are 54.629, 25.035, and 12.795%, respectively.

it has been demonstrated that standard LIBS, treated statistically, provides better results on the considered geological samples. This result is interesting, especially given applications of LIBS on natural materials with more than 99% of the correctly classified spectra.

Wang et al. [100] used LIBS combined with PCA and artificial neural network (ANN) to analyze and identify three kinds of Chinese herbal medicines from different habitats and different parts. The first three PCs of the cumulative interpretation rate are more than 92%, and the three-dimensional score plot is presented in Figure 7.25, each scatter representing a sample, which shows a good clustering effect. As can be seen from Figure 7.25, the samples of each origin are concentrated together and have clear intervals from other groups. The results show that the average classification accuracy of PCA is 99.89%, which is better than LDA and SVM learning method.

7.7 Soft Independent Modeling of Class Analogy

7.7.1 Introduction

Soft independent modeling of class analogy (SIMCA) is a subspace-based classification method, which is widely used in the classification of high-dimensional spectral data in chemometrics [101]. SIMCA can be used to classify the two categories into two stages. First, in the training stage, two class models are established for two classes. The class model in SIMCA is represented by PC class subspace by PCA. Second, in the test phase, according to the distance between the new samples and the two class subspaces, the new samples are classified by classification rules. There are usually two distances: the square of the orthogonal distance between the new sample and the class subspace (OD2) and the square of the score distance between the projection of the new sample to the center of the class subspace (SD2). SIMCA is designed for the "soft" assignment of a new sample, which means that a new sample can be assigned to a known class, including two known classes and one with no assignment. Therefore, SIMCA can be used as both a classifier and an outlier detector [102]. The theory of SIMCA is as follows in this section.

7.7.2 Theory

When SIMCA is used to classify two classes, PCA is used to establish two class subspaces for the two classes. Then, the classification rules based on OD2 and or SD2 are used to determine the class membership of the new samples. Although SIMCA is widely used in the classification of high-dimensional spectral data, it has the problem of establishing class subspace without considering the information between classes. Therefore, calculating the F value of each category separately may not be enough to distinguish a new sample. The solution is to find a more discriminative subspace than the original feature space before applying SIMCA, and the data should be projected to this subspace [103]. Compared with the projection of the original feature space, the projection from the sample to this discriminant subspace is expected to be more separate and easier to classify, as shown in Figure 7.26.

As a preprocessing method, GDS projection improves the subspace mutual subspace method (MSM) commonly used in image set-based target recognition. The purpose of GDS is to solve the problem of MSM: class subspace is generated independently by PCA according to class, so classification may not be very distinctive. This problem is actually the same as that of SIMCA. Therefore, we believe that GDS projection can also be used as a preprocessing method for SIMCA to improve its classification performance. GDS is

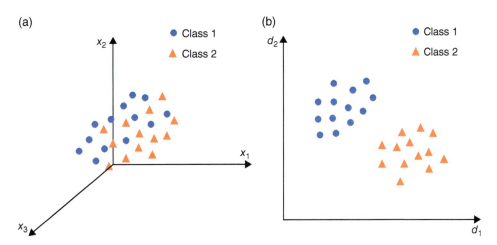

Figure 7.26 (a) Two classes of samples are mixed together in the original three-dimensional feature space. (b) The same groups of samples can be well separated when they are projected to a discriminative two-dimensional subspace.

a subspace that contains the difference information between classes, so it should have stronger discrimination ability than the original feature space. GDS is generated on the basis of generation matrix G_D, which is calculated from the projection matrix of two class subspaces and can provide inter class information. The feature vectors of G_D with smaller eigenvalues contain the difference information between class subspaces, while the eigenvectors of G_D with larger eigenvalues contain similar information between class subspaces. In order to make full use of the difference information, GDS projection only retains the feature vectors with smaller eigenvalues and discards the eigenvectors with larger eigenvalues.

The purpose of developing DOS is to solve the SIMCA problem. The discrimination ability of feature vector is measured by the classification accuracy of SIMCA to the samples projected on the eigenvector. The higher the classification accuracy, the stronger the recognition ability. Considering the simplicity and efficiency of the proposed method, as well as the uncorrelation and orthogonality of the candidate eigenvectors, this kind of filtering feature vector selection scheme was chosen. The effectiveness of DOS preprocessing SIMCA.

In the training phase of SIMCA, suppose $X_k \in R^{nk \times p}$ is the training set of class k ($k = 1, 2$), in that there are n_k training instances and each instance is represented by a p-dimensional data vector (i.e. in the original p-dimensional feature space). To build the PC subspace for each class, eigen decomposition was used to the covariance matrix of the kth class:

$$\text{Cov}(X_k) = \frac{1}{n_k - 1}(X_{k(c)})^T X_{k(c)} = V_k \sum_k V_k^T a \qquad (7.40)$$

In formula (7.40), $X_{k(c)}$ is the column centered X_k; the columns of $V_k \in R^{p \times qk}$ (qk = rank : (Cov(X_k))) denote the normalized eigenvectors, and $\sum_k$ is the eigenvalue of a diagonal matrix of $\{\delta_1 \geq \delta_2 \geq \cdots \geq \delta_{qk}\}$; the first$r_k$($r_k \leq q_k$) columns of V_k was selected as the basis vectors W_k that spans the kth class subspace φ_k, as the first set of base vectors of r_k($r_k \leq q_k$); W_k is the k-class subspace; it belongs to r_k space, which is r_k-dimensional.

It follows that the projection matrix $P_k \in R^{p \times p}$ of φ_k:

$$P_k = W_k W_k^T \qquad (7.41)$$

In formula (7.41), in the test phase, a new sample x_{new} is assigned according to the following two residuals. First of all, the kth class residuals in the training set are as follows:

$$E_k = X_{k(c)} - X_{k(c)} P_k \qquad (7.42)$$

In formula (7.42), the residuals of x_{new} projection to the kth subspace:

$$e^{k,\text{new}} = x_{(c)}^{k,\text{new}} - x_{(c)}^{k,\text{new}} P_k \qquad (7.43)$$

In formula (7.43), center on the average vector X_k. Then, x_{new} is assigned to the class with the smallest F-value [104] where the F-value is defined as

$$F = \frac{\|e^{k,\text{new}}\|_2^2}{\|E_k\|_2^2/(n_k - r_k - 1)} \qquad (7.44)$$

In formula (7.44), $\|\cdot\|_2$ is the Frobenius rule $\|e^{k,\text{new}}\|_2^2$ is OD^2.

Because the class subspace in SIMCA is constructed independently, SIMCA does not consider the information between classes, which limits the classification performance. In order to improve the performance of SIMCA, the aim is to find a subspace that is more different than the original feature space. When SIMCA is applied to the projection of samples in this subspace, better performance can be obtained due to the higher separation degree of samples in the subspace. The process of finding and projecting to the discriminating subspace can be used as a preprocessing step of SIMCA.

7.7.3 Application

Some works about the application of SIMCA coupled with LIBS in different fields were introduced in this section.

Schroder et al. [105] used LIBS to analyze a group of eight salts consisting of four chlorides and four sulfates with the same cation as pure pressed particles and in frozen saline solution. The spectral analysis of LIBS was performed by MVA method, PCA, SIMCA, and PLS-DA. The results show that LIBS combined with SIMCA method can correctly classify calcium, potassium, and sodium salts. There are no unclassified spectra and no misclassified samples. However, the correct classification spectral number of SIMCA in magnesium salt is close to those in wrong classification, so it cannot be correctly classified. As can be seen from the results in Table 7.14 (top) the calcium, potassium, and the sodium salts in this study could be correctly classified with no unclassified spectrum and moreover no falsely classified sample. Therefore, independent soft pattern recognition (SIMCA) is not successful for salt classification in frozen solutions. PLS-DA is better for LIBS spectral analysis.

Myakalwa et al. [106] used LIBS to detect the content of tablets and discussed the feasibility of its application in conventional classification. Two chemometrics algorithms, PCA and SIMCA, were used to mine the multivariable characteristics of LIBS data, which proved that LIBS has the potential to distinguish and identify tablets. SIMCA algorithm uses the excellent expected classification accuracy of supervised classification and shows its potential application in process analysis technology in the future, especially in the application of rapid online process control monitoring in pharmaceutical industry. Compared with PCA, SIMCA is a supervised classification technique. Figure 7.27 shows a bar plot visualization of SIMCA classifications for a representative set of 30 test samples (for a specific iteration). To obtain a more comprehensive sense of the results, we computed the average rate of unclassification, misclassification, and correct classification over 100 iterations (Table 7.15). The results show that the average accuracy of SIMCA is about 94%. Based on the results of this study, it is expected that the combination of LIBS and chemometrics can be successfully used for quality control and routine monitoring of tablets.

Aquino et al. [107] used LIBS to analyze the polymerized part of mobile phone waste and used KNN (K-nearest neighbor), SIMCA, and PLS-DA, respectively, to propose classification models using black and white polymers to identify manufacturers and countries of origin. The results showed the average correct classification and sensitivity: K-nearest neighbor 92.8 (93.8); SIMCA 82.4 (71.3); PLS-DA cross-validation 17.1 (29.5); and PLS-DA cross-validation 16.0 (27.3). As can be observed in Table 7.16, which summarizes the results obtained for each model, the KNN technique followed by SIMCA provided the best results for the prediction of the manufacturer and the origin of the samples analyzed in

Table 7.14 SIMCA prediction results for the pressed salt samples (top) and frozen salt solutions (bottom).

	CaCl$_2$	CaSO$_4$	KCl	K$_2$SO$_4$	MgCl$_2$	MgSO$_4$	NaCl	Na$_2$SO$_4$
Sample (pressed)								
CaCl$_2$	8	—	—	—	—	—	—	—
CaSO$_4$	—	8	—	—	—	—	—	—
KCl	—	—	8	—	—	—	—	—
K$_2$SO$_4$	—	—	—	8	—	—	—	—
MgCl$_2$	—	—	—	—	8	6	—	—
MgSO$_4$	—	—	—	—	8	8	—	—
NaCl	—	—	—	—	—	—	8	—
Na$_2$SO$_4$	—	—	—	—	—	—	—	8
Sample (ice)								
CaCl$_2$	8	1	—	—	1	—	—	—
CaSO$_4$	—	8	—	—	—	—	—	—
KCl	—	—	8	8	8	5	—	—
K$_2$SO$_4$	—	—	8	8	8	7	—	—
MgCl$_2$	2	—	4	3	8	4	—	—
MgSO$_4$	—	—	7	3	8	8	—	—
NaCl	—	—	—	—	—	—	8	8
Na$_2$SO$_4$	—	—	—	—	—	—	6	8

The samples listed in the column on the left are treated as "unknown" and are assigned to one or more of the classes listed in the top line. Stated are the total numbers of spectra assigned to the class in the first line.

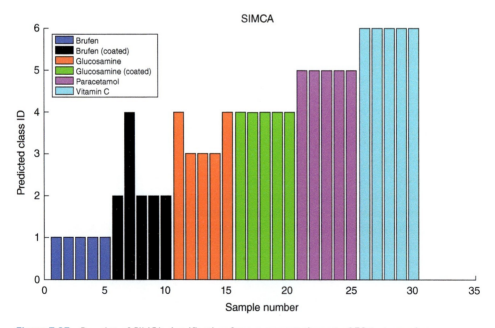

Figure 7.27 Bar plot of SIMCA classification for a representative set of 30 test samples.

7.7 Soft Independent Modeling of Class Analogy

Table 7.15 SIMCA classification results obtained from 30 test samples over 100 iterations.

	Correct classification (%)	Wrong classification (%)	Unclassification (%)
Brufen	90.8	7.6	1.6
Brufencoated	93.4	6.2	0.4
Glucosamine	90.4	8.8	0.8
Glucosaminecoated	94.0	0.2	5.8
Paracetamol	96.8	0	3.2
Vitamin C	98.8	0	1.2
Average	94.03	3.8	2.17

Table 7.16 Average percentage of correct classifications and sensitivity for the models built from white and black polymers.

Technique	Manufacturers Correct classification (%)	Sensitivity	Origin Correct classification (%)	Sensitivity
KNN (1 neighbor)	99.5	99.2	100	100
SIMCA	95.9	95.1	98.0	98.5
PLS-DA without cross-validation	73.2	78.7	98.8	98.0
PLS-DA with cross-validation (leave-one-out)	72.8	78.4	98.8	98.0
Black polymers models (variables selection B), Matrix = 2210 : 2367				
KNN (1 neighbor)	100	100	100	100
SIMCA	84.7	74.5	86.9	73.8
PLS-DA without cross-validation	15.0	25.2	33.5	81.3
PLS-DA with cross-validation (leave-one-out)	15.5	25.9	33.5	81.4
Black polymers models (variables selection C), Matrix¼ = 2210 : 2192				
KNN (1 neighbor)	92.8	93.8	100	100
SIMCA	82.4	71.3	86.8	76.0
PLS-DA without cross-validation	17.1	29.5	32.1	82.6
PLS-DA with cross-validation (leave-one-out)	16.0	27.3	32.2	82.7

terms of the average of the percentage of correct classifications and sensitivity. The results show that KNN and SIMCA classification models have good prediction ability, and PLS-DA has no corresponding analysis ability.

7.8 Conclusion and Expectation

LIBS combined with classification algorithm (distance-based method) has been introduced in this section. The brief introduction, theory, and introduction of seven classification algorithm (CA, ICA, K-NN, LDA, PLS-DA, PCA, and SIMCA) were summed up. The classification algorithm based on distance is not limited to the several mentioned in this section. This paper only introduces the more prominent ones here. There are also related reports on the improved algorithm of the methods mentioned above and other methods. Readers can make further inquiries according to their work needs. The characteristics of data or sample sets are different, of that some are noisy, some sparse, and some are highly correlated. The effect of classification is generally related to the characteristics of the data, and there is no algorithm that can accurately analyze all classification problems. Some suggestions to enhance the classification performance of the classification algorithm combined with LIBS technology are as follows:

1) Sample pretreatment. Although the development of spectral analysis technology has made some breakthroughs, there are still deficiencies in the analysis of complex systems. Therefore, the processing of some black box information cannot be completed only by classification algorithm. Sample separation or purification that should be considered by researchers can sometimes have an immediate effect on the progress of classification.
2) Spectral pretreatment. The sample is excited by a laser to generate plasma, so as to obtain the corresponding LIBS spectral data of this sample. It is well known that the instability of the laser signal will lead to large fluctuations in LIBS spectra of the obtained samples. In addition, due to the reason of the sample itself and the difference of experimental environment, LIBS spectrum will be affected. Therefore, when LIBS combined with classification algorithm to construct the classification model, it is necessary to preprocess LIBS spectral data.
3) Feature variable extraction. Generally, the LIBS spectral data of samples contain a lot of data information, including not only the information related to the classification problem solving but also a lot of interference information and redundant information. If the model is directly constructed based on full LIBS spectrum, the performance of the model will be seriously affected, such as the reduction of classification accuracy and the increase of the modeling time. Therefore, when using LIBS data for modeling and analysis, it will play an effective role in improving the performance of classification model by feature variable extraction method.
4) Data fusion. Data fusion strategy has been paid more and more attention in the analysis of complex system in the recent years. Data fusion uses data from multiple analytical instruments to analyze and process the same complex sample. It can be characterized from multidimensional to obtain more comprehensive sample information. Combining other analysis techniques with LIBS technology can improve the analysis performance.

References

1 K. S. Booksh, B. R. Kowalski, Theory of analytical chemistry, *Analytical Chemistry*, (1994), **66**, 782–791.
2 M. T. Jimare Benito, C. Bosch Ojeda, F. Sanchez Rojas, Process analytical chemistry: applications of near infrared spectrometry in environmental and food analysis: an overview, *Applied Spectroscopy Reviews*, (2008), **43**:5, 452–484.
3 S. Chigome, N. Torto, A review of opportunities for electrospun nanofibers in analytical chemistry, *Analytica Chimica Acta*, (2011), **706**, 25–36.
4 E. L. Willighagen, R. Wehrens, L. M. C. Buydens, Molecular chemometrics, *Critical Reviews in Analytical Chemistry*, (2006), **36**, 189–198.
5 Y. Roggo, P. Chalus, L. Maurer, C. Lema-Martinez, A. Edmond, N. Jent, A review of near infrared spectroscopy and chemometrics in pharmaceutical technologies, *Journal of Pharmaceutical and Biomedical Analysis*, (2007), **44**, 683–700.
6 A. Hoskuldsson, PLS regression methods, *Journal of Chemometrics*, (1988), **2**, 211–228.
7 A. M. Yehia, H. T. Elbalkiny, S. M. Riad, Y. S. Elsaharty, Chemometrics for resolving spectral data of cephalosporines and tracing their residue in waste water samples, *Spectrochimica Acta Part A: Molecular and Biomolecular Spectroscopy*, (2019), **219**, 436–443.
8 P. D. Wentzell, D. T. Andrews, D. C. Hamilton, K. Faber, B. R. Kowalski, Maximum likelihood principal component analysis, *Journal of Chemometrics*, (1997), **11**, 339–366.
9 K. Jetter, U. Depczynski, K. Molt, A. Niemöller, Principles and applications of wavelet transformation to chemometrics, *Analytica Chimica Acta*, (2000), **420**, 169–180.
10 J.-B. Sirven, B. Bousquet, L. Canioni, L. Sarger, Laser-induced breakdown spectroscopy of composite samples: comparison of advanced chemometrics methods, *Analytical Chemistry*, (2006), **78**, 1462–1469.
11 S. J. Hill, S. Chenery, J. B. Dawson, E. H. Evans, A. Fisher, W. J. Price, C. M. M. Smith, K. L. Suttonf, J. F. Tyson, Advances in atomic emission, absorption and fluorescence spectrometry, and related techniques, *Journal of Analytical Atomic Spectrometry*, (2000), **15**, 763–805.
12 L. Liu, D. Cozzolino, W. U. Cynkar, R. G. Dambergs, L. Janik, B. K. O'Neill, C. B. Colby, M. Gishen, Preliminary study on the application of visible-near infrared spectroscopy and chemometrics to classify Riesling wines from different countries, *Food Chemistry* (2008), **106**, 781–786.
13 R. M. Maggio, P. M. Castellano, T. S. Kaufman, A new principal component analysis-based approach for testing "similarity" of drug dissolution profiles, *European Journal of Pharmaceutical Sciences*, (2008), **34**, 66–77.
14 E. Bingham, A. Hyvarinen, A fast fixed-point algorithm for independent component analysis of complex valued signals, *International Journal of Neural Systems*, **10**, (2000), 1–8.
15 M. Li, J. Xue, Y. Du, T. Zhang, H. Li, Data fusion of Raman and near-infrared spectroscopies for the rapid quantitative analysis of methanol content in methanol-gasoline, *Energy & Fuels* (2019), **33**, 12286–1229.
16 G. Fu, D. Cao, Q. Xua, H. Li, Y. Liang, Combination of kernel PCA and linear support vector machine for modeling a nonlinear relationship between bioactivity and molecular descriptors, *Journal of Chemometrics*, (2011), **25**, 92–99.

17 J. Liang, M. Li, Y. Du, C. Yan, Y. Zhang, T. Zhang, X. Zheng, H. Li, Data fusion of laser induced breakdown spectroscopy (LIBS) and infrared spectroscopy (IR) coupled with random forest (RF) for the classification and discrimination of compound salvia miltiorrhiza, *Chemometrics and Intelligent Laboratory Systems*, (2020), **207**, 104179.

18 R. J. May, H. R. Maier, G. C. Dandy, T. M. Fernando, Non-linear variable selection for artificial neural networks using partial mutual information, *Environmental Modelling & Software*, (2008), **23**, 1312–1326.

19 J. A. Westerhuis, T. Kourti, J. F. MacGregor, Analysis of multiblock and hierarchical PCA and PLS models, *Journal of Chemometrics*, (1998), **12**, 301–321.

20 B. Mertens, M. Thompson, T. Fearn, Principal component outlier detection and SIMCA: a synthesis, *Analyst*, (1994), **119**, 2777–2784.

21 L. Breiman, Random forest, *Machine Learning*, (2001), **45**, 5–32.

22 M. R. E. Schapire, Y. Singer, BoosTexter: a boosting-based system for text categorization, *Machine Learning*, (2000), **39**, 135–168.

23 D. H. Laidlaw, Partial-volume Bayesian classification of material mixtures in MR volume data using voxel histograms, *IEEE Transactions on Medical Imaging*, (2002), **17**, 1094–1096.

24 M. A. Cazorla, F. Escolano, Two Bayesian methods for junction classification, *IEEE Transactions on Image Processing*, (2003), **12**(3), 317–327.

25 C. M. Rocco Sanseverino, A rule induction approach to improve Monte Carlo system reliability assessment, *Reliability Engineering & System Safety*, (2003), **82**, 85–92.

26 J. Zhang, R. S. Michalski, An integration of rule induction and exemplar-based learning for graded concepts, *Machine Learning*, (1995), **21**, 235–267.

27 T. Zhang, H. Tang, H. Li, Chemometrics in laser-induced breakdown spectroscopy, *Journal of Chemometrics*, (2003), **32**(11), e2983.

28 A. Fisher, P. S. Goodall, M. W. Hinds, S. M. Nelms, D. M. Penny, Atomic spectrometry update. Industrial analysis: metals, chemicals and advanced materials, *Journal of Analytical Atomic Spectrometry*, (2003), **18**, 1497–1528.

29 M. E. Essington, G. V. Melnichenko, M. A. Stewart, R. A. Hull, Soil metals analysis using laser-induced breakdown spectroscopy (LIBS), *Soil Science Society of America Journal*, (2009), **73**, 1469–1478.

30 D. L. Massart, L. Kaufman, The interpretation of analytical chemical data by the use of cluster analysis, *Chemical Analysis Series*, (1983), **65**, 273.

31 Y. Guo, Y. Tang, Y. Du, S. Tang, L. Guo, X. Li, Y. Lu, X. Zeng, Cluster analysis of polymers using laser-induced breakdown spectroscopy with K-means, *Plasma Science and Technology*, (2018), **20**, 103–107.

32 L. Brunnbauer, S. Larisegger, H. Lohninger, M. Nelhiebel, A. Limbeck, Spatially resolved polymer classification using laser induced breakdown spectroscopy (LIBS) and multivariate statistics, *Talanta*, (2020), **209**, 120572.

33 H. Peng, G. Chen, X. Chen, Z. Lu, S. Yao, Hybrid classification of coal and biomass by laser-induced breakdown spectroscopy combined with K-means and SVM, *Plasma Science and Technology*, (2019), **03**, 64–72.

34 X.H. Li, S.B. Yang, X. Chen, G.D. Yao, A.C. Liu, X. Yu, Multi-elemental imaging of breast cancer tissues using laser-induced breakdown spectroscopy, Conference on Lasers and Electro-Optics Europe & European Quantum Electronics Conference (CLEO/EUROPE-EQEC), Munich, Germany, June 23–27, 2019, (1999).

35 A. Jia, Survey on partitional clustering algorithms, *Electronic Design Engineering*, (2014), **22**, 38–41.
36 L. He, Q. Wang, Y. Zhao, L. Liu, P. Zhong, Study on cluster analysis used with laser-induced breakdown spectroscopy, *Plasma Science and Technology*, (2016), **18**, 647–653.
37 Q. Wang, L.A. He, Y. Zhao, Z. Peng, L. Liu, Study of cluster analysis used in explosives classification with laser-induced breakdown spectroscopy, *Laser Physics*, (2016), **26**(6), 065605.
38 J. Yadav, M. Sharma, A review of K-mean algorithm, *International Journal of Engineering Trends & Technology*, (2013), **4**, 2972–2974.
39 T. M. Kodinariya, D. P. R. Makwana, Review on determining number of cluster in K-means clustering, *International Journal of Advance Research in Computer Science and Management Studies*, (2013), **1**, 90–95.
40 B. Feng, W. Hao, G. Chen, D. Zhan, Optimization to K-means initial cluster centers, *Computer Engineering and Applications*, (2013), **49**, 182–185.
41 S. J. Haswell, *Practical Guide to Chemometrics*, Marcel Dekker, New York, (1992).
42 Y. Tang, Y. Guo, Q. Sun, S. Tang, J. Li, L. Guo, J. Duan, Industrial polymers classification using laser-induced breakdown spectroscopy combined with self-organizing maps and K-means algorithm, *Optik*, (2018), **165**, 179–185.
43 J. Herault and C. Jutten. Space or time adaptive signal processing by neural network models,John S. Denker (ed.), Neural Networks for Computing: AIP Conference Proceedings 151, Snowbird, UT, USA, 13–16 April 1986, Vol. 151, ISBN: 0-88318-351-X, (1986).
44 G. Burel, Blind separation of sources: a nonlinear neural algorithm. *Neural Networks*, (1992), **5**, 937–947.
45 J. P. Nadal, N. Parga, Non-linear neurons in the low noise limit: a factorial code maximizes information, *Network*, (1994), **5**, 565–581.
46 A. Cichocki, L. Moszczynski, A new learning algorithm for blind separation of sources, *Electronics Letters*, (1992), **28**, 1386–1387.
47 A. Cichocki, R. Unbehauen, E. Rummert, Robust learning algorithm for blind separation of signals, *Electronics Letters*, (1994), **30**, 1986–1987.
48 A. Cichocki, R. Unbehauen, Robust neural networks with on-line learning for blind identification and blind separation of sources, *IEEE Transactions on Circuits and Systems*, (1996), **43**, 894–906.
49 P. Common, Independent component analysis-a new concept, *Signal Processing*, (1994), **36**, 287–314.
50 A. Bell, T. Sejnowski, An information-maximization approach to blind separation and blind deconvolution, *Neural Computation*, (1995), **7**, 1129–1159.
51 S. I. Amari, A. Cichocki, H. Yang, A new learning algorithm for blind source separation, in Mozer, M. (ed.) *Advances in Neural Information Processing System*, MIT Press, Cambridge, MA, (1996), **8**, 757–763. DOI: US20050105644 A1.
52 A. Hyvirinen, E. Oja, A fast fixed-point algorithm for independent component analysis. *Neural Computation*, (1997), **9**, 1483–1492.
53 A. Hyvirinen, Fast and robust fixed-point algorithm for independent component analysis, *IEEE Transactions on Neural Networks*, (1999), **10**, 626–634.

54 A. Hyvarinen, Independent component analysis: algorithms and applications, *Neural Networks*, (2000), **13**, 411–430.
55 A. Hyvarinen, Fast and robust fixed-point algorithms for independent component analysis, *IEEE Transactions on Neural Networks*, (1999), **8**, 622–634.
56 T. Zhang, C. Yan, J. Qi, H. Tang, H. Li, Classification and discrimination of coal ash by laser-induced breakdown spectroscopy (LIBS) coupled with advanced chemometric methods, *Journal of Analytical Atomic Spectrometry*, (2017), **32**, 1960–1965.
57 O. Forni, S. Maurice, O. Gasnault, C. R. Wiens, A. Cousin, S. M. Clegg, J. Sirven, J. Lasue, Independent component analysis classification of laser induced breakdown spectroscopy spectra, *Spectrochimica Acta Part B: Atomic Spectroscopy*, (2013), **86**, 31–41.
58 T. Cover, P. Hart, Nearest neighbor pattern classification, *IEEE Transactions on Information Theory*, (1967), **13**, 21–27.
59 W. Shang, H. Huang, H. Zhu, Y. Lin, Z. Wang, Y. Qu, An improved KNN algorithm-fuzzy KNN, *Lecture Notes in Computer Science*, (2005), **3801**, 741–746.
60 K. Saetern, N. Eiamkanitchat, An ensemble K-nearest neighbor with neuro-fuzzy method for classification, *Advances in Intelligent Systems & Computing*, (2014), **265**, 43–51.
61 F. Jiang, C. Chu, Application of KNN improved algorithm in automatic classification of network public proposal cases, International Conference on Cloud Computing and Big Data Analysis (IEA), IEEE, Chengdu, People's Republic of China, 28–30 April 2017, (2017), 82–86.
62 G. Li, H. Chen, Y. Hu, J. Wang, Y. Guo, J. Liu, H. Li, R. Huang, H. Lu, J. Li, An improved decision tree-based fault diagnosis method for practical variable refrigerant flow system using virtual sensor-based fault indicators, *Applied Thermal Engineering*, (2018), **129**, 1292–1303.
63 J. C. Bezdek, S. K. Chuah, D. Leep, Generalized K-nearest neighbor rules, *Fuzzy Sets and Systems*, (1986), **18**, 237–256.
64 S. Cost, S. Salzberg, A weighted nearest neighbor algorithm for learning with symbolic features, *Machine Learning*, (1993), **10**, 57–78.
65 Y. Hu, B. Shi, Fast KNN test classification based on area division, *Computer Science*, (2012), **10**, 182–186.
66 Z. Zhang, Y. Huang, H. Wang, A new KNN classification approach, *Computer Science*, (2008), **35**, 170–172.
67 C. Huang, J. Chen, Chinese text classification based on improved K-nearest neighbor algorithm, *Journal of Shanghai Normal*, (2019), **48**, 96–101.
68 Y. Ge, Y. Zeng, C. Cheng, A novel nearest neighbor classification algorithm based on local mean and class mean, Proceedings of the 36th Chinese Control Conference, Dalian, People's Republic of China, 26–28 July 2017, (2017), 11059–11065.
69 Y. Yang, X. Hao, L. Zhang, L. Ren, Application of Scikit and Keras Libraries for the classification of iron ore data acquired by laser-induced breakdown spectroscopy (LIBS), *Sensors*, (2020), **20**, 1–11.
70 J. Alvarez, M. Velasquez, A. K. Myakalwar, C. Sandoval, R. Fuentes, R. Castillo, D. Sbarbarod, J. Yanez, Determination of copper-based mineral species by laser induced breakdown spectroscopy and chemometric methods, *Journal of Analytical Atomic Spectrometry*, (2019), **34**, 2459–2468.

71 P. R. Michael, P. M. Dirchwolf, T. V. Silva, R. N. Villafañe, J. A. G. Neto, R. G. Pellerano, E. C. Ferreira, Brown rice authenticity evaluation by spark discharge-laser-induced breakdown spectroscopy, *Food Chemistry*, (2019), **297**, 1–6.

72 F. Colao, R. Fantoni, P. Ortiz, M. A. Vazquez, J. M. Martin, R. Ortiz, N. Idris, Quarry identification of historical building materials by means of laser induced breakdown spectroscopy, X-ray fluorescence and chemometric analysis, *Spectrochimica Acta Part B: Atomic Spectroscopy*, (2010), **65**, 688–694.

73 G. S. Larios, G. Nicolodelli, G. S. Senesi, M. C. S. Ribeiro, A. A. P. Xavier, D. M. B. P. Milori, C. Z. Alves, B. S. Marangoni, C. Cena, Laser-induced breakdown spectroscopy as a powerful tool for distinguishing high- and low-vigor soybean seed lots, *Food Analytical Methods*, (2020), **13**, 1691–1698.

74 W. Li, Y. Zhu, X. Li, Z. Hao, L. Guo, X. Li, X. Zeng, Y. Lu, In situ classification of rocks using stand-off laser-induced breakdown spectroscopy with a compact spectrometer, *Journal of Analytical Atomic Spectrometry*, (2018), **33**, 461–467.

75 R. Gaudiuso, E. Ewusi-Annan, N. Melikechi, X. Sun, B. Liu, L. F. Campesato, T. Merghoub, Using LIBS to diagnose melanoma in biomedical fluids deposited on solid substrates: limits of direct spectral analysis and capability of machine learning, *Spectrochimica Acta Part B: Atomic Spectroscopy*, (2018), **146**, 106–114.

76 Z. Rokhbin, S. Z. Shoursheini, H. Shirvani-Mahdavi, Rapid quality assessment of isogams using laser plasma spectroscopy, *Applied Physics B*, (2020), **126**, 65.

77 J. Yan, S. Li, K. Liu, R. Zhou, W. Zhang, Z. Hao, X. Li, D. Wang, Q. Li, X. Zeng, An image features assisted line selection method in laser-induced breakdown spectroscopy, *Analytica Chimica Acta*, (2020), **1111**, 139–146.

78 G. Vítkova, K. Novotny, L. Prokes, A. Hrdlicka, J. Kaiser, J. Novotny, R. Malina, D. Prochazka, Fast identification of biominerals by means of stand-off laser-induced breakdown spectroscopy using linear discriminant analysis and artificial neural networks, *Spectrochimica Acta Part B: Atomic Spectroscopy*, (2012), **73**, 1–6.

79 F. Zhou, P. Ye, H. Fu, G. Ji, Recognition of heavy metal contamination using *Tegillarca granosa* LIBS data, *IOP Conference Series, Earth and Environmental Science*, (2019), **237**, 1–6.

80 L. H. Chiang, E. L. Russell, R. D. Braatz, Partial least squares, in: *Fault Detection and Diagnosis in Industrial Systems. Advanced Textbooks in Control and Signal Processing*, Springer, London, https://doi.org/10.1007/978-1-4471-0347-9_6, (2001).

81 X. Chen, Y. Xu, L. Meng, X. Chen, L. Yuan, Q. Cai, W. Shi, G. Huang, Non-parametric partial least squares-discriminant analysis model based on sum of ranking difference algorithm for tea grade identification using electronic tongue data, *Sensors and Actuators B-Chemical*, (2020), **311**, 127924.

82 J. J. Remus, R. S. Harmon, R. R Hark, G. Haverstock, D. Baron, I. K. Potter, S. K. Bristol, L. J. East, Advanced signal processing analysis of laser-induced breakdown spectroscopy data for the discrimination of obsidian sources, *Applied Optics*, (2012), **51**, 65–73.

83 J. P. Castro, E. R. Pereira-Filho, R. Bro, Laser-induced breakdown spectroscopy (LIBS) spectra interpretation and characterization using parallel factor analysis (PARAFAC): a new procedure for data and spectral interference processing fostering the waste electrical and electronic equipment (WEEE) recycling process, *Journal of Analytical Atomic Spectrometry*, (2020), **35**, 1115–1124.

84 F. C. De Lucia, J. L. Gottfried, C. A. Munson, A. W. Miziolek, Multivariate analysis of standoff laser-induced breakdown spectroscopy spectra for classification of explosive-containing residues, *Applied Optics*, (2008), **47**, 112–121.

85 R. R. Hark, J. J. Remus, L. J. East, R. S. Harmon, M. A. Wise, B. M. Tansi, K. M. Shughrue, K. S. Dunsin, C. Liu, Geographical analysis of "conflflict minerals" utilizing laser-induced breakdown spectroscopy, *Spectrochimica Acta Part B: Atomic Spectroscopy*, (2012), **74–75**, 131–136.

86 J. L. Gottfried, R. S. Harmon, F. C. DeLucia, A. W. Miziolek, Multivariate analysis of laser-induced breakdown spectroscopy chemical signatures for geomaterial classification, *Spectrochimica Acta Part B: Atomic Spectroscopy*, (2009), **64**,1009–1019.

87 G. Ji, P. Ye, Y. Shi, L. Yuan, X. Chen, M. Yuan, D. Zhu, X. Chen, X. Hu, J. Jiang, Laser-induced breakdown spectroscopy for rapid discrimination of heavy-metal-contaminated seafood *Tegillarca granosa*, *Sensors*, (2017), **17**, 1–11.

88 Y. Tian, Z. Wang, X. Han, H. Hou, R. Zheng, Comparative investigation of partial least squares discriminant analysis and support vector machines for geological cuttings identification using laser-induced breakdown spectroscopy, *Spectrochimica Acta Part B: Atomic Spectroscopy*, (2014), **102**, 52–57.

89 N. Porto, J. V. Roque, C. A. Wartha, W. Cardoso, L. A. Peternelli, M. H. P. Barbosa, R. F. Teofilo, Early prediction of sugarcane genotypes susceptible and resistant to Diatraea saccharalis using spectroscopies and classification techniques, *Spectrochimica Acta Part A: Molecular and Biomolecular Spectroscopy*, (2019), **218**, 69–75.

90 M. Gazmeh, M. Bahreini, S. H. Tavassoli, Discrimination of healthy and carious teeth using laser induced breakdown spectroscopy and partial least square discriminant analysis, *Applied Optics*, (2015), **54**,123–131.

91 Gemperline, Paul, *Practical Guide to Chemometrics* (Second Edition), CRC Press/Taylor & Francis, Boca Raton, FL, (2006).

92 P. Porízka, J. Klus, E. Képeš, D. Prochazka, D. W. Hahn, On the utilization of principal component analysis in laser-induced breakdown spectroscopy data analysis, a review, *Spectrochimica Acta Part B-Atomic Spectroscopy*, (2018), **148**, 65–82.

93 A. M. Martinez, A. C. Kak, PCA versus LDA, *IEEE Transactions On Pattern Analysis and Machine Intelligence*, (2002), **23**, 228–233.

94 J. Yang, D. Zhang, A. F. Frangi, J. Yang, Two-dimensional PCA:a new approach to appearance-based face representation and recognition, *IEEE Transactions on Pattern Analysis and Machine Intelligence*, (2004), **26**, 131–137.

95 Y. Yang, C. Li, S. Liu, H. Min, C. Yan, M. Yang, J. Yu, Classification and identification of brands of iron ores using laser-induced breakdown spectroscopy combined with principal component analysis and artificial neural networks, *Analytical Methods*, (2020), **12**, 1316–1323.

96 J. Wang, L. Li, P. Yang, Y. Chen, Y. Zhu, M. Tong, Z. Hao, X. Li, Identification of cervical cancer using laser-induced breakdown spectroscopy coupled with principal component analysis and support vector machine, *Lasers in Medical Science*, (2018), **33**, 1381–1386.

97 A. K. Pathak, A. Singh, R. Kuma, A. K. Rai, Laser-induced breakdown spectroscopy coupled with PCA study of human tooth, *National Academy Science Letters-India*, (2019), **42**, 87–90.

98 N. S. Hanasil, R. K. R. Ibrahim, M. Duralim, H. H. J. Sapingi, Discrimination of extracted animal fats using laser-induced breakdown spectroscopy assisted with principal component analysis, *Journal Teknologi-Sciences & Engineering*, (2020), **82**, 125–130.

99 R. H. El-Saeid, Z. A. Abdel-Salam, S. Pagnotta, V. Palleschi, M. A. Harith, Classification of sedimentary and igneous rocks by laser induced breakdown spectroscopy and nanoparticle-enhanced laser induced breakdown spectroscopy combined with principal component analysis and graph theory, *Spectrochimica Acta Part B-Atomic Spectroscopy*, (2019), **158**, 105622.

100 J. Wang, X. Liao, P. Zheng, S. Xue, R. Peng, Classification of Chinese herbal medicine by laser-induced breakdown spectroscopy with principal component analysis and artificial neural network, *Analytical Letters*, (2018), **51**, 575–586.

101 N. Kumar, A. Bansal, G. S. Sarma, R. K. Rawal, Chemometrics tools used in analytical chemistry: an overview, *Talanta*, (2014), **123**, 186–199.

102 J. Trygg, E. Holmes, T. Lundstedt, Chemometrics in metabonomics, *Journal of Proteome Reseearch*, (2007), **6**, 469–479.

103 S. M. Clegg, E. Sklute, M. D. Dyar, J. E. Barefield, R. C. Wiens, Multivariateanalysis of remote laser-induced breakdown spectroscopy spectra using partial least squares, principal component analysis, and related techniques, *Spectrochimica Actr Part B-Atomic Spectroscopy*, (2009), **64**, 79–88.

104 H. Claeys, W. Mertens, H. Verhaert, C. Vermylen, Evaluation of third-generation screening and confirmatory assays for HCV antibodies, *Vox Sanguinis*, (1994), **66**, 9–122.

105 S. Schroder, S. G. Pavlov, I. Rauschenbach, E. K. Jessberger, H.-W. Hubers, Detection and identification of salts and frozen salt solutions combining laser-induced breakdown spectroscopy and multivariate analysis methods: a study for future martian exploration, *Icarus*, (2013), **223**, 61–73.

106 A. K. Myakalwa, S. Sreedhar, I. Barman, N. C. Dingari, S. V. Rao, P. P. Kiran, S. P. Tewari, G. M. Kumar, Laser-induced breakdown spectroscopy-based investigation and classification of pharmaceutical tablets using multivariate chemometric analysis, *Talanta*, (2011), 53–59.

107 F. W. B. Aquino, E. R. Pereira-Filho, Analysis of the polymeric fractions of scrap from mobile phones using laser-induced breakdown spectroscopy: chemometric applications for better data interpretation, *Talanta*, (2015), **134**, 65–73.

8

Blind Source Separation in LIBS

Anna Tonazzini[1], Emanuele Salerno[1], and Stefano Pagnotta[2]

[1] *Institute of Information Science and Technologies, National Research Council of Italy, Pisa, Italy*
[2] *Department of Earth Sciences, University of Pisa, Pisa, Italy*

8.1 Introduction

A laser-induced emission spectrum is a complex object from the point of view of both the physical meaning of the emission peaks and the amount of information contained within it. Simplifying, a spectrum is a sum of different signals to which noise, depending on different factors, is added. There are several variables that can be extracted from an emission spectrum, whose formal description is referred to previous chapters in this book.

The analyte signal (S) in a laser-induced breakdown spectrum is the light radiation emitted by an analyte at a specific wavelength and collected by a detector. The plasma also emits a signal referred to as continuum background (B) composed of bremsstrahlung radiation from free electrons and recombination emission. With the plasma decay, the emission lines begin to appear and, therefore, to optimize the signal to noise ratio (SNR), the plasma light is collected by a time-resolved detector at times that are generally between a few hundred nanoseconds and a few microseconds (Sun and Yu 2009). Therefore, the detector response (x) at a given wavelength in a single-analyte spectrum is the sum of S, B, and the noise (D) introduced by the detector (dark current, stray light, etc.) (Hahn and Omenetto 2012; Tognoni and Cristoforetti 2016):

$$x = S + B + D \tag{8.1}$$

The subtraction of the continuum background (B) to the analyte signal results in a net analyte signal (Hahn and Omenetto 2012; Tognoni and Cristoforetti 2016):

$$x_{net} = x - B \tag{8.2}$$

This formula is valid for only one analyte. In the most frequent case, the signal (S) will be the sum of the signals of all the analytes present in the material plus their backgrounds. The elimination or correction of the different contributions of these backgrounds is not simple because their intensity depends on the emission wavelengths, the surface characteristics of

Chemometrics and Numerical Methods in LIBS, First Edition. Edited by Vincenzo Palleschi.
© 2023 John Wiley & Sons Ltd. Published 2023 by John Wiley & Sons Ltd.

the sample and its matrix, and, finally, on the conditions of plasma formation (Tognoni et al. 2002; Xu et al. 1997). This means that these parameters are not always fixed, also due to fluctuations in the laser power and differences due to the inhomogeneity of the sample and leads us to estimate the continuum background spectrum by spectrum and sample by sample, as proposed by many authors (Sun and Yu 2009; Yaroshchyk and Eberhardt 2014), or using the msbackadj function of the Matlab bioinformatics toolbox©.

Besides the continuum background, random noise also affects the laser-induced breakdown spectroscopy (LIBS) signal. Mermet et al. (2008) recognize four main factors contributing to random noise in LIBS spectra: source noise, shot noise, detector noise, and thermal drift (Table 8.1).

Bulajic et al. (2002) simulate shot and detector noise in typical LIBS spectra as Gaussian variables, as also done by Motto-Ros et al. (2019), who remark that, when the line intensities are weak (i.e. close to the background value), the noise can be considered independent of the intensity. Tognoni and Cristoforetti (2016), furthermore, come to the conclusion that the detector noise can be considered negligible. In this chapter, we follow these authors, considering the random noise as Gaussian and signal-independent. A simple correction of the thermal drift in the spectrum can be done using a calibrated lamp.

Let us now consider a 2D map with P pixels (Figure 8.1), whose generic pixel p is equipped with an entire LIBS spectrum observed at M discrete wavelengths. At each wavelength, the observed spectrum for an N-analyte material has the form

$$x_p(\lambda) = \sum_{i=1}^{N} a_{p,i} s_i(\lambda) + n_p(\lambda) \qquad \lambda = \lambda_1, \lambda_2, ..., \lambda_M \qquad p = 1, 2, ..., P \qquad (8.3)$$

where $s_i(\lambda)$ is the ideal spectrum of the ith analyte, $a_{p,i}$ is its abundance at point p, and n_p is the associated noise. Hence, at each pixel, the measured spectrum is a weighted

Table 8.1 Four main factors that cause noise in LIBS signals.

Noise type	Description	Intensity	Reference
Source noise	Shot-to-shot fluctuations in laser energy, rate of ablation, laser-plasma coupling, and plasma characteristics.	Standard deviation proportional to the signal.	(Bauer et al. 1998; Danzer 1984; Kempenaers et al. 2002; Sallé et al. 2006)
Shot noise	Caused by photons that arrives on the detector (described by Poisson statistics).	Standard deviation proportional to the square root of the signal.	(Mermet et al. 2008)
Detector noise	Fluctuation caused mainly by shot noise plus negligible effects due to the characteristics of the detector	Same as shot noise	(Mermet et al. 2008)
Thermal drift	Caused by heating of the laser components and the optical path	Spectrum (thermal) drift	(Mermet et al. 2008)

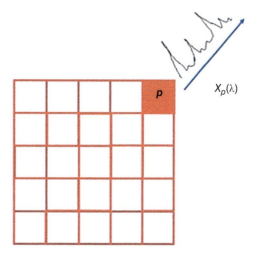

Figure 8.1 Schematic drawing of a LIBS 2D map.

superposition of the spectra of the fundamental elements, where neither the mixing coefficients nor the element spectra are known.

The first problem to be faced before processing the signals, whether they come from a punctual analysis or from an elementary map, is the elimination or reduction not only of background noise but other types of noise too. Many authors (Alexander et al. 2012; Bauer et al. 1998; Fu et al. 2019; Mermet et al. 2008; Schlenke et al. 2012; Zhang et al. 2013, 2015; Zou et al. 2014) have faced the problem in different ways, some using purely mathematical methods and some trying to give a physical meaning to the operations performed.

A typical LIBS spectrum is formed by the sum of the emissions of the neutral (I) and first ionization (II) atomic species present in the plasma light captured by the spectrometer. Furthermore, these are associated with the presence of different kinds of interferences caused by various factors due to the instrument, measurement conditions, plasma physics, etc. The most influential is the continuum background emitted, as said earlier, by the plasma (see Figure 8.2).

The simplest and most used method to analyze a LIBS spectrum is to select the wavelengths which the neutral (I) and first ionization (II) analyte emissions correspond to in a supervised manner (Boué-Bigne 2008; Busser et al. 2018; Carvalho et al. 2015; Casado-Gavalda et al. 2017; Dixit et al. 2017; Fabre et al. 2018; Fortes et al. 2015; Gimenez et al. 2016; Häkkänen and Korppi-Tommola 1995; Hausmann et al. 2017; Hoehse et al. 2011; Hong et al. 2014; Kim et al. 1998; Le Guével et al. 2018; Lednev et al. 2017; Lefebvre et al. 2016; Li et al. 2017; López-López et al. 2017; Lopez-Quintas et al. 2012; Motto-Ros et al. 2012; Noll et al. 2001; Pagnotta et al. 2017; Peng et al. 2016; Rifai et al. 2018; Romero and Laserna 1997; Schiavo et al. 2016; Sheta et al. 2015; Škarková et al. 2017; Sperança et al. 2017; St-Onge et al. 2002; Taparli et al. 2018; Trichard et al. 2018; Wiggins et al. 2013; Xu et al. 2016). This procedure is valid when we know the elemental composition of the analysed material or, in any case, when we have a clear idea of its qualitative composition. In the problematic cases of very noisy signals or presence of elementary peaks that show

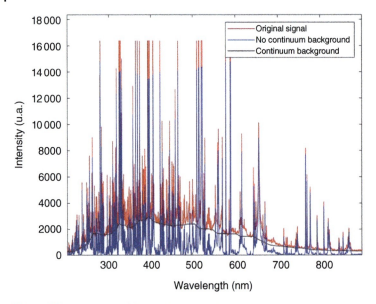

Figure 8.2 A typical LIBS signal, as output from the spectrometer and corrected for the continuum background.

self-absorption, this method brilliantly bypasses each of these problems by making an *a priori* selection of the best peak lines.

In the analysis of completely unknown samples, consisting of many analytes with highly variable proportions, the supervised selection of analyte peaks attributable to elemental qualitative composition begins to show its negative sides. Specifically, it will be necessary to start considering various other factors such as the SNR, signals that overlap each other and create sums of peaks, various types of noise and, finally, self-absorption.

The continuum background at each spectral line is typically estimated from the signal values measured at the rising and dropping slopes of that line. Usually, the analyte signal is characterized through the intensities of its peaks. These can normally be derived in three ways: the intensity at the central wavelength of each peak, the integral of the signal over the line width or, less frequently, the sum of the signal values along each spectral line.

When the number and type of analytes to be detected in a LIBS spectrum is completely unknown, the use of source separation methods could be of great help: these methods can separate the sources of a signal without any knowledge of its components or how they are mixed. In the literature, there are several methods and examples of algorithms for separating the components of a signal (Choi et al. 2005; Jain and Rai 2012; Pal et al. 2013); among these, the most interesting are the Principal Component Analysis (PCA) (Karhunen et al. 1995; Karhunen et al. 1998; Karhunen and Joutsensalo 1994; Oja 1995; Oja and Plumbley 2003; Pajunen and Karhunen 1997; Zhu et al. 2006), the minimum noise fraction (MNF) (Dabiri and Lang 2018; Gao et al. 2017; Harris et al. 2005; Luo et al. 2016) and the fast-independent component analysis (FastICA) (Hyvarinen 1999a, 1999c; Hyvärinen 2001; Lin et al. 2007; Sun 2005).

8.2 Data Model

We consider the emission spectra at the P pixels as a multivariate signal resulting from the sum of individual signals, each describing the characteristics of the LIBS emission of a single element. We can rewrite Eq. 8.3 in matrix form:

$$x(\lambda) = As(\lambda) + n(\lambda) \qquad \lambda = \lambda_1, \lambda_2, ..., \lambda_M \tag{8.4}$$

where the P-vector x contains all the measured spectra at each wavelength, the N-vector s contains all the elemental spectra, A is a $P \times N$ mixing matrix, and the P-vector n contains the noise contributions at each wavelength.

We have said that the analysis of a LIBS map consists in identifying the elements that compose the material under investigation, and their relative abundances at each inspected point. This means estimating s and A from knowledge of x. Even in the case of $P \geq N$ and zero noise, if no additional assumption is made, this problem is clearly underdetermined, since any full-rank choice for A can give an estimate of s that accounts for the evidence x. Hence, Eq. 8.4 is a particular instance of a problem of blind source separation (BSS) (Cichocki and Amari 2002). Here, the sources are the spectra of the N elements, and the mixtures are the spectra at the P measurement points.

8.3 Analyzing LIBS Data via Blind Source Separation

8.3.1 Second-order BSS

Even if no specific information is available, statistical assumptions can often be made on the sources. In our case, it can be assumed that the spectra of the elements are mutually uncorrelated, and it is intuitively clear why one could decorrelate the input data to try to extract the individual spectra. This amounts to applying second-order statistical techniques to estimate A and s from x.

Let us first pose the case where the noise in Eq. (8.4) is zero and assume that the data vectors are zero-centered by preprocessing. We seek for a linear transformation $y(\lambda) = Wx(\lambda)$ such that $\langle y_i^T y_j \rangle = 0, \forall i, j = 1, ..., P, i \neq j$, where W is a $P \times P$ matrix and the notation $\langle \cdot \rangle$ means expectation. In other words, the components of the transformed data vector y are orthogonal. This operation is not unique, since, given an orthonormal basis of a subspace, any further rigid rotation still yields an orthonormal basis of the same subspace. Our data covariance matrix is the $P \times P$ matrix:

$$\text{Cov}_x = \langle xx^T \rangle \approx \frac{1}{M} \sum_{\lambda = \lambda_1}^{\lambda_M} x(\lambda) x^T(\lambda) \tag{8.5}$$

Since the data are usually correlated, matrix Cov_x will be nondiagonal. The covariance matrix of vector y is

$$\text{Cov}_y = \langle Wxx^T W^T \rangle = W \text{Cov}_x W^T \tag{8.6}$$

To obtain an output y with mutually orthogonal components, Cov_y should be diagonal. Let us perform the eigenvalue decomposition of matrix Cov_x, and call V_x the matrix of its eigenvectors, and Λ_x the diagonal matrix of its eigenvalues, sorted in decreasing order. It is easy to verify that both the following choices for W (among possible others) yield a diagonal Cov_y:

$$W_{PCA} = V_x^T \tag{8.7}$$

$$W_W = \Lambda_x^{-(1/2)} V_x^T \tag{8.8}$$

Matrix W_{PCA} produces a set of vectors y_i that are orthogonal to each other and whose squared Euclidean norms (i.e. their variances) are equal to the eigenvalues of the data covariance matrix. This is the principal components transform, or principal component analysis (PCA) (Cichocki and Amari 2002). The output principal components are sorted by decreasing values of variance. By using matrix W_w, we obtain a set of orthogonal vectors of unit norms, i.e. mutually orthogonal vectors located on a spherical surface (whitening, or Mahalanobis transform). Other choices for the data decorrelating matrix W can be taken via the multiplication from the left of any whitening matrix by any orthogonal matrix. Being $P \geq N$, and assuming that matrix A is full-rank, all the eigenvalues of order larger than N vanish, and in principle matrix W_w cannot be computed, and the $P \times P$ matrix W_{PCA} has no meaning. Moreover, if $P \gg N$, the problem could also become computationally unmanageable. Equations (8.7) and (8.8) maintain their validity if only the first N eigenvalues and eigenvectors of Cov_x are selected, and the size of the PCA and the whitening matrices becomes $N \times P$. Alternatively, the user could choose to guess a number N' of significant analytes and only compute the first N' eigenvalues and eigenvectors. Successively, possible components associated with near-zero eigenvalues can be neglected as non-significant. In the noiseless case, thus, the BSS problem under uncorrelation assumption simply becomes a problem of eigenvalue analysis. As its solution is not unique, however, the correspondence of the output vectors y_i and the original sources s_i is not ensured theoretically.

8.3.2 Maximum Noise Fraction

The second-order Blind Source Separation (BSS) techniques described earlier have been derived under the noiseless assumption. In practice, the measured data are always more or less noisy. One possibility to face this problem is to try to denoise the data before separation. In remote-sensed multispectral data, it has been observed that PCA is often able to sort the decorrelated data based on quality besides variance. Hence, the most significant extracted components are also somehow less affected by noise. This fact could depend on the high cross-correlation that often exists in multispectral data, which can lead to a compression of information into the low-order principal components. This compression is manifested as a steadily decreasing signal-to-noise ratio as the order of the component increases, i.e. the related eigenvalue decreases. However, there are numerous examples, especially among aircraft scanner data, where this is not the case (Green et al. 1988). Indeed, when the noise variances are unequal in different bands, it may happen that noise contributes significantly to the variance of one of the first components, so that this might actually contain less useful information than another component with lower variance. This is because the principal component transform is sensitive to the scaling of the data to which is applied. To some extent, the scaling (or weighting) of the multispectral bands is arbitrary, and usually equal weighting is applied. A more natural strategy, conversely, is to weight the bands so that the noise level in each of them is the same. Green et al. (1988) exploit this principle to derive a transform based on maximization of SNR, so that the transformed components are ranked by SNR rather than variance as done in PCA. They show that this transform, the maximum noise fraction (MNF) transform, produces uncorrelated components that maximize their noise fraction (or, equivalently, their SNR if taken in reverse order). Then, MNF produces components ordered by image quality. Lee et al. (1990) show that the MNF transform is equivalent to a two stage transformation in which the data are first transformed so that the noise is whitened, i.e. it has unit variance in all bands, and is uncorrelated across the bands, and the second stage is a PCA. This transform has been called noise adjusted principal component (NAPC) transform. The first stage of NAPC provides a natural weighting where the noise in each band is equal in magnitude and is uncorrelated with the noise in any other. Therefore, in the second stage, maximizing the noise-whitened multispectral data variance with PCA results in maximizing their corresponding SNR. In Roger (1994), a fast computation for NAPC was also proposed. Assuming that the noise and signal components of the data are uncorrelated, the NAPC transform has been formally derived by Lee et al. (1990, p. 297), by using arguments similar to those used in the derivation of the principal component transform, and tracing back the variational problem of maximizing SNR at each subsequent component to that of maximizing variance. The practical NAPC (or MNF) algorithm requires the knowledge of the symmetric, positive definite covariance matrix Cov_n of the additive noise affecting the data, and consists of the following steps:

1) From Cov_n, compute its eigenvector matrix, V_n, and the diagonal matrix of its eigenvalues Λ_n, such that

$$V_n Cov_n V_n^T = \Lambda_n \tag{8.10}$$

2) Define a noise-whitening matrix $W_{wn} = \Lambda_n^{-(1/2)} V_n^T$ for which

$$W_{wn} Cov_n W_{wn}^T = I \tag{8.11}$$

where I is the identity matrix

3) Transform the data covariance matrix, Cov_x, by W_{wn} to obtain the noise-adjusted data covariance matrix, Cov_x^{adj}:

$$Cov_x^{adj} = W_{wn} Cov_x W_{wn}^T \tag{8.12}$$

4) From Cov_x^{adj}, compute its eigenvector matrix, U_x, and the diagonal matrix of its eigenvalues, Λ_x^{adj}, such that

$$U_x Cov_x^{adj} U_x^T = \Lambda_x^{adj} \tag{8.13}$$

5) The MNF components are given by the set of vectors y, produced from the original data vectors x by the transform:

$$y = W_{MNF} x \tag{8.14}$$

where W_{MNF}, the NAPC-MNF transform matrix, is given by $W_{MNF} = U_x^T \Lambda_n^{-(1/2)} V_n^T$.

Green et al. (1988) state without proof that the MNF components are the left-hand eigenvectors of $\text{Cov}_n \text{Cov}_x^{-1}$, and that the eigenvalues of Λ_x^{adj} are the corresponding noise fraction values. In Roger (1994), it is formally shown that matrix W_{MNF} simultaneously diagonalizes Cov_x and Cov_n, that is:

$$W_{\text{MNF}} \text{Cov}_x W_{\text{MNF}}^T = \Lambda_x^{\text{adj}} \tag{8.15}$$

and

$$W_{\text{MNF}} \text{Cov}_n W_{\text{MNF}}^T = 1 \tag{8.16}$$

which implies that the transformed data are uncorrelated and ordered by their variance, i.e. by their SNR. In many situations the covariance matrix of the noise is available. If not, it may be estimated from the covariance of the signal first-order differences Green et al. (1988). This approximation is more accurate the more regular and smooth the data signals are. Still in Green et al. (1988), MNF has been analysed in different situations of noise. It is apparent that MNF reduces to PCA for noiseless data (i.e. when Cov_n is taken identically null). However, MNF performs exactly like PCA also in other circumstances. For instance, when noise is uncorrelated with equal variance in all bands, it is straightforward to see that the two procedures produce the same set of eigenvectors. Finally, it is worth noting that MNF can be used for noise reduction. Indeed, once the data have been transformed into components with ordered SNR, it is logical to spatially filter the noisiest components and subsequently to transform back to the original coordinate system. As the components that will be filtered by this procedure contain a reduced signal component, the resulting signal degradation will be much less than if the same smoothing were performed on the untransformed data. This procedure should allow much stronger smoothing to be applied, without severe signal degradation. However, since the signal content of even the noisiest MNF component is rarely so low to be overlooked, virtually every MNF component needs to be filtered before re-transformation, which leads back to the typical over-smoothing problems of denoising. In any case, denoising is not the feature of NAPCA-MNF that we are interested in here, where we aim at exploiting its signal separation capabilities.

8.3.3 Independent Component Analysis

An effective technique used to solve a BSS problem like the LIBS problem of Eq. 8.4 could be the so-called independent component analysis (ICA) technique (Hyvärinen et al. 2001), which assumes full statistical independence of the sources and has been first derived for noise-free data to be then extended to the realistic noisy case. Unlike the above-reported second-order approaches, the ICA assumptions guarantee the uniqueness of the solution, ensuring that its outputs reproduce the original sources up to arbitrary scaling and permutation.

If the prior distribution for each source is known, independence is equivalent to assume a factorized form for the joint prior distribution of s:

$$\Psi(s(\lambda)) = \prod_{i=1}^{N} \Psi_i(s_i(\lambda)) \qquad \forall \lambda \tag{8.17}$$

The separation problem can be formulated as the maximization of function (Eq. 8.17) in A and s, subject to the constraint $x = As$. This is equivalent to the search for a matrix $W = (w_1, w_2, ..., w_N)^T$, such that, when applied to the data $x = (x_1, x_2, ..., x_p)$, produces the set of vectors $w_i^T x$ that are maximally independent, and whose distributions are given by the Ψ_i. By taking the logarithm of Eq. 8.17, the problem solved by ICA algorithms is then:

$$\hat{W} = \text{argmax}_w \sum_\lambda \sum_i \log \Psi_i \left(w_i^T x(\lambda) \right) + M \log |\det(W)| \tag{8.18}$$

In the square case $P = N$, matrix $\hat{W}$ is an estimate of A^{-1} up to arbitrary scale factors and permutations of the columns. Hence, each vector $\hat{s}_i = \hat{w}_i^T x$ is one of the original source vectors up to a scale factor. Besides independence, to make separation possible a necessary extra condition for the sources is that they all, but at most one, must be non-Gaussian. To enforce non-Gaussianity, generic super-Gaussian or sub-Gaussian distributions can be used as priors for the sources. These have proven to give very good estimates for the mixing matrix and for the sources as well, no matter of the true source distributions, which, on the other hand, are usually unknown (Bell and Sejnowski 1995a, 1995b).

8.3.4 ICA for Noisy Data

In Hyvärinen and Oja (1997), a fast-fixed-point algorithm (FastICA) is developed that takes some measure of non-Gaussianity and then finds projections in which this is locally maximized for whitened data. Projections in such directions give consistent estimates of the independent components if the measure of non-Gaussianity is well chosen. In Hyvarinen, (1999b), this approach was taken for deriving a noisy version of the algorithm, based on the availability of measures of non-Gaussianity that are insensitive to Gaussian noise on the data. Specifically, a modification of the fixed-point FastICA algorithm was introduced where Gaussian moments, simply estimated from noisy observations, are used as contrast functions for data affected by Gaussian noise. As for the NAPC transform, also this "noise-adjusted ICA" algorithm assumes knowledge of the covariance matrix of the noise. In the noise-free ICA algorithm, a first step entails the whitening of the data. In the noisy version of the algorithm, the noise is considered by replacing the ordinary whitening, based on the eigenvalue decomposition of the data covariance matrix Cov_x (see Eq. 8.8), with a "quasi-whitening" operation, based on the covariance matrix of the ideal, noise-free mixture, which is given by $\text{Cov}_x - \text{Cov}_n$. The quasi-whitened data follow a noisy ICA model as well, with an orthogonal mixing matrix, and easily derivable noise covariance matrix. The quasi-whitened data are then fed into a fixed-point algorithm with Gaussian moments, which uses the covariance matrix of the transformed noise to avoid bias (Hyvarinen 1999b).

8.4 Numerical Examples

To demonstrate the potential effectiveness of BSS methods in isolating elemental spectra from LIBS measurements, we rely on simulated spectra built from known data and a simplified noise model. In this section, we show and compare the results obtained by MNF and ICA for different signal-to-noise ratios.

Table 8.2 Simulated relative abundances in four measurement points of a nonhomogeneous bronze sample.

Measurement point	%Cu	%Sn	%Zn	%Pb
#1	83.6	1.9	11.4	3.1
#2	87.1	5.0	5.5	2.4
#3	90.9	0.3	7.8	1.0
#4	84.1	1.4	10.5	4.0

We start from the typical relative abundances (see Table 8.2) found in a bronze alloy containing copper, tin, zinc, and lead. By small random perturbations of these abundances, we simulated the LIBS spectra in 40 measurement points of a nonhomogeneous bronze sample. From the NIST LIBS database, (https://physics.nist.gov/PhysRefData/ASD/LIBS/), we retrieved the spectra of the four analytes in the ionization statuses likely to appear when acquired by a spectrometer, that is, Cu I, Sn II, Zn II, Pb I, and Pb II (hereafter, the sum of the latter two is denoted by Pb Sum). The chosen spectral range is 200–400 nm in air, with 2500 wavelengths per spectrum, electron temperature $T_e = 1eV$, and electron number density $n_e = 10^{17}$ cm^{-3}. Figure 8.3 shows these spectra with the line intensities in arbitrary units.

The elemental spectra are then composed using the random abundances mentioned above and corrupted by white Gaussian noise with variable signal-to-noise ratio. These simulated spectra would be the ones obtained ideally from a real measurement in the hypothesis of being able to perfectly subtract the continuum background B.

In Figures 8.4 and 8.5, we show example MNF outputs, against noisy data with SNR = 18 dB and SNR = 42 dB, respectively. We cannot compare these results point by point with the original spectra, as the outputs are copies of the source functions up to scaling factors. For this reason, we rely on correlation coefficients. Since the LIBS spectra are strictly non-negative, whereas the algorithm outputs can also assume negative values, we compute the correlation coefficients, r, between the original spectra and the positive values of the reconstructed ones.

In the case of 18 dB SNR, MNF produces 40 components, but only the first three correlate significantly with the ideal element spectra, while tin is not recovered at all ($r < 0.1$). Moreover, from the plot, it is well noticeable that the estimated copper spectrum contains residual components from the zinc. As we added Gaussian noise with equal variance to the spectra of all the 40 points, from a theoretical point of view PCA performs identically to MNF. Indeed, with PCA, copper and zinc are both recovered with correlation values similar to those of MNF, copper exhibits the same residual components from the zinc, tin is not recovered, and the estimated lead spectrum is even worse than the one obtained by MNF ($r = 0.192$).

For an SNR of 42 dB, the elemental spectra are recovered much better, and the noisiest component, the fourth one, this time has a correlation coefficient $r = 0.416$ with the tin. However, copper still exhibits contaminations from the zinc, which, conversely, is recovered almost perfectly ($r = 1$).

The corresponding results from ICA are shown in Figures 8.6 and 8.7. From Figure 8.7, we note that the reconstructed tin spectrum has a very low correlation coefficient even with

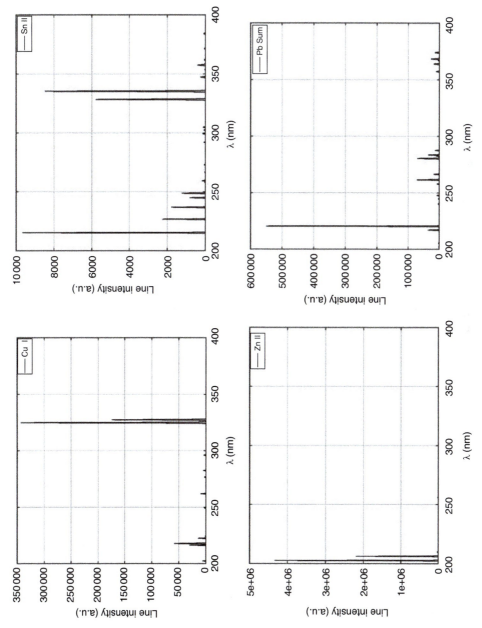

Figure 8.3 Elemental LIBS spectra as retrieved from the NIST database.

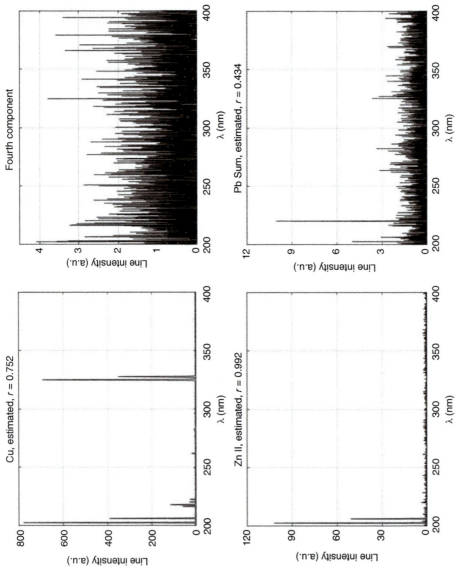

Figure 8.4 Elemental spectra estimated by MNF from data with Gaussian noise of 18 dB SNR.

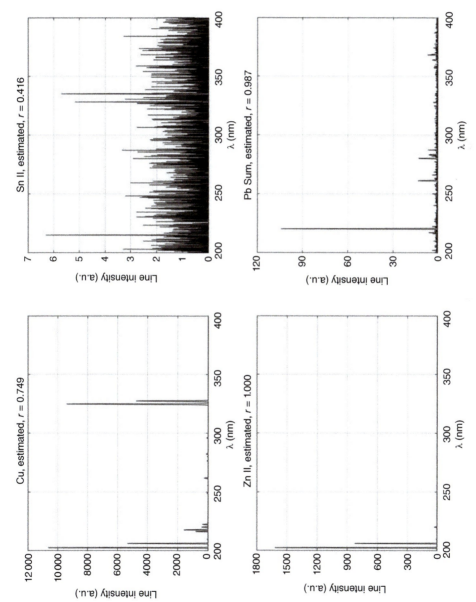

Figure 8.5 Elemental spectra estimated by MNF from data with Gaussian noise of 42 dB SNR.

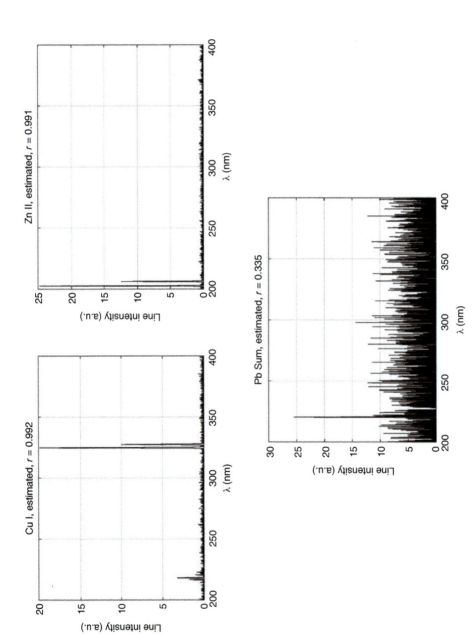

Figure 8.6 Elemental spectra estimated by ICA with Gaussian noise, 18 dB SNR. Only three outputs are computed, as the fourth eigenvalue of the data covariance matrix is equal or less than zero.

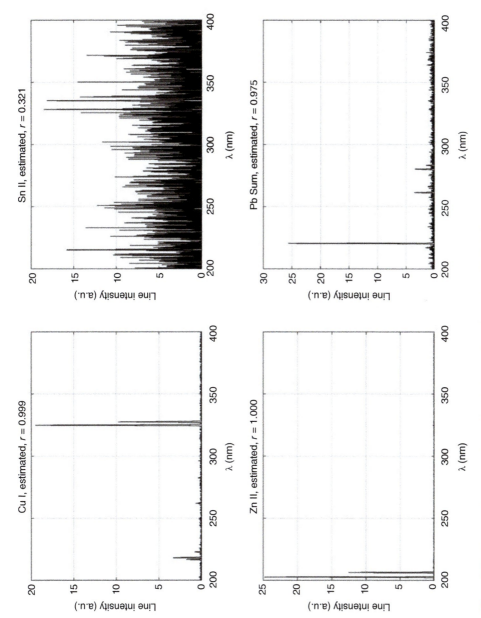

Figure 8.7 Elemental spectra estimated by ICA with Gaussian noise, 42 dB SNR.

such a high SNR, so in practice tin is not extracted from the measured spectra. A possible reason for this is the small abundance of this analyte in the alloy under examination, as can be seen from the sample values in Table 8.2. In the case of Figure 8.6, the ICA algorithm does not even evaluate the fourth component, since only three significant principal components are found (see later), and the correlation coefficient for the reconstructed lead spectrum is also quite low. Conversely, the reconstructed spectra of copper and zinc are still strongly correlated with the corresponding originals.

To study the behavior of ICA with different noise levels, we let them range from 60 to 12 dB in steps of 6 dB. The results are averaged over 10 runs of the algorithm to consider the variability of the noise realization and the fact that the initial guess for the solution of the separation problem is random. Table 8.3 reports the resulting correlation coefficients for all the reconstructed spectra, including for comparison, the case of zero noise.

Considering as well recovered sources that correlate for more than an 80% with the corresponding outputs, it is easy to note that the copper and the zinc spectra are always recovered correctly, whereas the tin spectrum is very sensitive to noise and the lead spectrum is recovered correctly if the SNR is higher than 30 dB.

Following a suggestion by Hyvarinen (1999b), saying that the noisy version of ICA can also work with non-Gaussian noise, we also tested this algorithm with data corrupted by multiplicative lognormal-distributed noise, as this is a natural choice for strictly positive measurements. We thus have colored, signal-dependent noise.

Also, in this case, although the data model used to design the separation algorithm includes Gaussian signal-independent noise, useful results are obtained. From Table 8.4, we see that copper and zinc are still reconstructed correctly for all the SNRs, and the lead

Table 8.3 Mean correlation coefficients between the elemental original and reconstructed spectra, with signals corrupted by white Gaussian additive noise.

	Average correlation coefficients (10 runs)			
	Noisy ICA	White, signal-independent Gaussian noise		
SNR (dB)	Cu I	Sn II	Zn II	Pb Sum
∞	0.99994	0.99996	0.99995	0.99992
60	0.99991	0.95533	0.99993	0.99961
54	0.99985	0.84880	0.99992	0.99861
48	0.99977	0.61573	0.99936	0.99454
42	0.99102	0.37058	0.99172	0.97925
36	0.89172	0.15477	0.95968	0.89317
30	0.9995960	−0.0024554	0.9994730	0.7874864
24	0.9984934	−0.0044540	0.9971295	0.5355158
18	0.9892834	0.0015834	0.9894578	0.3302940
12	0.9643751	0.0061326	0.9634158	0.2050567

Table 8.4 Mean correlation coefficients between the elemental original and reconstructed spectra, with signals corrupted by multiplicative lognormal noise.

		Average correlation coefficients (10 runs)		
	Noisy ICA	Multiplicative, lognormal-distributed noise		
SNR (dB)	Cu I	Sn II	Zn II	Pb Sum
∞	0.99994	0.99996	0.99995	0.99992
60	0.99734	0.96766	0.99888	0.99950
54	0.9999407	−0.0040305	0.9999168	0.9978723
48	0.9998225	−0.0045019	0.9998574	0.9924288
42	0.9998953	−0.0053434	0.9996803	0.9707330
36	0.9990394	−0.0062307	0.9985958	0.9257757
30	0.9994727	−0.0064664	0.9989289	0.0208064
24	0.9980711	−0.0067057	0.9973147	0.0197596
18	0.9928784	−0.0069262	0.9894208	0.0197565
12	0.9740855	−0.0075006	0.9688107	0.0166150

spectrum is accurate for SNR>30 dB. The tin spectrum, conversely, is only recovered for SNR = 60 dB.

Note that the ICA outputs reproduce the sources slightly better than the ones produced by MNF. Furthermore, the ICA outputs that correlate more than 0.8 with the corresponding elemental spectra do not present the spurious lines appearing in Figures 8.4 and 8.5. From Tables 8.3 and 8.4, it is possible to note a different behavior of the ICA outputs with data affected by either signal-independent or signal-dependent noise. In the latter case, we have, for tin and lead, an abrupt transition between correlations well above 0.8 and very small values. This kind of values also appear in the results with Gaussian noise, except that the transitions are a little bit smoother. Following Hyvarinen (1999a, 1999b), in our ICA algorithm, the data are first quasi-whitened and then rotated to maximize their mutual independence. In the presence of noise, the quasi-whitening matrix has the same form as in Eq. (8.8), but it is even possible that some eigenvalue of order lower than N is zero. In this situation, ICA only produces less than N independent components. In our case ($N = 4$), the estimation quality index for a given analyte is computed as the maximum correlation coefficient between its original spectrum and all the outputs. When less than four outputs are available, some original spectra (Sn II, Pb Sum, or both, in our case) will have all very small, meaningless, correlation values with all the outputs. Thus, the number of independent components computed as a function of the SNR can also be derived by just inspecting Tables 8.3 and 8.4. With Gaussian noise, ICA extracts only three independent components for SNR ≤ 30 dB, whereas for lognormal noise, ICA extracts only three components with SNR between 36 dB and 54 dB and only two components for SNR ≤ 30 dB.

8.5 Final Remarks

This short account of BSS methods for LIBS spectroscopy is mainly aimed at enabling the reader to understand the potentialities and the advantages offered by a correct isolation of the different elemental spectra in terms of "readability" of the results. Indeed, the analysis by visual inspection of the individual spectra would be facilitated for the obtained separation of nearby lines belonging to different analytes and the consequent suppression of possible distortions due to line superposition. Also, some possibilities for an automatic spectrum analysis would be opened up, for example, if a dictionary of ideal elemental spectra were available, by computing their correlation coefficients with the algorithm outputs or, as often done with remote-sensed imagery, through the spectral angle mapping (SAM) technique (Kruse et al. 1993)

To our knowledge, this topic has not been explored thoroughly so far. Its complexity and promises, however, should encourage the research community to take it into account as a concrete possibility of improving the performance of LIBS analysis in many application fields. Many issues deserve to be addressed. First of all, an appropriate noise model would be helpful in not only analyzing the behavior of existing methods in front of different signal and noise situations but also to try to develop separation methods that are specific to data affected by physically plausible noise and/or interference. Obviously, any algorithm would yield satisfactory results if the SNR is the least possible. This condition can be approached by either data preprocessing or accurate data capture, such as instrument calibration against systematic effects and the suppression of plasma-physics-associated phenomena.

As far as the study of theoretical/simulated performance is concerned, Tables 8.3 and 8.4 above suggest the possibility to establish separability bounds as functions of the SNR and the composition of the material being examined, once a separability threshold is fixed (for example, $r = 0.8$, as above). Intuitively, an element whose relative abundance is small should need a very high SNR to be separated from the embedding spectra. Concrete results, however, let us conjecture that the number of independent measurement points also affects the separability, as well as the shape of an individual spectrum, the so-called "skyline." In our simulated experiments with complex metal alloys, the most abundant elements can be separated for any reasonably assumed SNR value, whereas the minor elements are not always separable.

While the results on simulated spectra seem promising, real experiments with measured LIBS spectra are still needed to assess the applicability of BSS methods to elemental spectra separation. In particular, the behavior of these algorithms in the presence of interferences due to the physics of plasmas, such as self-absorption, is not yet known. This means that an extensive experimentation is still needed to assess the possible added value provided by these techniques. It is to be noted, however, that BSS does not need to replace other well-established approaches. The results from this kind of technique could indeed support other results in particularly complicated cases, for example, when no preliminary idea is available on the composition of the samples analyzed, thus helping a confident search for the significant spectral lines.

References

Alexander, D. R., Anderson, T., & Bruce, J. C., III (2012). *Laser Induced Breakdown Spectroscopy Having Enhanced Signal-to-Noise Ratio*. United States Patent Application Publication no. US 2012/0314214 A1, Dec. 13, 2012.

Bauer, H. E., Leis, F., & Niemax, K. (1998). Laser induced breakdown spectrometry with an echelle spectrometer and intensified charge coupled device detection. *Spectrochimica Acta Part B: Atomic Spectroscopy*, **53**(13), 1815–1825.

Bell, A. J., & Sejnowski, T. J. (1995a). An information-maximization approach to blind separation and blind deconvolution. *Neural Computation*, **7**(6), 1129–1159.

Bell, A. J., & Sejnowski, T. J. (1995b). Fast blind separation based on information theory. *Proc. International Symposium on Nonlinear Theory and Applications, Las Vegas*.

Boué-Bigne, F. (2008). Laser-induced breakdown spectroscopy applications in the steel industry: Rapid analysis of segregation and decarburization. *Spectrochimica Acta – Part B Atomic Spectroscopy*, **63**(10), 1122–1129. https://doi.org/10.1016/j.sab.2008.08.014

Bulajic, D., Corsi, M., Cristoforetti, G., Legnaioli, S., Palleschi, V., Salvetti, A., & Tognoni, E. (2002). A procedure for correcting self-absorption in calibration free-laser induced breakdown spectroscopy. *Spectrochimica Acta Part B: Atomic Spectroscopy*, **57**(2), 339–353.

Busser, B., Moncayo, S., Coll, J.-L., Sancey, L., & Motto-Ros, V. (2018). Elemental imaging using laser-induced breakdown spectroscopy: A new and promising approach for biological and medical applications. *Coordination Chemistry Reviews*, **358**, 70–79. https://doi.org/10.1016/j.ccr.2017.12.006

Carvalho, R. R. V, Coelho, J. A. O., Santos, J. M., Aquino, F. W. B., Carneiro, R. L., & PereiraFilho, E. R. (2015). Laser-induced breakdown spectroscopy (LIBS) combined with hyperspectral imaging for the evaluation of printed circuit board composition. *Talanta*, **134**, 278–283. https://doi.org/10.1016/j.talanta.2014.11.019

Casado-Gavalda, M. P., Dixit, Y., Geulen, D., Cama-Moncunill, R., Cama-Moncunill, X., Markiewicz-Keszycka, M., Cullen, P. J., & Sullivan, C. (2017). Quantification of copper content with laser induced breakdown spectroscopy as a potential indicator of offal adulteration in beef. *Talanta*, **169**, 123–129. https://doi.org/10.1016/j.talanta.2017.03.071

Choi, S., Cichocki, A., Park, H.-M., & Lee, S.-Y. (2005). Blind source separation and independent component analysis: A review. *Neural Information Processing-Letters and Reviews*, **6**(1), 1–57.

Cichocki, A., & Amari, S. (2002). *Adaptive blind signal and image processing: Learning algorithms and applications*. New York, USA: John Wiley & Sons.

Dabiri, Z., & Lang, S. (2018). Comparison of independent component analysis, principal component analysis, and minimum noise fraction transformation for tree species classification using APEX hyperspectral imagery. *ISPRS International Journal of Geo-Information*, **7**(12), 488.

Danzer, K. (1984). Comparison of sampling procedures for investigation of chemical homogeneity of solids by spark emission spectrography. *Spectrochimica Acta Part B: Atomic Spectroscopy*, **39**(8), 949–954.

Dixit, Y., Casado-Gavalda, M. P., Cama-Moncunill, R., Cama-Moncunill, X., MarkiewiczKeszycka, M., Cullen, P. J., & Sullivan, C. (2017). Laser induced breakdown spectroscopy for quantification of sodium and potassium in minced beef: A potential technique for detecting beef kidney adulteration. *Analytical Methods*, **9**(22), 3314–3322. https://doi.org/10.1039/c7ay00757d

Fabre, C., Devismes, D., Moncayo, S., Pelascini, F., Trichard, F., Lecomte, A., Bousquet, B., Cauzid, J., & Motto-Ros, V. (2018). Elemental imaging by laser-induced breakdown spectroscopy for the geological characterization of minerals. *Journal of Analytical Atomic Spectrometry*, **33**(8), 1345–1353. https://doi.org/10.1039/c8ja00048d

Fortes, F. J., Perez-Carceles, M. D., Sibon, A., Luna, A., & Laserna, J. J. (2015). Spatial distribution analysis of strontium in human teeth by laser-induced breakdown spectroscopy: Application to diagnosis of seawater drowning. *International Journal of Legal Medicine*, **129**(4), 807–813. https://doi.org/10.1007/s00414-014-1131-9

Fu, Y., Hou, Z., Li, T., Li, Z., & Wang, Z. (2019). Investigation of intrinsic origins of the signal uncertainty for laser-induced breakdown spectroscopy. *Spectrochimica Acta Part B: Atomic Spectroscopy*, **155**, 67–78.

Gao, L., Zhao, B., Jia, X., Liao, W., & Zhang, B. (2017). Optimized kernel minimum noise fraction transformation for hyperspectral image classification. *Remote Sensing*, **9**(6), 548.

Gimenez, Y., Busser, B., Trichard, F., Kulesza, A., Laurent, J. M., Zaun, V., Lux, F., Benoit, J. M., Panczer, G., Dugourd, P., Tillement, O., Pelascini, F., Sancey, L., & Motto-Ros, V. (2016). 3D Imaging of Nanoparticle Distribution in Biological Tissue by Laser-Induced Breakdown Spectroscopy. *Scientific Reports*, **6**. https://doi.org/10.1038/srep29936

Green, A. A., Berman, M., Switzer, P., & Craig, M. D. (1988). A transformation for ordering multispectral data in terms of image quality with implications for noise removal. *IEEE Transactions on Geoscience and Remote Sensing*, **26**(1), 65–74.

Hahn, D.W. and Omenetto, N. (2012) Laser-Induced Breakdown Spectroscopy (LIBS), part II: review of instrumental and methodological approaches to material analysis and applications to different fields. *Applied Spectroscopy*, **66**, 347–419. http://dx.doi.org/10.1366/11-06574

Häkkänen, H. J., & Korppi-Tommola, J. E. I. (1995). UV-laser plasma study of elemental distributions of paper coatings. *Appl. Spectrosc.*, **49**(12), 1721–1728.

Harris, J. R., Rogge, D., Hitchcock, R., Ijewliw, O., & Wright, D. (2005). Mapping lithology in Canada's Arctic: Application of hyperspectral data using the minimum noise fraction transformation and matched filtering. *Canadian Journal of Earth Sciences*, **42**(12), 2173–2193.

Hausmann, N., Siozos, P., Lemonis, A., Colonese, A. C., Robson, H. K., & Anglos, D. (2017). Elemental mapping of Mg/Ca intensity ratios in marine mollusc shells using laser-induced breakdown spectroscopy. *Journal of Analytical Atomic Spectrometry*, **32**(8), 1467–1472. https://doi.org/10.1039/c7ja00131b

Hoehse, M., Gornushkin, I., Merk, S., & Panne, U. (2011). Assessment of suitability of diode pumped solid state lasers for laser induced breakdown and Raman spectroscopy. *Journal of Analytical Atomic Spectrometry*, **26**(2), 414–424. https://doi.org/10.1039/c0ja00038h

Hong, S., Lai, W. W.-L., Wilsch, G., Helmerich, R., Helmerich, R., Günther, T., & Wiggenhauser, H. (2014). Periodic mapping of reinforcement corrosion in intrusive chloride contaminated concrete with GPR. *Construction and Building Materials*, **66**, 671–684. https://doi.org/10.1016/j.conbuildmat.2014.06.019

Hyvarinen, A. (1999a). Fast and robust fixed-point algorithms for independent component analysis. *IEEE Transactions on Neural Networks*, **10**(3), 626–634.

Hyvarinen, A. (1999b). Gaussian moments for noisy independent component analysis. *IEEE Signal Processing Letters*, **6**(6), 145–147.

Hyvarinen, A. (1999c). Fast ICA for noisy data using Gaussian moments. *1999 IEEE International Symposium on Circuits and Systems (ISCAS)*, **5**, 57–61.

Hyvärinen, A. (2001). Fast ICA by a fixed-point algorithm that maximizes non-Gaussianity. In: Stephen Roberts and Richard Everson (Editors) *Independent component analysis: Principles and practice*, 71–94. Cambridge University Press. doi: https://doi.org/10.1017/CBO9780511624148.003

Hyvärinen, A, Karhunen, J., & Oja, E. (2001). *Independent component analysis*. John Wiley and Sons. Inc., New York.

Hyvärinen, A, & Oja, E. (1997). A fast fixed-point algorithm for independent component analysis. *Neural Computation*, **9**(7), 1483–1492.

Jain, S. N., & Rai, C. (2012). Blind source separation and ICA techniques: A review. *IJEST*, **4**(4), 1490–1503.

Karhunen, J., & Joutsensalo, J. (1994). Representation and separation of signals using nonlinear PCA type learning. *Neural Networks*, **7**(1), 113–127.

Karhunen, J., Pajunen, P., & Oja, E. (1998). The nonlinear PCA criterion in blind source separation: Relations with other approaches. *Neurocomputing*, **22**(1–3), 5–20.

Karhunen, J., Wang, L., & Vigario, R. (1995). Nonlinear PCA type approaches for source separation and independent component analysis. *Proceedings of ICNN'95-International Conference on Neural Networks*, 2, 995–1000.

Kempenaers, L., Janssens, K., Vincze, L., Vekemans, B., Somogyi, A., Drakopoulos, M., Simionovici, A., & Adams, F. (2002). A Monte Carlo model for studying the microheterogeneity of trace elements in reference materials by means of synchrotron microscopic X-ray fluorescence. *Analytical Chemistry*, **74**(19), 5017–5026.

Kim, T., Lin, C. T., & Yoon, Y. (1998). Compositional mapping by laser-induced breakdown spectroscopy. *Journal of Physical Chemistry B*, **102**(22), 4284–4287.

Kruse, F. A., Lefkoff, A. B., Boardman, J. W., Heidebrecht, K. B., Shapiro, A. T., Barloon, P. J., & Goetz, A. F. H. (1993). The spectral image processing system (SIPS)-interactive visualization and analysis of imaging spectrometer data. *AIP Conference Proceedings*, **283**(1), 192–201.

Lednev, V. N., Sdvizhenskii, P. A., Grishin, M. Y. A., Cheverikin, V. V, Stavertiy, A. Y. A., Tretyakov, R. S., Taksanc, M. V, & Pershin, S. M. (2017). Laser-induced breakdown spectroscopy for three-dimensional elemental mapping of composite materials synthesized by additive technologies. *Applied Optics*, **56**(35), 9698–9705. https://doi.org/10.1364/AO.56.009698

Lee, J. B., Woodyatt, A. S., & Berman, M. (1990). Enhancement of high spectral resolution remote-sensing data by a noise-adjusted principal components transform. *IEEE Transactions on Geoscience and Remote Sensing*, **28**(3), 295–304.

Lefebvre, C., Catalá-Espí, A., Sobron, P., Koujelev, A., & Léveillé, R. (2016). Depth-resolved chemical mapping of rock coatings using Laser-Induced Breakdown Spectroscopy: Implications for geochemical investigations on Mars. *Planetary and Space Science*, **126**, 24–33. https://doi.org/10.1016/j.pss.2016.04.003

Le Guével, X., Henry, M., Motto-Ros, V., Longo, E., Montañez, M. I., Pelascini, F., De La Rochefoucauld, O., Zeitoun, P., Coll, J.-L., Josserand, V., & Sancey, L. (2018). Elemental and optical imaging evaluation of zwitterionic gold nanoclusters in glioblastoma mouse models. *Nanoscale*, **10**(39), 18657–18664. https://doi.org/10.1039/c8nr05299a

Li, J., Hao, Z., Zhao, N., Zhou, R., Yi, R., Tang, S., Guo, L., Li, X., Zeng, X., & Lu, Y. (2017). Spatially selective excitation in laser-induced breakdown spectroscopy combined with laser-induced fluorescence. *Optics Express*, **25**(5), 4945–4951. https://doi.org/10.1364/OE.25.004945

Lin, Q.-H., Zheng, Y.-R., Yin, F.-L., Liang, H., & Calhoun, V. D. (2007). A fast algorithm for one-unit ICA-R. *Information Sciences*, **177**(5), 1265–1275.

López-López, M., Alvarez-Llamas, C., Pisonero, J., García-Ruiz, C., & Bordel, N. (2017). An exploratory study of the potential of LIBS for visualizing gunshot residue patterns. *Forensic Science International*, **273**. https://doi.org/10.1016/j.forsciint.2017.02.012

Lopez-Quintas, I., Mateo, M. P., Piñon, V., Yañez, A., & Nicolas, G. (2012). Mapping of mechanical specimens by laser induced breakdown spectroscopy method: Application to an engine valve. *Spectrochimica Acta – Part B Atomic Spectroscopy*, **74–75**, 109–114. https://doi.org/10.1016/j.sab.2012.06.035

Luo, G., Chen, G., Tian, L., Qin, K., & Qian, S.-E. (2016). Minimum noise fraction versus principal component analysis as a preprocessing step for hyperspectral imagery denoising. *Canadian Journal of Remote Sensing*, **42**(2), 106–116.

Mermet, J. M., Mauchien, P., & Lacour, J. L. (2008). Processing of shot-to-shot raw data to improve precision in laser-induced breakdown spectrometry microprobe. *Spectrochimica Acta Part B: Atomic Spectroscopy*, **63**(10), 999–1005.

Motto-Ros, V., Moncayo, S., Trichard, F., & Pelascini, F. (2019). Investigation of signal extraction in the frame of laser induced breakdown spectroscopy imaging. *Spectrochimica Acta Part B: Atomic Spectroscopy*, **155**, 127–133.

Motto-Ros, V., Sancey, L., Ma, Q. L., Lux, F., Bai, X. S., Wang, X. C., Yu, J., Panczer, G., & Tillement, O. (2012). Mapping of native inorganic elements and injected nanoparticles in a biological organ with laser-induced plasma. *Applied Physics Letters*, **101**(22). https://doi.org/10.1063/1.4768777

Noll, R., Bette, H., Brysch, A., Kraushaar, M., Mönch, I., Peter, L., & Sturm, V. (2001). Laser-induced breakdown spectrometry - Applications for production control and quality assurance in the steel industry. *Spectrochimica Acta - Part B Atomic Spectroscopy*, **56**(6), 637–649. https://doi.org/10.1016/S0584-8547(01)00214-2

Oja, E. (1995). *The nonlinear PCA learning rule and signal separation: Mathematical analysis*. Helsinki University of Technology Espoo, Finland.

Oja, E., & Plumbley, M. (2003). Blind separation of positive sources using non-negative PCA. *4th International Symposium on Independent Component Analysis and Blind Signal Separation*.

Pagnotta, S., Lezzerini, M., Ripoll-Seguer, L., Hidalgo, M., Grifoni, E., Legnaioli, S., Lorenzetti, G., Poggialini, F., & Palleschi, V. (2017). Micro-laser-induced breakdown spectroscopy (Micro-LIBS) study on ancient roman mortars. *Applied Spectroscopy*, **71**(4), 721–727.

Pajunen, P., & Karhunen, J. (1997). Least-squares methods for blind source separation based on nonlinear PCA. *International Journal of Neural Systems*, **8**(05n06), 601–612.

Pal, M., Roy, R., Basu, J., & Bepari, M. S. (2013). Blind source separation: A review and analysis. *2013 International Conference Oriental COCOSDA Held Jointly with 2013 Conference on Asian Spoken Language Research and Evaluation (O-COCOSDA/CASLRE)*, 1–5

Peng, J., Liu, F., Zhou, F., Song, K., Zhang, C., Ye, L., & He, Y. (2016). Challenging applications for multi-element analysis by laser-induced breakdown spectroscopy in agriculture: A review. *TrAC – Trends in Analytical Chemistry*, **85**, 260–272. https://doi.org/10.1016/j.trac.2016.08.015

Rifai, K., Doucet, F., Özcan, L., & Vidal, F. (2018). LIBS core imaging at kHz speed: Paving the way for real-time geochemical applications. *Spectrochimica Acta – Part B Atomic Spectroscopy*, **150**, 43–48. https://doi.org/10.1016/j.sab.2018.10.007

Roger, R. E. (1994). A faster way to compute the noise-adjusted principal components transform matrix. *IEEE Transactions on Geoscience and Remote Sensing*, **32**(6), 1194–1196.

Romero, D., & Laserna, J. J. (1997). Multielemental chemical imaging using laser-induced breakdown spectrometry. *Analytical Chemistry*, **69**(15), 2871–2876. https://doi.org/10.1021/ac9703111

Sallé, B., Lacour, J.-L., Mauchien, P., Fichet, P., Maurice, S., & Manhes, G. (2006). Comparative study of different methodologies for quantitative rock analysis by laser-induced breakdown spectroscopy in a simulated Martian atmosphere. *Spectrochimica Acta Part B: Atomic Spectroscopy*, **61**(3), 301–313.

Schiavo, C., Menichetti, L., Grifoni, E., Legnaioli, S., Lorenzetti, G., Poggialini, F., Pagnotta, S., & Palleschi, V. (2016). High-resolution three-dimensional compositional imaging by doublepulse laser-induced breakdown spectroscopy. *Journal of Instrumentation*, **11**(08), c8002.

Schlenke, J., Hildebrand, L., Moros, J., & Laserna, J. J. (2012). Adaptive approach for variable noise suppression on laser-induced breakdown spectroscopy responses using stationary wavelet transform. *Analytica Chimica Acta*, **754**, 8–19.

Sheta, S. A., Di Carlo, G., Ingo, G. M., & Harith, M. A. (2015). Surface heterogeneity study of some reference Cu-Ag alloys using laser-induced breakdown spectroscopy. *Surface and Interface Analysis*, **47**(4), 514–522. https://doi.org/10.1002/sia.5741

Škarková, P., Novotný, K., Lubal, P., Jebavá, A., Pořízka, P., Klus, J., Farka, Z., Hrdlička, A., & Kaiser, J. (2017). 2d distribution mapping of quantum dots injected onto filtration paper by laser-induced breakdown spectroscopy. *Spectrochimica Acta – Part B Atomic Spectroscopy*, **131**, 107–114. https://doi.org/10.1016/j.sab.2017.03.016

Sperança, M. A., de Aquino, F. W. B., Fernandes, M. A., Lopez-Castillo, A., Carneiro, R. L., & Pereira-Filho, E. R. (2017). Application of Laser-Induced Breakdown Spectroscopy and Hyperspectral Images for Direct Evaluation of Chemical Elemental Profiles of Coprolites. *Geostandards and Geoanalytical Research*, **41**(2), 273–282. https://doi.org/10.1111/ggr.12155

St-Onge, L., Detalle, V., & Sabsabi, M. (2002). Enhanced laser-induced breakdown spectroscopy using the combination of fourth-harmonic and fundamental Nd: YAG laser pulses. *Spectrochimica Acta Part B: Atomic Spectroscopy*, **57**(1), 121–135.

Sun, L., & Yu, H. (2009). Automatic estimation of varying continuum background emission in laser-induced breakdown spectroscopy. *Spectrochimica Acta Part B: Atomic Spectroscopy*, **64**(3), 278–287.

Sun, Y. (2005). The study and test of ICA algorithms. *Proceedings. 2005 International Conference on Wireless Communications, Networking and Mobile Computing, 2005*, **1**, 602–605.

Taparli, U. A., Jacobsen, L., Griesche, A., Michalik, K., Mory, D., & Kannengiesser, T. (2018). In situ laser-induced breakdown spectroscopy measurements of chemical compositions in stainless steels during tungsten inert gas welding. *Spectrochimica Acta – Part B Atomic Spectroscopy*, **139**, 50–56. https://doi.org/10.1016/j.sab.2017.11.012

Tognoni, E., & Cristoforetti, G. (2016). Signal and noise in laser induced breakdown spectroscopy: An introductory review. *Optics & Laser Technology*, **79**, 164–172.

Tognoni, E., Palleschi, V., Corsi, M., & Cristoforetti, G. (2002). Quantitative micro-analysis by laser-induced breakdown spectroscopy: A review of the experimental approaches. *Spectrochimica Acta Part B: Atomic Spectroscopy*, **57**(7), 1115–1130.

Trichard, F., Gaulier, F., Barbier, J., Espinat, D., Guichard, B., Lienemann, C.-P., Sorbier, L., Levitz, P., & Motto-Ros, V. (2018). Imaging of alumina supports by laser-induced breakdown spectroscopy: A new tool to understand the diffusion of trace metal impurities. *Journal of Catalysis*, **363**, 183–190. https://doi.org/10.1016/j.jcat.2018.04.013

Wiggins, B., Tupitsyn, E., Bhattacharya, P., Rowe, E., Lukosi, E., Chvala, O., Burger, A., & Stowe, A. (2013). Investigation of non-uniformity and inclusions in 6LiInSe 2 utilizing laser induced breakdown spectroscopy (LIBS). *Proceedings of SPIE – The International Society for Optical Engineering*, **8852**. https://doi.org/10.1117/12.2024125

Xu, L., Bulatov, V., Gridin, V. V, & Schechter, I. (1997). Absolute analysis of particulate materials by laser-induced breakdown spectroscopy. *Analytical Chemistry*, **69**(11), 2103–2108.

Xu, T., Liu, J., Shi, Q., He, Y., Niu, G., & Duan, Y. (2016). Multi-elemental surface mapping and analysis of carbonaceous shale by laser-induced breakdown spectroscopy. *Spectrochimica Acta – Part B Atomic Spectroscopy*, **115**, 31–39. https://doi.org/10.1016/j.sab.2015.10.008

Yaroshchyk, P., & Eberhardt, J. E. (2014). Automatic correction of continuum background in Laser-induced Breakdown Spectroscopy using a model-free algorithm. *Spectrochimica Acta Part B: Atomic Spectroscopy*, **99**, 138–149. https://doi.org/10.1016/j.sab.2014.06.020

Zhang, B., Sun, L., Yu, H., Xin, Y., & Cong, Z. (2013). Wavelet denoising method for laser-induced breakdown spectroscopy. *Journal of Analytical Atomic Spectrometry*, **28**(12), 1884–1893.

Zhang, B., Sun, L., Yu, H., Xin, Y., & Cong, Z. (2015). A method for improving wavelet threshold denoising in laser-induced breakdown spectroscopy. *Spectrochimica Acta Part B: Atomic Spectroscopy*, **107**, 32–44.

Zhu, X.-L., Zhang, X.-D., Ding, Z.-Z., & Jia, Y. (2006). Adaptive nonlinear PCA algorithms for blind source separation without prewhitening. *IEEE Transactions on Circuits and Systems I: Regular Papers*, **53**(3), 745–753.

Zou, X. H., Guo, L. B., Shen, M., Li, X. Y., Hao, Z. Q., Zeng, Q. D., Lu, Y. F., Wang, Z. M., & Zeng, X. Y. (2014). Accuracy improvement of quantitative analysis in laser-induced breakdown spectroscopy using modified wavelet transform. *Optics Express*, **22**(9), 10233–10238.

9

Artificial Neural Networks for Classification

Jakub Vrábel[1,2], Erik Képeš[1,2], Pavel Pořízka[1,2], and Jozef Kaiser[1,2]

[1] CEITEC BUT, Brno University of Technology, Brno, Czech Republic
[2] Institute of Physical Engineering, Brno University of Technology, Brno, Czech Republic

9.1 Introduction and Scope

Like in other scientific revolutions, almost every aspect of human endeavor was affected and changed by modern machine learning (ML) approaches. Their usage ranges from the simple personalized advertisement, image recognition to more advanced autonomous driving systems, game playing (e.g. Go [1] and chess), and, of course, scientific applications [2]. Most of these tasks can be now automatized with an unprecedented performance.

The present success and interest in the ML community is based on artificial neural networks (ANNs). While ANNs are definitely not a novel topic in the field of computer science, their development has been far from straightforward (see Figure 9.1). The renaissance of ANNs (as well as the huge growth of the ML field) was sparked off around the year 2006 by a "simple" rebranding of the topic to *deep learning* [3]. The key to the success of deep learning was based on several "good old ideas" combined with new computing possibilities (e.g. parallelization and GPU computation).

Naturally, the success of deep learning in countless fields triggered the spectroscopists' interest as well. The possibility to carry out robust analyses without the need of extensive hypothesis testing and programming transformed many aspects of spectroscopic analysis. Namely, LIBS has struggled with the nonlinear response of spectroscopic signal since its inception. This nonlinearity can be favorably addressed by applying ANNs, which are general function approximators [4, 5]. Thus, ANNs are able to reflect the complex signal response of laser-induced plasma emission. However, the use of ANNs comes with a major drawback; ANNs are mainly employed as black-box models without a deeper understanding of their function. This undermines their full potential and hinders the users' ability to effectively troubleshoot issues that arise during the development of ANN models.

This chapter is intended to provide interested readers with a guideline and general good practices for using ANNs. We believe that this necessitates a basic **understanding of the "inner-mechanisms" of ANNs**. Thus, apart from describing the basic ANN architectures in Section 9.2, this chapter also briefly introduces the **training of ANNs** in general

Chemometrics and Numerical Methods in LIBS, First Edition. Edited by Vincenzo Palleschi.
© 2023 John Wiley & Sons Ltd. Published 2023 by John Wiley & Sons Ltd.

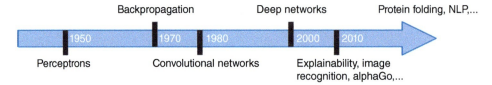

Figure 9.1 History of artificial neural networks.

(Section 9.3), **the backpropagation algorithm** (the workhorse of training ANNs, Section 9.4), **convolutions** (Section 9.5), the evaluation of ANNs (Section 9.6), and their regularization (Section 9.7). Lastly, in Section 9.8, we build on the established knowledge to comment on the complex applications of ANNs to specific classification problems.

9.2 Artificial Neural Networks (ANNs)

In general, ANNs are introduced as ML models that are inspired by the human brain. However, it should be emphasized that the current understanding of the brain function reveals the existence of more complex relations and behavior than a relatively simple ANN mechanism. In this chapter, we focus on two major ANN architectures: the **fully connected feed-forward** ANN (also commonly referred to as **multilayer perceptron – MLP**), and **convolutional neural networks** (CNN). Both are described later in this chapter. Note that owing to their lack of spectroscopic applications, ANN architectures such as recurrent ANNs, long short-term memory ANNs, transformer ANNs, and many others are not included in this chapter [6].

The fundamental building block of ANNs is the artificial neuron (or node), which is schematically shown in Figure 9.2a. Such a neuron has weighted input connections $x_i w_i$ for ($i = 1, 2, 3, \ldots$) and a bias $x_0 w_0$ (described in Section 9.4). Inside a neuron, all inputs are summed and passed to a nonlinear **activation function** $\sigma(\cdot)$. Historically, the most frequently used activation functions were the sigmoid $\left(\sigma(x) = \dfrac{1}{1 + e^{-x}}\right)$ and the hyperbolic tangent ($\sigma(x) = \tanh(x)$), which were intended to imitate the firing of real neurons in the brain. However, in most state-of-the-art applications, the more computationally efficient rectified linear unit (ReLU, defined as $\sigma(x) = \max(0, x)$) activation function replaced the former two. The shapes of these three activation functions and their derivatives[1] are shown in Figure 9.3. A complete review of the existing activation functions is beyond the scope of this chapter. A more exhaustive list of activation functions can be found in recent review and comparative works [7, 8]. The value yielded by the activation function is the neuron's output value y, which is passed to the subsequent layer[2].

An artificial neural network is constructed by connecting multiple artificial neurons into a hierarchical structure (see Figure 9.2b). All data enter the network through the **input layer** and pass through the **hidden layers** reaching the **output layer** and finally yielding

---

1 The importance of the derivative of the activation function will become evident in Section 9.4.
2 Note that y is a scalar value for MLP, and a tensor for CNN (Section 9.5).

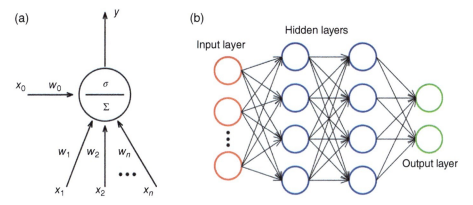

Figure 9.2 (a) Structure of an artificial neuron. (b) Schematic structure of a ANN. The ANN comprises an input layer, through which the data are introduced to the model, several hidden layers (two in this particular case), and an output layer.

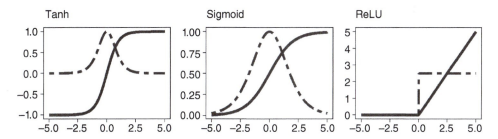

Figure 9.3 Shape of three common activation functions (full lines) and their derivatives (dashed lines). The values of the derivatives were scaled up.

the output of the ANN. If there are more hidden layers than just one, the network is generally referred to as a deep ANN (hence the term **"deep learning"**). This basic architecture is called the **feed-forward network** and contains only **fully connected layers**. In feed-forward networks, each neuron is connected to each and every neuron in both adjacent layers (except for the input and output layers[3]), but not connected to other neurons in the same layer. By far, fully connected ANNs are the most common architecture used by the LIBS community (Table 9.1). A common variation of fully connected ANNs present in the LIBS literature is the radial basis function ANN (RBFNN). Note that the RBFNN differs from the regular fully connected ANN in the use of the radial basis function as the activation function of the neurons in its hidden layer. The radial basis function, unlike the sigmoid and ReLU functions, is a parametric function. Thus, using the radial basis function provides further trainable parameters [9].

Note that by defining the ANN architecture we only define its structure, i.e. the number of layers, number of neurons in the individual layers, and the activation function of the neurons and the related trainable parameters. These are collectively referred to as the

---

3 The input and output layers have a single adjacent layer.

hyperparameters of the network. Thus, at this stage, the ANN would yield random output values for any input. Finally, the network has to be rigorously trained prior to its application to classification tasks (Section 9.3). A more comprehensive review of the application of ANNs for the classification of LIBS spectra is provided in Section 9.8.

9.3 Cost Functions and Training

The training of ANNs is the process of adjusting the model parameters based on the feedback about the network performance. Namely, during the training, data are propagated through the network, and an output value is obtained for each input sample. These computed output values are then compared with the expected output values (the latter is also referred to as the **ground truth**). In terms of LIBS classification, the input would be, e.g. a spectrum, set of emission line intensities, or principal component analysis (PCA) [10] scores. The output would describe the class dependency of the input spectrum.

Prior to the parameter optimization, the subject of the optimization must be defined. Consequently, the so-called **cost function**[4] is introduced. In the context of supervised learning, the cost function represents the discrepancy between the ground-truth value (e.g. categorical variable for classification, continuous value for regression) and the value predicted by the model. Thus, the cost function is mathematically expressed as $C(y_n, g(x_n, \omega))$, where y_n is the ground-truth value corresponding to the n-th observation, the function $g(...)$ represents the model prediction based on the input data and model parameters, x_n is a vector of the n-th observation, and ω represents the model parameters. As a simplified example, for a simple binary classifier designed to distinguish between images of cats and dogs, x_n would be an image that was labeled as y_n (a ground-truth value was attached to it) by a human expert prior to the classification. For the following text, we define the error $E(\omega)$ as a shorthand notation for the cost function: $E(\omega) \equiv C(y_n, g(x_n, \omega))$.

The class dependency is commonly given in the form of **one-hot encoding**. One-hot encoding transforms the initial class labels (which are commonly given either as strings or integers) into vectors of length m, where m is the number of classes being considered in the classification task. Each vector created by one-hot encoding consists of $m - 1$ zeros and a single value equal to 1: For an observation belonging to the q-th class, the q-th element of the encoding vector is equal to 1. The correctness of the computed output, i.e. the similarity between the computed and expected output values, is quantified in terms of the cost function. The gradient of the cost function, which is computed with respect to (w.r.t.) each parameter of the network (weights and biases), is used for parameter adjustments. The efficient implementation of the algorithm for computing the gradients is called backpropagation, Section 9.4.

A standard choice of the cost function for **classification** tasks is the **categorical cross-entropy** [11]. For the one-hot encoded categorical variable y, the categorical cross-entropy is defined as

---

4 In the literature, the cost function is also frequently referred to as error or loss. These terms are used interchangeably.

$$E(\omega) = -\sum_{n=1}^{N}\sum_{m=1}^{M} y_{mn} \log g_m(x_n, \omega), \qquad (9.1)$$

where N is the number of observations (spectra), M is the number of dimensions, n and m are the corresponding indices, and $g(...)$ is the model prediction (as described earlier).

For pedagogical purposes, it is useful to mention the **mean squared error (MSE)** cost function that the reader is familiar with in the context of (linear) regression. The formula for the MSE is

$$E(\omega) = \frac{1}{N}\sum_{n=1}^{N}\sum_{m=1}^{M} \|y_{mn} - g_m(x_n, \omega)\|^2. \qquad (9.2)$$

The choice of the cost function is a crucial step, completely defining the functionality of the model, and the tasks that the model is suitable for.

Given a dataset X, the cost function can be imagined as a generalized hypersurface over a hyperplane defined by the model parameters, commonly referred to as the cost-function landscape or **loss landscape** (Figure 9.4). Each point on the plane (representing a unique combination of parameters) has a corresponding value of the cost function. Typically, the parameter space is very high-dimensional, which leads to a rough cost-function landscape with an arbitrary number of local extrema. The optimization of the model parameters starts by computing the generalized gradient of the cost function. Thus, a computationally efficient and robust algorithm has to be utilized. By following the negative generalized gradient, we

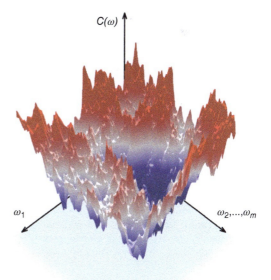

Figure 9.4 Illustration of the cost function-landscape (or loss-landscape). Embedding diagram of a loss-function in two dimensions, illustrating the complexity (surface roughness) of the optimization problem (many local extrema).

move toward a local minimum[5]. It is worth mentioning that most of the critical points in the loss landscape are not local minima or maxima, but saddle points. However, the proof of this rigorous fact is beyond the scope of this chapter and for the sake of providing intuition, we will only consider local minima.

The **gradient descent** (GD) is an iterative optimization algorithm used for minimizing the cost function in ML models. The fundamental idea behind the GD is the following: GD determines the direction and size of the next step in the high-dimensional parameter space. This is done by first computing the negative gradient of the cost function w.r.t. each parameter of the model. Subsequently, the gradient is multiplied by a constant (or variable) **learning rate** (LR) η:

$$\delta\omega_m = -\eta \frac{\partial E(\omega)}{\partial \omega_m}, \tag{9.3}$$

where $\delta\omega_m$ is the change of the m-th parameter that determines the size of the step in the negative gradient's direction. The LR plays a crucial role in the optimization as very small values lead to a slow convergence. Moreover, small LR may result in the optimization getting stuck in a local minimum (see Figure 9.5a). On the contrary, a very high η value might lead to the training process completely missing the important valleys (minima) in the loss landscape. In addition, the standard definition of the cost function incorporates the whole data set. Thus, the GD is an average over all possible inputs, making the parameter optimization extremely costly. Consequently, several expansions of the GD algorithm were developed to achieve an improved performance [12].

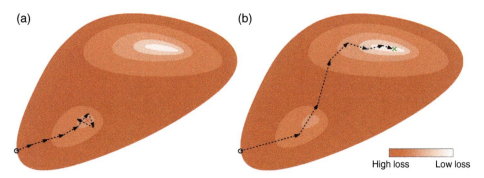

Figure 9.5 Gradient descent optimization algorithm. (a) Gradient descent with a small, constant learning rate (LR) step. This leads to getting stuck inside a local minimum, where the algorithm will oscillate around the extremum point. (b) Gradient descent with adaptive learning rate and momentum. When the algorithm enters first (local) minimum, it slightly lowers the LR step (size of the arrow). In the presented case, it jumped out from the minima (can be caused by the momentum) where gradients are smaller, leading to an increase of LR and tending to the second minimum. In the area of the second minimum, gradients are high, leading to lowering of the learning rate and reaching the global minimum.

---

5 This is an iterative process. The number of iterations carried out during training is referred to as the number of epochs.

The first to mention is **stochastic gradient descent** (SGD), where the approximation of the GD is computed as an average over a mini-batch (usually a small fraction of the full dataset). An obvious advantage is a considerable reduction of the computational burden. Moreover, SGD incorporates a degree of stochasticity following from the approximate direction of the gradient step. Consequently, SGD provides a degree of regularization to the network (see Section 9.7) [13].

A further expansion of GD can be achieved by reflecting the actual position in the loss-landscape and adjusting the value of the LR. In locations with small gradients, the LR is increased to speed up the optimization. On the contrary, at points with large gradients, the LR is lowered in order to adjust for the fine structure of the parameter space (see the Figure 9.5b). This principle is utilized, e.g. in the **adaptive gradient algorithm** (**AdaGrad**) [14].

Lastly, GD can be expanded by assigning a **momentum** to the movement in the parameter space leading to the optimization to exhibit inertia, i.e. the proportion of the optimization steps will depend on the previous steps. The use of a momentum term smooths out the optimization trajectory [15].

In practice, a combination of all mentioned expansions is used [16]. For example, all the following advanced optimizers were used in the context of LIBS classification: adaptive moment estimation (Adam) [17, 18], conjugate gradient method BP [19–23], and scaled conjugate gradient BP [24–26]. Lastly, we note a few niche optimizers, which found applications in the LIBS literature, such as the Broyden–Fletcher–Goldfarb–Shanno [27–29] and the Levenberg–Marquadt [30, 31] algorithms. Unfortunately, a complete description of advanced algorithms is beyond the scope of this chapter. Hence, the reader is advised to study the referenced works.

9.4 Backpropagation

To complete the discussion of the ANNs' training process, we address the **backpropagation**, which is the practical and computationally efficient method for a gradient computation in each step of the GD algorithm. Before the adoption of the backpropagation, the scalability of neural networks was challenging because of the huge computation cost of the numerical differentiation over all parameters to compute the gradients. The idea for backpropagation originates from the well-known chain rule of calculus. Note that backpropagation in a certain form was a well-known idea in computer science even prior to the recent spread of ML. Nevertheless, the (re-)discovery of backpropagation is often attributed to Hinton and his colleagues [32].

The GD starts with feeding an input data vector (representing one spectrum) to a deep neural network with all parameters (weights) selected randomly in the range $-1 \leq w_i \leq 1$ and propagating the input through the network. As we deal with the supervised learning here, we know what the desired output is (specific value or category). After passing data through the network, the network output is compared with the desired output by evaluating the error using a chosen cost function. At this point, we would like to adjust the model parameters in order to lower the cost function and consequently to improve the performance of model predictions. However, the explicit computation of the gradient would be extremely costly, given the huge number of parameters present in the network. Therefore, we exploit the structure of the network and propagate the error in the backward direction to quantify the effect of each parameter to the final value. In other words, we use just a chain rule of calculus to obtain an algebraic relation that represents the whole function of the model and serves to obtain a gradient vector.

We need to compute the partial derivative of the cost function w.r.t. any parameter ω^* (weight or bias) in the network: $\frac{\partial C}{\partial \omega^*}$. First, let us introduce a proper notation to make the algorithm clear. A weight that is connecting the k-th neuron of the layer $(l-1)$ to j-th neuron of the layer[6] (l) is $\omega_{jk}^{(l)}$. The bias term is represented by $\omega_{j0}^{(l)}$ with the corresponding activation $a_0^{(l-1)} = 1$.[7] The activation (or the output) of the j-th neuron in the (l)-th layer is

$$a_j^{(l)} = \sigma\left(\sum_k \omega_{jk}^{(l)} a_k^{(l-1)}\right), \tag{9.4}$$

where the index k is going from zero to the number of neurons in the layer $(l-1)$ and σ is the activation function (e.g. sigmoid). It is convenient to denote the sum inside the brackets by a custom symbol:

$$z_j^{(l)} = \sum_k \omega_{jk}^{(l)} a_k^{(l-1)}. \tag{9.5}$$

Lastly, we need to specify the cost function, which was selected for the demonstration as

$$C(\omega) = \frac{1}{2} \sum_j \left(a_j^{(L)} - y_j\right)^2, \tag{9.6}$$

where (L) is the last layer of the network (or the output layer).

By a formal differentiation of the cost function w.r.t. a selected parameter in the network, we obtain

$$\frac{\partial C}{\partial \omega^*} = \sum_j \left(a_j^{(L)} - y_j\right) \frac{\partial a_j^{(L)}}{\partial \omega^*} = \sum_j \left(a_j^L - y_j\right) \sigma'\left(z_j^L\right) \underline{\frac{\partial z_j^{(L)}}{\partial \omega^*}}. \tag{9.7}$$

Eq. 9.7 was obtained just by applying the chain rule and using the definition of variables present in the cost function. See that the parameter ω^* affects the cost function through activations of neurons in previous layer that are entering the sum $z_j^{(L)}$. Now, to extend the underlined term in Eq. 9.7, we apply the chain rule again to obtain

$$\frac{\partial z_j^L}{\partial \omega^*} = \sum_k \frac{\partial z_j^{(L)}}{\partial a_k^{(L-1)}} \frac{\partial a_k^{(L-1)}}{\partial \omega^*} = \sum_k \omega_{jk}^{(L)} \sigma'\left(z_k^{(L-1)}\right) \frac{\partial z_k^{(L-1)}}{\partial \omega^*}. \tag{9.8}$$

As you may notice, by repeating the same procedure again and again, we can propagate backwards through the network to enumerate the dependence of the cost function on any parameter. For the general case, we have a recursive formula

---

6 Notice that the layer indices are denoted by brackets to improve the readability of the formulas.
7 This parameter was introduced in Figure 9.2a as "x_0" for demonstrative purposes. At this point, it was changed to match the more general notation required for describing the backpropagation algorithm.

$$\frac{\partial z_j^{(l)}}{\partial \omega^*} = \sum_k w_{jk}^{(l)} \sigma'\left(z_k^{(l-1)}\right) \frac{\partial z_k^{(l-1)}}{\partial \omega^*} = \sum_k M_{jk}^{(l)} \frac{\partial z_k^{(l-1)}}{\partial \omega^*}, \tag{9.9}$$

for any two layers (l) and ($l-1$). Each pair of layers will contribute to the relation with a matrix $M_{jk}^{(l)}$. So finally, we can write

$$\frac{\partial z_j^{(l)}}{\partial \omega^*} = \sum_{k,l,\ldots,u,v} M_{jk}^{(l)} M_{kl}^{(l-1)} \ldots M_{uv}^{(\text{target layer}+1)} \frac{\partial z_v^{(\text{target layer})}}{\partial \omega^*} \tag{9.10}$$

where the target layer is the layer from which ω^* connects the following layer (in the forward direction).

This is all the "magic" standing behind the backpropagation and therefore behind the neural networks as well. Especially, the simplicity of the approach should be appreciated. All necessary values for the gradient computation are obtained just with a single forward pass of the data through the network and by propagating the error back. In this explanation, we emphasized the intuitive computation from the cost function (adapting some ideas from the Marquardt's course [33]). A more formal (and detailed) explanation of the backpropagation algorithm can be found in the textbooks by Nielsen [34] and Bishop [35].

9.5 Convolutional Neural Networks

Our discussion has so far focused on multilayer perceptrons (MLPs). However, MLPs are not the only ANN architecture that have been successfully applied for the analysis of spectroscopic data. Namely, CNNs have been gaining attention in the past few years among spectroscopists [36–38]. While CNNs were introduced in 1997 [39], they have been popularized only recently [40]. CNNs were intended to solve the issue of translational invariance in image data. This was achieved by looking for patterns in the data rather than strictly learning the distribution function of each individual feature (e.g. the intensity recorded at each individual wavelength in the context of spectra). Most importantly, CNNs were able to learn these patterns automatically from labeled data without the need of manually defining them. The ability of CNNs to learn local patterns motivated the shift from MLPs to CNNs in various fields.

MLPs and CNNs differ fundamentally, while MLPs treat each input feature individually, CNNs aim at modeling the relationship between neighboring features. Thus, CNNs learn **intrinsic patterns** present in the data. In the context of spectroscopic data (i.e. spectra), this can be roughly interpreted as working with emission lines rather than the individual intensity values comprising the emission lines. CNNs achieve this by introducing the **convolutional layer** (hence the name).

Note that CNNs still contain fully connected layers. However, completely convolutional networks have been recently demonstrated, too. These are yet to be applied to spectroscopic applications and are omitted from the present work. Thus, in this section, we describe the fundamentals of the convolutional layer, its hyperparameters, and how changing these hyperparameters is expected to affect the data processing. Moreover, since spectroscopic data are generally treated individually (i.e. as vectors), we limit our discussion to 1D convolutions. Nevertheless, each discussed step can be generalized to n dimensions, and references to relevant literature are provided [41]. Lastly, since spectroscopic data are generally digitized, we only consider the **discrete convolution operations**.

The principles of CNNs are summarized in Figure 9.6. Namely, compared to MLPs, CNNs do not take a weighted sum of every feature in the preceding layer. Instead, CNNs carry out various sliding window operations. Hence, each neuron in the convolutional layer is exposed only to a subset of the features at any given time. Naturally, this subset consists of adjacent features (intensity recorded at adjacent wavelength values).

The most important sliding window operation carried out by the convolutional layer is the (discrete) convolution between the input features and a **convolutional filter**. The convolutional filter is defined by its trainable weights (i.e. its shape) and a set of hyperparameters. These **hyperparameters** are specified by the user prior to training the model and include the **number of filters** (i.e. number of neurons in the convolutional layer, since each neuron is represented by a unique filter), **filter dimensions** (the length of the filters in the case of 1D data), **stride**, and **padding**. The role and expected impact of each hyperparameter will be described later. However, first, let us address the convolution itself.

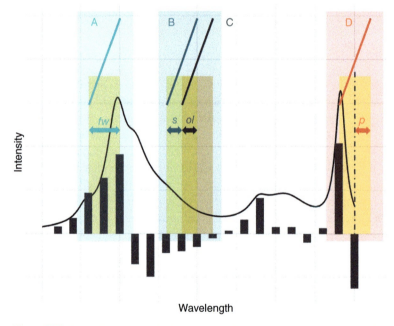

Figure 9.6 Principles of convolutional layers. The black line shows a recorded spectrum. The yellow (and the orange) areas signify the portion of the spectrum exposed to the convolution operation. The tilted lines show the shape of the applied convolutional filter at four distinct positions marked with A, B, C, and D, respectively. The bars show the values yielded by the convolution at multiple positions. The highlighted regions showcase the following (from left to right): convolution with a linear filter with a positive slope yields positive values that increase with the increasing steepness of the emission line; the same convolutional filter yields values close to zero for approximately constant spectral regions; and lastly, padding must be used for the same convolutional filter to be applicable to the last portion of the spectrum. f_w – filter width, s – stride, ol – overlap between two subsequent applications of the convolutional filter, p – padding.

The discrete convolution operation (henceforth simply convolution) is rather straightforward. At each position, the convolution is the weighted sum of the selected features, where the weights are determined by the **shape of the filter**. The result of the convolution operation is then fed into a nonlinear activation function (Section 9.2). Thus, each neuron in the convolutional layer yields a vector (or tensor, in the more general case), whose length (dimensions) is determined by the hyperparameters of the filter. Consequently, the next crucial difference between fully connected (dense) layers and the convolutional layer is the shape of their outputs. Namely, a neuron in a dense layer propagates its output to every neuron in the proceeding (dense) layer. Thus, the output of a dense layer can be understood as a set of weighted vectors where the weights are determined by the connection weights between neurons of adjacent layers. Moreover, the dimension of these outputs is also determined by the number of neurons in the preceeding layer. On the contrary, while a neuron in a convolutional layer yields a vector as well, these vectors are generally stacked into a tensor with an increased rank (a tensor with an additional axis) compared to the input tensor. This has two consequences. A convolutional layer preceded by another convolutional layer can apply a higher dimensional filter. Alternatively, the higher rank tensor can be flattened before further processing. This is generally done before a dense layer. Flattening reduces the rank of the feature tensor while increasing its dimension along a single axis. For example, the flattening of a 2D matrix with dimensions of $n \times m$ would result in a vector of length $n \cdot m$. This transformation must maintain the number of features held by the tensor.

The convolutional layer is often followed by a layer carrying out a different type of sliding window operation, pooling. **Pooling** refers to the selection of a subset of the values yielded by the convolutional filters. In other words, pooling is a sliding window operation where the windows do not overlap. The pooling operation is generally rather flexible, allowing the user to define any unique criteria. Nevertheless, the most common pooling operations are **max-pooling** and **mean-pooling** (see the Figure 9.7). These refer to the rolling maxima and rolling average operations, respectively.

Pooling was designed to address small possible shifts in the input features, providing limited translational invariance. As such, pooling could be crucial while dealing with spectra exhibiting wavelength shifts or partially resolved emission lines resulting in shifting maxima. Moreover, pooling has been shown to increase the training and inference speed and to help with regularization (see Section 9.7). Note that pooling results in the reduction of the feature tensor's dimensions. Thus, extensive pooling might lead to a substantial loss of information available to subsequent layers and negatively affect the model's performance.

To fully understand the structure of a CNN, there are only a few remaining details to be discussed, namely the hyperparameters of the convolutional layer. As mentioned, the filters

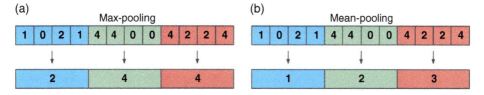

Figure 9.7 Principles of the pooling operation: (a) max-pooling; (b) mean-pooling.

are gradually exposed to parts of the input features. Consequently, the distance between subsequent expositions must be defined. This is given by the parameter called **stride**. In general, the stride should be smaller than the filter size in order to ensure a partial overlap between the consequent applications of the convolutional filter. Naturally, using a disproportionately large stride will result in a considerable loss of information and poor network performance. Nevertheless, optimizing the stride might help with the network's ability to generalize. Moreover, as the stride reduces the size of the feature vectors, it speeds up training and inference. As with other hyperparameters, stride should be optimized via cross-validation (CV) (see Section 9.6).

The last hyperparameter characterizing convolutional layers is **padding**. As already mentioned, CNN learn intrinsic patterns present in the data. This is done via fitting the filters to the patterns. However, features learnt by the filters might be only partially present in the data. An example would be an emission line at the boundary of the recorded spectral range and consequently only partially recorded. In this case, the convolution operation between the partial feature and the corresponding filter might underestimate the match. As a possible solution, padding can be used. Padding sticks additional values to the input feature vector, artificially lengthening them. This enables the partial match between the filter and the data, as shown in Figure 9.6.

The last step in using CNNs is understanding its **training** process, i.e. optimizing its filters' shapes. As mentioned earlier, the shape of the individual filters is numerically given by a weight vector. Moreover, the convolution operation is followed by the application of a nonlinear activation function $\sigma(\cdot)$. Thus, the final output of these two operations can be written as

$$a_k^{(l)} = \sigma\left(\sum_{i=1}^{f_w} a_{(k-1)s+i+1}^{(l-1)} \cdot f_i\right) \tag{9.11}$$

where a_k^l is the k-th element of the output vector in layer l, s is the stride, f_w is the filter width, and $f_i^{(l)}$ is the i-th weight of a filter in the l-th layer. Note that $\sigma(.)$ can be any of the activation functions introduced in Section 9.2. Consequently, the model can be trained via gradient descent with backpropagation (Section 9.4) by applying the **chain rule**, similarly to MLPs.

9.6 Evaluation and Tuning of ANNs

Up to this point, we have described the most used ANN architectures and provided some insights into their training. Both aspects of using ANNs involve a wide range of hyperparameters that must be tuned for optimal performance. Moreover, the final performance of ANNs must be clearly determined before it can be used. Thus, this section provides a list of **performance metrics** commonly used for classification models. Subsequently, we discuss how ANN performance is and should be evaluated at various stages of model development; namely, during training, hyperparameter tuning, and testing. In addition, we discuss why **dataset splitting** should be carried out. We cover two data-set splitting approaches for specific classification tasks (in-sample and out-of-sample) that are relevant to the processing of spectroscopic data.

In the context of classification, all common **figures of merit (FOMs)** boil down to the number of true positives (TP), true negatives (TN), false positives (FP), and false negatives (FN) among the predicted class labels. TP and FP refer to the correctly and incorrectly assigned class labels. TN and FN refer to the number of observations that were correctly or incorrectly rejected as representatives of the considered class. Moreover, using TP, TN, FP, and FN, additional FOMs can be defined: **Accuracy, sensitivity** (also referred to as true positive rate or recall), **specificity** (also referred to as selectivity or true-negative rate), and **precision** (also referred to as positive predictive value). Some of the more in-depth FOMs include the **receiver operating characteristic (ROC),** and the closely related **area under the curve (AUC)**. The definition of these quantities is readily available in the literature (either LIBS, ML, or statistics) [42, 43]. At this point, the attentive reader might ask why we need two sets of metrics to evaluate ANN performance, the loss (or cost) function on one hand and the FOMs listed earlier on the other. The answer is straightforward; the loss function is derivable and, hence, can be used for backpropagation. Meanwhile, FOMs are user-friendly performance metrics, but their derivation w.r.t. the model parameters does not exist.

A seemingly minor detail, but one with potentially large consequences, is how the ANN performance is evaluated. In other words, what data are used to determine the FOMs of a trained ANN. ANNs can **approximate any function with arbitrary accuracy** [4, 5]. Moreover, it was shown that such an approximation can be achieved even with shallow networks (with a suitable number of nodes). Consequently, ANNs are expected (with a properly tuned architecture) to learn the distribution of the training data with an arbitrary accuracy. On the contrary, in practical applications, trained ANNs are applied to data which were not available for training. Thus, **performance metrics achieved on the training data** are **not directly implicative of the real performance** of the model. Consequently, exclusive datasets should be used for training and testing.

It is important to emphasize that the data used for testing should be only used for the final evaluation of the model. Thus, a model should be trained on one dataset and evaluated on an exclusive dataset. However, the training only includes the optimization of the model weights (and biases) but does not include the model's hyperparameters. Consequently, an additional splitting of the available dataset is required, which mimics the final testing. This third dataset is commonly referred to as the **validation dataset**. The roles of the three datasets can be briefly summarized as follows: First, the training dataset is used to optimize the weights of the model via an optimizer. Second, the validation dataset is used to tune the architecture of the model, generally via manual adjustments. Lastly, the testing dataset is used for the final evaluation of the model.

Thus, prior to the construction of ANN models, the initial dataset should be split into the three subsets: **training, validation**, and **test**. The first rule is that all the **subsets** must be **disjoint**. The goal of the rigorous dataset splitting is to achieve good **generalizability** of the model (network). In other words, models are built on a selected set of input data (spectra), but they should be able to generalize over these observations to perform well on yet unseen spectra as well. This principle can be easily demonstrated by an example taken from image analysis; imagine an ML model that was trained on images of red sport cars. However, ultimately, the model is to be employed to classify images of standard personal cars of various colors. A model with good generalizability will rely on features that are shared by every car, regardless of their type or color (e.g. all cars have wheels, windows, and a shape prolonged along a single direction). On the contrary, an overtrained model can completely ignore car-like features and may focus only on color, resulting in the classification of random red-colored images as a car. To translate this idea of generalizability to spectroscopy, we will distinguish between two classification tasks: "in-sample" and "out-of-sample" and corresponding approaches to dataset splitting.

In the case of **in-sample classification,** the dataset is split following the random sampling of the available spectra according to a user-defined ratio (e.g. 60% for the training, 20% for validation, and 20% for the testing). Even though this setup fulfills the basic rules of dataset splitting (disjoint training, validation, and testing subsets), it does not reflect the need for models that are capable of generalizing. This is, of course, implied by the nature of spectroscopic (LIBS) data. Spectra obtained from different shots (measurements), but the same homogeneous sample (material) are highly similar, exhibiting only a small fluctuation in intensity. Therefore, trying to classify test spectra, using the model trained on the spectra obtained from the same specimen is **not a challenging task**. Again, using parallels from image processing, you may imagine classifying the same image with different lighting conditions/color by the model trained on the original image.

On the contrary, **out-of-sample** classification tasks put high requirements on generalizability. In such a setup, it is believed that a correctly trained model can be used for the classification of spectra obtained from completely different physical samples, which are somehow related to the samples used for training the model. This relation is based on the similarity in the elemental composition of the samples, varying only in reasonable intervals. Compared to in-sample classification, out-of-sample classification tasks are very challenging and usually require large amounts of data from a variety of samples (e.g. to address the complex variations caused by matrix effects). An out-of-sample classification task was the subject of the recent EMSLIBS 2019 classification contest [44], where an extensive dataset comprising soil-sample spectra was analyzed [45]. This was the fundamental idea of the contest: A single category (or class) comprised several physical samples of mineral powders (e.g. hematite) with slightly varying compositions defined by the category's geological classification. Subsequently, only spectra from a selected set of specimens were available to the contestants for training. Thus, the contest presented an out-of-sample classification task.

Lastly, we point out that the idea of dataset splitting is not foreign to LIBS. A common strategy is to split the available data into three parts, with the most data available for training and a comparable amount of data used for model validation and testing. For example, a common splitting ratio is 70-15-15 [23, 24, 46–50]. Note that this splitting generally samples the spectra obtained from the same specimens for the three datasets. A more robust approach is to use spectra obtained from distinct specimens for testing [16, 20, 21, 25, 27, 30, 51]. This approach forces the networks to learn more general features characterizing the classes rather than picking up measurement artefacts unique to the individual specimens. An alternative approach is to collect the training, validation, and testing data on different days [52], which, to a certain degree, simulates the real-life scenario of shifting instrument performance on an unchanging task. Naturally, the correct validation and testing of the models depend on the intended application.

9.7 Regularization

As repeatedly mentioned, **overtraining** (also frequently referred to as **overfitting**) of ANN models is a serious issue and requires further attention. The aim of this section is to provide the reader a set of tools that were designed to help training models which are capable of **generalizing** to hitherto unseen data. This process is commonly referred to as **regularization** and is commonly applied even in the context of linear regression with large numbers of predictive features. In fact, the above-mentioned dataset splitting, if carried out with care, is a form of regularization.

Throughout the chapter, it has been emphasized that modern ANNs are designed to work with large datasets. The main reason for this limitation is the need to prevent the overfitting of the model. Large and deep network architectures that are standardly used have millions of parameters. It leads to a huge overparameterization of the model (comparable to fitting a LIBS calibration curve with a 100-th order polynomial), whereas having a lower number of training observations than parameters risks overfitting. If the difference between the number of parameters and available training data is dramatic (several orders of magnitude), the network will easily incorporate the exact structure of the training data (i.e. the statistical distribution of each predictive variable), leading to a very poor **generalization.** The importance of this topic cannot be understated as correct regularization often means the difference between the successful application of ANNs and the completely unusable model. However, there are no strict rules for selecting the right regularization technique (or their combination). Therefore, regularization is most commonly chosen heuristically. In the following paragraphs, we describe a few well-established techniques, some of which have already been utilized in LIBS-related research. Nevertheless, the serious utilization of ANNs necessitates a more thorough study of regularization techniques. This is reflected by the fact that every modern textbook dedicated to ANNs and deep learning broadly covers the topic of regularization.

The most prominent form of regularization is CV. During CV, the training data are repeatedly randomly divided into training and validation sets. The number of iterations of the CV procedure is commonly presented as n-fold CV, which refers to carrying out the train-validation split n times. Owing to its general utility, CV is by far the most common regularization technique used in LIBS analysis [17, 52–61]. For a more detailed discussion of the various CV strategies, the reader is encouraged to consult one of the many review papers addressing the subject [62, 63].

Nevertheless, CV might be often computationally prohibitive. As such, various regularization techniques have been directly implemented into advanced optimization strategies. Namely, the SGD and its expansions (momentum, AdaGrad, Adam [64]) were already discussed in a different context. However, they also represent a form of regularization. It is the stochasticity that prevents the model from overtraining. The gradient step is computed as an average from a **mini-batch** that contains a selected number of input observations (e.g. $N = 50$), thus it cannot focus on specific features dominantly present in a training dataset. Of course, spectra in each mini-batch should exhibit a certain distinctiveness (i.e. representing a significant portion of the variance present in the training data).

Apart from the use of more advanced optimizers, the training process itself can be modified to achieve some level of regularization in another way. In fact, one of the simplest, yet efficient, regularization techniques is **early stopping** [65]. Early stopping refers to stopping of the iterative training process before the overtraining is reached. As described in Sections 9.3 and 9.4, during a network's training, its weights are iteratively updated according to the information extracted from the cost function. In practice, each individual observation (spectrum) present in the training dataset is propagated through the network multiple times (determined by the number of **epochs**). However, too many epochs might lead to the significant **overfitting** of the model. Thus, the validation subset of the data is used to monitor the network's performance during learning and to consequently optimize the number of epochs. This is done according to the following observation: An overfitted model will generally exhibit (nearly) perfect training performance, but a poor validation performance as is shown in Figure 9.8. Hence, to summarize, early stopping compares the training and validation performances to stop the training process when the two start to diverge. Early stopping is the second most frequently used form of regularization in LIBS (after CV) [16, 19, 20, 51, 66].

Regularization is not limited to the data preparation and alteration of the learning process. In fact, the network's architecture can be modified to incorporate regularization. For example, **dropout** is a simple method that randomly sets selected weights in the network equal to zero with a specified frequency. This is repeated during each training iteration and usually is specified for each layer separately. It is not uncommon to set the frequency up to 0.6 (meaning that 60% of weights are set equal to zero). In ML, the dropout has been introduced relatively recently with a great success by Hinton et al. [67, 68]. Similarly, the dropout has been successfully applied in the context of LIBS as well [18, 69].

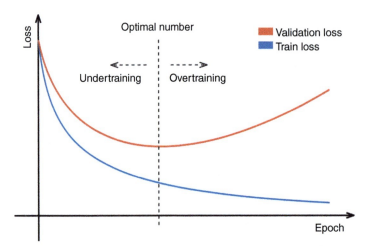

Figure 9.8 Generalization performance of the network during the training procedure. Training loss (error) goes to zero with an increasing number of epochs, but the validation error starts to rise after reaching minimal value.

Lastly, regularization can be enforced by modifying the cost function. Namely, the well-established L1 and L2 regularization techniques can be implemented into ANNs as

$$C_{L1} = \frac{1}{n}\sum_{i=1}^{n} C(\mathbf{y}_n, g(\mathbf{x}_n, \boldsymbol{\omega})) + \frac{\lambda}{n}\sum_{l=1}^{L} \left\|\mathbf{w}^{(l)}\right\|^1 \quad (9.12)$$

and

$$C_{L2} = \frac{1}{n}\sum_{i=1}^{n} C(\mathbf{y}_n, g(\mathbf{x}_n, \boldsymbol{\omega})) + \frac{\lambda}{n}\sum_{l=1}^{L} \left\|\mathbf{w}^{(l)}\right\|_F^2 \quad (9.13)$$

respectively, where $C(\mathbf{y}_n, g(\mathbf{x}_n, \boldsymbol{\omega}))$ is the general form of any loss function (Sections 9.3 and 9.4), n is the mini-batch size applied, considering the use of SGD, w_j are the weight parameters of the network, and λ is the regularization rate. Note that owing to the nature of the derivative of the added terms ($\lambda\|\mathbf{w}\|^1$ and $\lambda\|\mathbf{w}\|_F^2$, respectively), L1 regularization pushes the weights toward 0. Contrarily, L2 regularization leads to a weight decay. Naturally, L1 and L2 regularizations can be applied simultaneously with distinct λ values.

In summary, the use of regularization is strongly encouraged. A good rule of thumb is to maximize the applied regularization until a significant drop in the training accuracy is observed. This increases the likelihood that the trained model is capable of analyzing hitherto unseen data with reasonable performance. Lastly, note that the best approach to regularization is to collect more data covering a wider range of physical samples.

9.8 State-of-the-art LIBS Classification Using ANNs

The previous sections discussed the fundamentals necessary to construct and train ANNs for LIBS classification tasks. The aim of this section is to provide an overview of the existing applications of ANNs for LIBS classification. Some notable works are described in detail, while all the currently available peer-reviewed English articles on the topic are listed in Table 9.1.

The first application of ANNs to the classification of LIBS spectra dates back to 1998, when 21 plastic samples were classified into 5 classes by Sattman et al. [70]. Unfortunately, the work does not describe the applied architecture in sufficient detail. This is an issue regularly present in LIBS articles, likely due to the application of ANNs as black box models. However, it must be emphasized that this is often related to the employed software, which does not provide the user with extensive control over the ANN design.

On the contrary, several recent works present a complete description of the data collection and preparation processes and the applied ANN architecture and training as well. For example, Chen et al. [69] used a CNN with a single convolutional layer and a max pooling layer followed by a fully connected layer. The authors also employed regularization in the form of dropout and used SGD for training with a mini-batch size of 64. Similarly, He et al. [18] used a CNN with two convolutional layers separated by a max pooling layer. The second convolutional layer was followed by two fully connected layers. The authors also applied dropout as regularization and batch normalization to stabilize the training process. Lastly, the authors used the Adam optimizer for a more reliable convergence. The

Table 9.1 Summary of LIBS classification using ANNs.

Authors	Applications	Architecture[a]	Regularization	FOM
Alli et al. [72]	Pharmaceutics	CNN	NA	Relative accuracy
Alvarez et al. [66]	Geology	MLP (50) [sigmoid]	Early stopping	Accuracy, precision, sensitivity
Berto et al. [23]	Archeology and forensics	MLP	Early stopping	Accuracy
Bohling et al. [73]	Metals	NA	NA	Accuracy
Boueri et al. [52]	Polymers	MLP (1)	Cross-validation	Identification rate (correctly, misidentified, unidentified)
Caceres et al. [22]	Biology	MLP (1)	NA	Accuracy, AUC
Campanella et al. [74]	Metals	MLP (1)	NA	Accuracy
Chen et al. [75]	Polymers	NA	NA	Accuracy
Chen et al. [69]	Geology	CNN (conv + max pooling) [tanh]	Dropout	Accuracy
Cui et al. [76]	Biology	MLP (1) [radial basis function]	NA	Accuracy
Diego-Vallejo et al. [77]	Miscellaneous applications	MLP (1)	NA	Accuracy, precision, sensitivity, specificity
Diego-Vallejo et al. [78]	Miscellaneous applications	MLP (1)	NA	Accuracy, precision, sensitivity, specificity
Farhadian et al. [49]	Explosives and chemicals, nuclear materials	MLP (1)	NA	Accuracy
He et al. [18]	Biology	CNN (conv, max pooling, conv, dense, dense, softmax) [relu]	Batch normalization, dropout	Accuracy
He et al. [79]	Archeology and forensics	NA	NA	Accuracy
Junjuri et al. [46]	Explosives and chemicals, nuclear materials	MLP (1)	NA	Accuracy
Junjuri et al. [47]	Polymers	MLP (1)	NA	Accuracy
Junjuri et al. [48]	Polymers	MLP (1)	NA	Accuracy
Kong et al. [80]	Metals	MLP	NA	Accuracy

Table 9.1 (Continued)

Authors	Applications	Architecture[a]	Regularization	FOM
Kong et al. [81]	Metals	MLP (1)	NA	accuracy
Koujelev et al. [82]	Geology	MLP (1)	NA	Identification rate as function of classification threshold
Li et al. [83]	Medical	Generalized regression neural network [radial basis function]	NA	Accuracy, precision, sensitivity, specificity, ROC, AUC
Liu et al. [55]	Biology	MLP (1) [radial basis function]	Cross-validation	Accuracy
Liu et al. [84]	Biology	MLP (1) [radial basis function]	NA	Accuracy, sensitivity, specificity
Lui et al. [51]	Geology	MLP (1) [log-sigmoid]	Early stopping	Accuracy
Manzoor et al. [21]	Biology	MLP (1) [tanh]	NA	Spectral correlation
Manzoor et al. [85]	Biology	MLP (1)	NA	Spectral correlation
Marcos-Martinez et al. [86]	Biology	MLP (1)	NA	Accuracy, precision, sensitivity, AUC, robustness
Menkinghoggart et al. [54]	Forensics	MLP (1) [tanh]	Cross-validation	Accuracy, TP, FP, TN, FN
Messaud aberkane et al. [50]	Metals	MLP (1) [tanh]	NA	Accuracy
Moncayo et al. [25]	Biology	MLP (1)	NA	Spectral correlation
Moncayo et al. [20]	Biology	MLP (2)	Early stopping	Accuracy
Moncayo et al. [19]	Biology	MLP (1)	Early stopping	Accuracy
Moncayo et al. [71]	Biology	MLP	NA	Accuracy
Mordmueller et al. [87]	Biology	MLP (0)	NA	Accuracy
Moros et al. [30]	Explosives and chemicals, nuclear materials	MLP	NA	Accuracy, precision, sensitivity, specificity, FP rate, FN rate
Pokrajac et al. [88]	Biology	MLP (1) [tanh]	NA	Accuracy

(*Continued*)

Table 9.1 (Continued)

Authors	Applications	Architecture[a]	Regularization	FOM
Ramil et al. [16]	Archeology and forensics	MLP (1) [log-sigmoid]	Early stopping	Accuracy
Roh et al. [56]	Polymers	Fuzzy rules radial basis feedforward NN [radial basis function]	Cross-validation	Correct classification rate
Roldan et al. [89]	Biology	NA	NA	Spectral correlation
Rzecki et al. [90]	Miscellaneous applications	Generalized regression neural network, probabilistic NN, MLP	NA	Accuracy, Sensitivity, specificity, TP, FP, TN, FN
Sattmann et al. [70]	Polymers	NA	NA	Correct/incorrect classification
Snyder et al. [91]	Biology	MLP (1)	Overfit penalty	FP, FN
Ukwatta et al. [92]	Metals	MLP (1)	NA	TP rate, FP rate, accuracy
Vitkova et al. [27]	Biology	MLP (1)	NA	Confusion matrix
Vrabel et al. [44]	Geology	MLP (1) [relu]	Dropout	Accuracy
Wang et al. [58]	Biology	MLP (1) [sigmoid]	Cross-validation	Accuracy
Yalemale et al. [53]	Geology	MLP (4-100)	Cross-validation	Accuracy, sensitivity, specificity
Yang et al. [24]	Metals	MLP (1) [sigmoid + softmax]	NA	Accuracy
Yang et al. [17]	Metals	MLP (1) [relu + softmax]	Cross-validation	Accuracy
Yoshino et al. [93]	Geology	MLP (3)	NA	F-measure
Yoshino et al. [61]	Geology	MLP (3) [sigmoid]	Cross-validation	Accuracy
Zdunek et al. [60]	Metals	Probabilistic NN	Cross-validation	Misclassification rate
Zhang et al. [57]	Biology	RBFNN	Cross-validation	Accuracy
Zhang et al. [59]	Geology	Wavelet neural network	Cross-validation	Accuracy

[a] The values given in parentheses refer to the number of hidden layers. The employed activation functions are given in square brackets.

descriptive approach of these works not only allows to test the reproducibility of the achieved results but also serves as a solid reference for an initial experimentation with ANN.

Exhaustive comparisons of ANNs with other classification approaches have been published by several authors. Zdunek et al. [60] classified five solder metals by ANN, LDA, kNN, PLS-DA, SIMCA, Naïve Bayes, and SVM. The authors achieved comparable results using ANNs to those obtained by using linear classifiers such as SIMCA. This emphasizes that ANNs are not always necessary. Roh et al. [56] designed a rather unique ANN classifier based on fuzzy logic using the parametric nature of the radial basis function, which was used as the activation function in their network. The proposed ANN outperformed several classifiers, such as Naïve Bayes Net, kNN, SVM, IBK (an implementation of kNN in the Weka software suite), decision trees, pruned decision trees, random forest, and random trees. Moncayo et al. [71] showed that ANNs can outperform CART (classification and regression tree), BLR (binary logistic regression), SVM, PLS-DA, SIMCA, and LDA.

Similar efforts were recently made by the above-mentioned work of Chen et al. [69], who compared the classification performance of CNNs with HA-ANN, kNN, PCA-kNN, PLS-DA, SVM, and PCA-SVM. More importantly, their work also presents a comparison between 1D and 2D convolutional networks. For the latter to be applicable to LIBS, the authors proposed a methodology of transforming the 1D LIBS spectra into 2D matrices. The spectra were first segmented, followed by the stacking of the obtained segments.

Lastly, we comment on the tasks that are commonly addressed using ANNs. Unfortunately, the use of ANNs is often presented as a novelty, rather than a necessity. Consequently, most classification tasks are grossly overfitted by ANNs. Nevertheless, a few works are worth emphasizing. By far, the most challenging classification tasks (considering the number of specimens and the general design of experiment) have been carried out by Chen et al. [69] and by several contestants of the mentioned EMSLIBS 2019 classification contest [44]. The former classified a total of 119 rock samples into five geological classes. A distinct set of physical samples was used for training, validating (hyperparameter tuning), and testing their ANN models. Meanwhile, the EMSLIBS classification contest consisted of classifying a total 138 soil samples into 12 geological classes. From the 138 samples, only 100 samples were available for training, while the final challenge was to classify the remaining 38 (test) samples. This task was addressed using, among others, ANNs by multiple groups, who employed a wide range of preprocessing and feature engineering strategies.

9.9 Summary

In this chapter, we aspired to provide the reader with the basic knowledge about the fundamental terms necessary to use ANNs. Some more general topics were studied thoroughly (cost function, GD, backpropagation, MLP, and CNN), while other topics were covered only marginally. We described the recommended practices and pointed out common mistakes. Also, several troubleshooting approaches were provided. For a deeper understanding of the comprehensive topic of ANNs, we suggest to study following materials [35, 41, 94] that provide a broader view in the ML context. Lastly, we summarized the existing literature relevant for LIBS.

Acknowledgments

JK acknowledge the support of the Faculty of Mechanical Engineering, Brno University of Technology under grant FSI-S-20-6353. EK is grateful for the financial support received under the CEITEC VUT-J-20-6482 project. JV is grateful for the financial support received under the project CEITEC VUT-J-20-6528 and CEITEC-K-21-6978.

References

1 D. Silver, J. Schrittwieser, K. Simonyan, I. Antonoglou, A. Huang, A. Guez, T. Hubert, L. Baker, M. Lai, A. Bolton, Y. Chen, T. Lillicrap, F. Hui, L. Sifre, G. van den Driessche, T. Graepel, D. Hassabis, Mastering the game of Go without human knowledge, *Nature* **550** (2017), 354–359. https://doi.org/10.1038/nature24270.
2 Y. LeCun, Y. Bengio, G. Hinton, Deep learning, *Nature* **521** (2015), 436–444. https://doi.org/10.1038/nature14539.
3 G.E. Hinton, Reducing the dimensionality of data with neural networks, *Science* (**80-**.) 313 (2006), 504–507. https://doi.org/10.1126/science.1127647.
4 G. Cybenko, Approximation by superpositions of a sigmoidal function, *Math. Control Signals Syst.* **2** (1989), 303–314. https://doi.org/10.1007/BF02551274.
5 K. Hornik, M. Stinchcombe, H. White, Multilayer feedforward networks are universal approximators, *Neural Netw.* **2** (1989), 359–366. https://doi.org/10.1016/0893-6080(89)90020-8.
6 F. Van Veen, S. Leijnen, The Neural Network Zoo, (2019),. https://www.asimovinstitute.org/neural-network-zoo/ (accessed November 29, 2020).
7 C. Nwankpa, W. Ijomah, A. Gachagan, S. Marshall, Activation Functions: Comparison of trends in Practice and Research for Deep Learning, (2018),. http://arxiv.org/abs/1811.03378.
8 T. Szandała, Review and comparison of commonly used activation functions for deep neural networks, in: *Bio-inspired Neurocomputing. Studies in Computational Intelligence* **903** 2021 pp. 203–224. https://doi.org/10.1007/978-981-15-5495-7_11.
9 J. Ghosh, A. Nag, An overview of radial basis function networks, in: *Radial Basis Function Networks*, **67** 2001: pp. 1–36. https://doi.org/10.1007/978-3-7908-1826-0_1.
10 P. Pořízka, J. Klus, E. Képeš, D. Prochazka, D.W. Hahn, J. Kaiser, On the utilization of principal component analysis in laser-induced breakdown spectroscopy data analysis, a review, *Spectrochim. Acta - Part B At. Spectrosc.* **148** (2018), 65–82. https://linkinghub.elsevier.com/retrieve/pii/S0584854718301526 (accessed February 5, 2019).
11 K.P. Murphy, *Machine Learning: A Probabilistic Perspective*, 1st ed., MIT Press, Cambridge, 2012.
12 Y. LeCun, L. Bottou, G.B. Orr, K.-R. Müller, Efficient backprop, in: *Neural Networks: Tricks of the Trade. Lecture Notes in Computer Science* **1524** 1998: pp. 9–50. https://doi.org/10.1007/3-540-49430-8_2.
13 L. Bottou, F.E. Curtis, J. Nocedal, Optimization Methods for Large-Scale Machine Learning, (2016),. http://arxiv.org/abs/1606.04838.

14 J. Duchi, E. Hazan, Y. Singer, Adaptive subgradient methods for online learning and stochastic optimization, *J. Mach. Learn. Res.* **12** (2011), 2121–2159. https://doi.org/10.5555/1953048.2021068.

15 N. Qian, On the momentum term in gradient descent learning algorithms, *Neural Netw.* **12** (1999), 145–151. https://doi.org/10.1016/S0893-6080(98)00116-6.

16 A. Ramil, A.J. López, A. Yáñez, Application of artificial neural networks for the rapid classification of archaeological ceramics by means of laser induced breakdown spectroscopy (LIBS), *Appl. Phys. A Mater. Sci. Process.* **92** (2008), 197–202. https://doi.org/10.1007/s00339-008-4481-7.

17 Y. Yang, X. Hao, L. Zhang, L. Ren, Application of scikit and keras libraries for the classification of iron ore data acquired by laser-induced breakdown spectroscopy (LIBS), *Sensors* **20** (2020), 1393. https://doi.org/10.3390/s20051393.

18 Y. He, Y. Zhao, C. Zhang, Y. Li, Y. Bao, F. Liu, Discrimination of grape seeds using laser-induced breakdown spectroscopy in combination with region selection and supervised classification methods, *Foods* **9** (2020), 199. https://doi.org/10.3390/foods9020199.

19 S. Moncayo, S. Manzoor, T. Ugidos, F. Navarro-Villoslada, J.O. Caceres, Discrimination of human bodies from bones and teeth remains by aser induced breakdown spectroscopy and neural networks, *Spectrochim. Acta Part B At. Spectrosc.* **101** (2014), 21–25. https://doi.org/10.1016/j.sab.2014.07.008.

20 S. Moncayo, J.D. Rosales, R. Izquierdo-Hornillos, J. Anzano, J.O. Caceres, Classification of red wine based on its protected designation of origin (PDO) using laser-induced breakdown spectroscopy (LIBS), *Talanta* **158** (2016), 185–191. https://doi.org/10.1016/j.talanta.2016.05.059.

21 S. Manzoor, S. Moncayo, F. Navarro-Villoslada, J.A. Ayala, R. Izquierdo-Hornillos, F.J.M. de Villena, J.O. Caceres, Rapid identification and discrimination of bacterial strains by laser induced breakdown spectroscopy and neural networks, *Talanta* **121** (2014), 65–70. https://doi.org/10.1016/j.talanta.2013.12.057.

22 J.O. Caceres, S. Moncayo, J.D. Rosales, F.J.M. de Villena, F.C. Alvira, G.M. Bilmes, Application of laser-induced breakdown spectroscopy (LIBS) and neural networks to olive oils analysis, *Appl. Spectrosc.* **67** (2013), 1064–1072. https://doi.org/10.1366/12-06916.

23 T.M. Berto, M.C. Santos, F.M.V. Pereira, É.R. Filletti, Artificial neural networks applied to the classification of hair samples according to pigment and sex using non-invasive analytical techniques, *X-Ray Spectrom.* **49** (2020), 632–641. https://doi.org/10.1002/xrs.3163.

24 Y. Yang, C. Li, S. Liu, H. Min, C. Yan, M. Yang, J. Yu, Classification and identification of brands of iron ores using laser-induced breakdown spectroscopy combined with principal component analysis and artificial neural networks, *Anal. Methods* **12** (2020), 1316–1323. https://doi.org/10.1039/C9AY02443C.

25 S. Moncayo, S. Manzoor, J.D. Rosales, J. Anzano, J.O. Caceres, Qualitative and quantitative analysis of milk for the detection of adulteration by laser induced breakdown spectroscopy (LIBS), *Food Chem.* **232** (2017), 322–328. https://doi.org/10.1016/j.foodchem.2017.04.017.

26 M.F. Møller, A scaled conjugate gradient algorithm for fast supervised learning, *Neural Netw.* **6** (1993), 525–533. https://doi.org/10.1016/S0893-6080(05)80056-5.

27 G. Vítková, K. Novotný, L. Prokeš, A. Hrdlička, J. Kaiser, J. Novotný, R. Malina, D. Prochazka, Fast identification of biominerals by means of stand-off laser-induced breakdown

spectroscopy using linear discriminant analysis and artificial neural networks, *Spectrochim. Acta Part B At. Spectrosc.* **73** (2012), 1–6. https://doi.org/10.1016/j.sab.2012.05.010.
28 J.J. Hatch, T.R. McJunkin, C. Hanson, J.R. Scott, Automated interpretation of LIBS spectra using a fuzzy logic inference engine, *Appl. Opt.* **51** (2012), B155. https://doi.org/10.1364/AO.51.00B155.
29 N. Nawi, M. Ransing, R. Ransing, An improved learning algorithm based on The Broyden-Fletcher-Goldfarb-Shanno (BFGS) method for back propagation neural networks, in: *Sixth International Conference on Intelligent Systems Design and Applications*, IEEE, 2006: pp. 152–157. https://doi.org/10.1109/ISDA.2006.95.
30 J. Moros, J. Serrano, F.J. Gallego, J. Macías, J.J. Laserna, Recognition of explosives fingerprints on objects for courier services using machine learning methods and laser-induced breakdown spectroscopy, *Talanta* **110** (2013), 108–117. https://doi.org/10.1016/j.talanta.2013.02.026.
31 M.T. Hagan, M.B. Menhaj, Training feedforward networks with the Marquardt algorithm, *IEEE Trans. Neural Netw.* **5** (1994), 989–993. https://doi.org/10.1109/72.329697.
32 D.E. Rumelhart, G.E. Hinton, R.J. Williams, Learning representations by back-propagating errors, *Nature* **323** (1986), 533–536. https://doi.org/10.1038/323533a0.
33 F. Marquardt, Machine learning and quantum devices, *SciPost Phys. Lect. Notes*, 2021. https://doi.org/10.21468/SciPostPhysLectNotes.29.
34 M.A. Nielsen, *Neural Networks and Deep Learning*, 1st ed., Determination Press, San Francisco, CA 2015. https://books.google.cz/books/about/Neural_Networks_and_Deep_Learning.html?id=STDBswEACAAJ&redir_esc=y (accessed November 27, 2020).
35 C. Bishop, *Pattern Recognition and Machine Learning*, 1st ed., Springer-Verlag New York, 2006. https://www.springer.com/gp/book/9780387310732 (accessed November 27, 2020).
36 J. Liu, M. Osadchy, L. Ashton, M. Foster, C.J. Solomon, S.J. Gibson, Deep convolutional neural networks for Raman spectrum recognition: a unified solution, *Analyst* **142** (2017), 4067–4074. https://doi.org/10.1039/C7AN01371J.
37 J. Acquarelli, T. van Laarhoven, J. Gerretzen, T.N. Tran, L.M.C. Buydens, E. Marchiori, Convolutional neural networks for vibrational spectroscopic data analysis, *Anal. Chim. Acta* **954** (2017), 22–31. https://doi.org/10.1016/j.aca.2016.12.010.
38 C. Carey, T. Boucher, S. Mahadevan, P. Bartholomew, M.D. Dyar, Machine learning tools for mineral recognition and classification from Raman spectroscopy, *J. Raman Spectrosc.* **46** (2015), 894–903. https://doi.org/10.1002/jrs.4757.
39 Yann Le Cun, L. Bottou, Y. Bengio, Reading checks with multilayer graph transformer networks, in: *International Conference on Acoustics, Speech, and Signal Processing (ICASSP)*, IEEE, 1997: pp. 151–154. https://doi.org/10.1109/ICASSP.1997.599580.
40 A. Krizhevsky, I. Sutskever, G.E. Hinton, ImageNet classification with deep convolutional neural networks, *Commun. ACM* **60** (2017), 84–90. https://doi.org/10.1145/3065386.
41 I. Goodfellow, Y. Bengio, A. Courville, *Deep Learning*, 1st ed., MIT Press, Cambridge, MA, 2016. https://mitpress.mit.edu/books/deep-learning.
42 J. El Haddad, L. Canioni, B. Bousquet, Good practices in LIBS analysis: review and advices, *Spectrochim. Acta - Part B At. Spectrosc.* **101** (2014), 171–182. https://doi.org/10.1016/j.sab.2014.08.039.

43 T. Hastie, R. Tibshirani, J. Friedman, *The Elements of Statistical Learning*, Springer New York, New York, NY, 2009. https://doi.org/10.1007/978-0-387-84858-7.

44 J. Vrábel, E. Képeš, L. Duponchel, V. Motto-Ros, C. Fabre, S. Connemann, F. Schreckenberg, P. Prasse, D. Riebe, R. Junjuri, M.K. Gundawar, X. Tan, P. Pořízka, J. Kaiser, Classification of challenging laser-induced breakdown spectroscopy soil sample data - EMSLIBS contest, *Spectrochim. Acta Part B At. Spectrosc.* **169** (2020), 105872. https://doi.org/10.1016/j.sab.2020.105872.

45 E. Képeš, J. Vrábel, S. Střítežská, P. Pořízka, J. Kaiser, Benchmark classification dataset for laser-induced breakdown spectroscopy, *Sci. Data.* **7** (2020), 53. https://doi.org/10.1038/s41597-020-0396-8.

46 R. Junjuri, A. Prakash Gummadi, M. Kumar Gundawar, Single-shot compact spectrometer based standoff LIBS configuration for explosive detection using artificial neural networks, *Optik (Stuttg).* **204** (2020), 163946. https://doi.org/10.1016/j.ijleo.2019.163946.

47 R. Junjuri, M.K. Gundawar, Femtosecond laser-induced breakdown spectroscopy studies for the identification of plastics, *J. Anal. At. Spectrom.* **34** (2019), 1683–1692. https://doi.org/10.1039/C9JA00102F.

48 R. Junjuri, M.K. Gundawar, A low-cost LIBS detection system combined with chemometrics for rapid identification of plastic waste, *Waste Manag.* **117** (2020), 48–57. https://doi.org/10.1016/j.wasman.2020.07.046.

49 A.H. Farhadian, M.K. Tehrani, M.H. Keshavarz, S.M.R. Darbani, Energetic materials identification by laser-induced breakdown spectroscopy combined with artificial neural network, *Appl. Opt.* **56** (2017), 3372. https://doi.org/10.1364/AO.56.003372.

50 S. Messaoud Aberkane, M. Abdelhamid, F. Mokdad, K. Yahiaoui, S. Abdelli-Messaci, M.A. Harith, Sorting zamak alloys via chemometric analysis of their LIBS spectra, *Anal. Methods* **9** (2017), 3696–3703. https://doi.org/10.1039/C7AY01138E.

51 S.-L. Lui, A. Koujelev, Accurate identification of geological samples using artificial neural network processing of laser-induced breakdown spectroscopy data, *J. Anal. At. Spectrom.* **26** (2011), 2419. https://doi.org/10.1039/c1ja10093a.

52 M. Boueri, V. Motto-Ros, W.-Q. Lei, Q.-L. Ma, L.-J. Zheng, H.-P. Zeng, J. Yu, Identification of polymer materials using laser-induced breakdown spectroscopy combined with artificial neural networks, *Appl. Spectrosc.* **65** (2011), 307–314.

53 M. Yelameli, B. Thornton, T. Takahashi, T. Weerakoon, K. Ishii, Classification and statistical analysis of hydrothermal seafloor rocks measured underwater using laser-induced breakdown spectroscopy, *J. Chemom.* **33** (2019), e3092. https://doi.org/10.1002/cem.3092.

54 K. Menking-Hoggatt, L. Arroyo, J. Curran, T. Trejos, Novel LIBS method for micro-spatial chemical analysis of inorganic gunshot residues, *J. Chemom.* (2019),. https://doi.org/10.1002/cem.3208.

55 X. Liu, X. Feng, Y. He, Rapid discrimination of the categories of the biomass pellets using laser-induced breakdown spectroscopy, *Renew. Energy* **143** (2019), 176–182. https://doi.org/10.1016/j.renene.2019.04.137.

56 S.-B. Roh, S.-B. Park, S.-K. Oh, E.-K. Park, W.Z. Choi, Development of intelligent sorting system realized with the aid of laser-induced breakdown spectroscopy and hybrid preprocessing algorithm-based radial basis function neural networks for recycling black plastic wastes, *J. Mater. Cycles Waste Manag.* **20** (2018), 1934–1949. https://doi.org/10.1007/s10163-018-0701-1.

57 C. Zhang, T. Shen, F. Liu, Y. He, Identification of coffee varieties using laser-induced breakdown spectroscopy and chemometrics, *Sensors* **18** (2017), 95. https://doi.org/10.3390/s18010095.

58 J. Wang, X. Liao, P. Zheng, S. Xue, R. Peng, Classification of Chinese herbal medicine by laser-induced breakdown spectroscopy with principal component analysis and artificial neural network, *Anal. Lett.* **51** (2018), 575–586. https://doi.org/10.1080/00032719.2017.1340949.

59 T. Zhang, C. Yan, J. Qi, H. Tang, H. Li, Classification and discrimination of coal ash by laser-induced breakdown spectroscopy (LIBS) coupled with advanced chemometric methods, *J. Anal. At. Spectrom.* **32** (2017), 1960–1965. https://doi.org/10.1039/C7JA00218A.

60 R. Zdunek, M. Nowak, E. Plinski, Statistical classification of soft solder alloys by laser-induced breakdown spectroscopy: review of methods, *J. Eur. Opt. Soc. Rapid Publ.* **11** (2016), 16006i. https://doi.org/10.2971/jeos.2016.16006i.

61 S. Yoshino, B. Thornton, T. Takahashi, Y. Takaya, T. Nozaki, Signal preprocessing of deep-sea laser-induced plasma spectra for identification of pelletized hydrothermal deposits using artificial neural networks, *Spectrochim. Acta Part B At. Spectrosc.* **145** (2018), 1–7. https://doi.org/10.1016/j.sab.2018.03.015.

62 S. Arlot, A. Celisse, A survey of cross-validation procedures for model selection, *Stat. Surv.* **4** (2010), 40–79. https://doi.org/10.1214/09-SS054.

63 B. Ghojogh, M. Crowley, The Theory Behind Overfitting, Cross Validation, Regularization, Bagging, and Boosting: Tutorial, (2019),. https://arxiv.org/abs/1905.12787.

64 D.P. Kingma, J. Ba, Adam: A Method for Stochastic Optimization, (2014),. http://arxiv.org/abs/1412.6980.

65 Y. Yao, L. Rosasco, A. Caponnetto, On early stopping in gradient descent learning, *Constr. Approx.* **26** (2007), 289–315. https://doi.org/10.1007/s00365-006-0663-2.

66 J. Álvarez, M. Velásquez, A.K. Myakalwar, C. Sandoval, R. Fuentes, R. Castillo, D. Sbarbaro, J. Yáñez, Determination of copper-based mineral species by laser induced breakdown spectroscopy and chemometric methods, *J. Anal. At. Spectrom.* **34** (2019), 2459–2468. https://doi.org/10.1039/C9JA00271E.

67 G.E. Hinton, N. Srivastava, A. Krizhevsky, I. Sutskever, R.R. Salakhutdinov, Improving neural networks by preventing co-adaptation of feature detectors, (2012),. http://arxiv.org/abs/1207.0580.

68 N. Srivastava, G. Hinton, A. Krizhevsky, R. Salakhutdinov, Dropout: A Simple Way to Prevent Neural Networks from Overfitting, *Journal of Machine Learning Research* **15** 2014. https://dl.acm.org/doi/10.5555/2627435.2670313.

69 J. Chen, J. Pisonero, S. Chen, X. Wang, Q. Fan, Y. Duan, Convolutional neural network as a novel classification approach for laser-induced breakdown spectroscopy applications in lithological recognition, *Spectrochim. Acta Part B At. Spectrosc.* **166** (2020), 105801. https://doi.org/10.1016/j.sab.2020.105801.

70 R. Sattmann, I. Mönch, H. Krause, R. Noll, S. Couris, A. Hatziapostolou, A. Mavromanolakis, C. Fotakis, E. Larrauri, R. Miguel, Laser-induced breakdown spectroscopy for polymer identification, *Appl. Spectrosc.* **52** (1998), 456–461. https://doi.org/10.1366/0003702981943680.

71 S. Moncayo, S. Manzoor, F. Navarro-Villoslada, J.O. Caceres, Evaluation of supervised chemometric methods for sample classification by laser induced breakdown spectroscopy,

Chemom. Intell. Lab. Syst. **146** (2015), 354–364. https://doi.org/10.1016/j.chemolab.2015.06.004.

72 M.B. Alli, D. Szwarcman, D.S. Civitarese, P. Hayden, Vacuum ultraviolet laser induced breakdown spectroscopy (VUV-LIBS) with machine learning for pharmaceutical analysis, *J. Phys. Conf. Ser.* **1289** (2019), 012031. https://doi.org/10.1088/1742-6596/1289/1/012031.

73 C. Bohling, K. Hohmann, D. Scheel, C. Bauer, W. Schippers, J. Burgmeier, U. Willer, G. Holl, W. Schade, All-fiber-coupled laser-induced breakdown spectroscopy sensor for hazardous materials analysis, *Spectrochim. Acta Part B At. Spectrosc.* **62** (2007), 1519–1527. https://doi.org/10.1016/j.sab.2007.10.038.

74 B. Campanella, E. Grifoni, S. Legnaioli, G. Lorenzetti, S. Pagnotta, F. Sorrentino, V. Palleschi, Classification of wrought aluminum alloys by Artificial Neural Networks evaluation of laser induced breakdown spectroscopy spectra from aluminum scrap samples, *Spectrochim. Acta Part B At. Spectrosc.* **134** (2017), 52–57. https://doi.org/10.1016/j.sab.2017.06.003.

75 P. Chen, X. Wang, X. Li, Q. Lyu, N. Wang, Z. Jia, A quick classifying method for tracking and erosion resistance of HTV silicone rubber material via laser-induced breakdown spectroscopy, *Sensors* **19** (2019), 1087. https://doi.org/10.3390/s19051087.

76 X. Cui, Q. Wang, Y. Zhao, X. Qiao, G. Teng, Laser-induced breakdown spectroscopy (LIBS) for classification of wood species integrated with artificial neural network (ANN), *Appl. Phys. B Lasers Opt.* **125** (2019), 56. https://doi.org/10.1007/s00340-019-7166-3.

77 D. Diego-Vallejo, D. Ashkenasi, A. Lemke, H.J. Eichler, Selective ablation of Copper-Indium-Diselenide solar cells monitored by laser-induced breakdown spectroscopy and classification methods, *Spectrochim. Acta Part B At. Spectrosc.* **87** (2013), 92–99. https://doi.org/10.1016/j.sab.2013.06.012.

78 D. Diego-Vallejo, H.J. Eichler, D. Ashkenasi, Inspection of thin-film solar cell processing by laser-induced breakdown spectroscopy and neural networks, in: *International Symposium on Optomechatronic Technology (ISOT 2012)*, IEEE, 2012: pp. 1–2. https://doi.org/10.1109/ISOT.2012.6403264.

79 J. He, Y. Liu, C. Pan, X. Du, Identifying ancient ceramics using laser-induced breakdown spectroscopy combined with a back propagation neural network, *Appl. Spectrosc.* **73** (2019), 1201–1207. https://doi.org/10.1177/0003702819861576.

80 H.Y. Kong, L.X. Sun, J.T. Hu, Y. Xin, Z.B. Cong, A comparative study of two data reduction methods for steel classification based on LIBS, *Appl. Mech. Mater.* **644–650** (2014), 4722–4725. https://doi.org/10.4028/www.scientific.net/AMM.644-650.4722.

81 H. Kong, L. Sun, J. Hu, Y. Xin, Z. Cong, Selection of spectral data for classification of steels using laser-induced breakdown spectroscopy, *Plasma Sci. Technol.* **17** (2015), 964–970. https://doi.org/10.1088/1009-0630/17/11/14.

82 A. Koujelev, M. Sabsabi, V. Motto-Ros, S. Laville, S.L. Lui, Laser-induced breakdown spectroscopy with artificial neural network processing for material identification, *Planet. Space Sci.* **58** (2010), 682–690. https://doi.org/10.1016/j.pss.2009.06.022.

83 J. Li, F. Chen, G. Huang, S. Zhang, W. Wang, Y. Tang, Y. Chu, J. Yao, L. Guo, F. Jiang, Identification of Graves' ophthalmology by laser-induced breakdown spectroscopy combined with machine learning method, *Front. Optoelectron.* (2020),. https://doi.org/10.1007/s12200-020-0978-2.

84 F. Liu, T. Shen, J. Wang, Y. He, C. Zhang, W. Zhou, F. Liu, T. Shen, J. Wang, Y. He, C. Zhang, W. Zhou, Detection of sclerotinia stem rot on oilseed rape (*Brassica napus* L.) based on laser-

induced breakdown spectroscopy, *Trans. ASABE* **62** (2019), 123–130. https://doi.org/10.13031/trans.12206.

85 S. Manzoor, L. Ugena, J. Tornero-Lopéz, H. Martín, M. Molina, J.J. Camacho, J.O. Cáceres, Laser induced breakdown spectroscopy for the discrimination of Candida strains, *Talanta* **155** (2016), 101–106. https://doi.org/10.1016/j.talanta.2016.04.030.

86 D. Marcos-Martinez, J.A. Ayala, R.C. Izquierdo-Hornillos, F.J.M. de Villena, J.O. Caceres, Identification and discrimination of bacterial strains by laser induced breakdown spectroscopy and neural networks, *Talanta* **84** (2011), 730–737. https://doi.org/10.1016/j.talanta.2011.01.069.

87 M. Mordmueller, C. Bohling, A. John, W. Schade, Rapid test for the detection of hazardous microbiological material, in: *Proceedings of SPIE 7484, Optically Based Biological and Chemical Detection for Defence V* 2009. https://doi.org/10.1117/12.830200.

88 D. Pokrajac, A. Lazarevic, V. Kecman, A. Marcano, Y. Markushin, T. Vance, N. Reljin, S. McDaniel, N. Melikechi, Automatic classification of laser-induced breakdown spectroscopy (LIBS) data of protein biomarker solutions, *Appl. Spectrosc.* **68** (2014), 1067–1075. https://doi.org/10.1366/14-07488.

89 A. Marín Roldán, V. Dwivedi, J. Yravedra Sainz de los Terreros, P. Veis, Laser-Induced breakdown spectroscopy (LIBS) for the analyses of faunal bones: assembling of individuals and elemental quantification, *Optik (Stuttg).* **218** (2020), 164992. https://doi.org/10.1016/j.ijleo.2020.164992.

90 K. Rzecki, T. Sośnicki, M. Baran, M. Niedźwiecki, M. Król, T. Łojewski, U. Acharya, Ö. Yildirim, P. Pławiak, Application of computational intelligence methods for the automated identification of paper-ink samples based on LIBS, *Sensors* **18** (2018), 3670. https://doi.org/10.3390/s18113670.

91 E.G. Snyder, C.A. Munson, J.L. Gottfried, F.C. De Lucia, Jr., B. Gullett, A. Miziolek, Laser-induced breakdown spectroscopy for the classification of unknown powders, *Appl. Opt.* **47** (2008), G80. https://doi.org/10.1364/AO.47.000G80.

92 E. Ukwatta, J. Samarabandu, Vision based metal spectral analysis using multi-label classification, in: *2009 Canadian Conference on Computer and Robot Vision*, IEEE, 2009: pp. 132–139. https://doi.org/10.1109/CRV.2009.42.

93 S. Yoshino, T. Takahashi, B. Thornton, Towards in-situ chemical classification of seafloor deposits: application of neural networks to underwater laser-induced breakdown spectroscopy, in: *Ocean. 2017 - Aberdeen*, IEEE, 2017: pp. 1–5. https://doi.org/10.1109/OCEANSE.2017.8084734.

94 P. Mehta, M. Bukov, C.-H. Wang, A.G.R. Day, C. Richardson, C.K. Fisher, D.J. Schwab, A high-bias, low-variance introduction to machine learning for physicists, *Phys. Rep.* **810** (2019), 1–124. https://doi.org/10.1016/j.physrep.2019.03.001.

10

Data Fusion

LIBS + Raman

Beatrice Campanella and Stefano Legnaioli

Applied and Laser Spectroscopy Laboratory, Institute of Chemistry of Organometallic Compounds, Research Area of the National Research Council, Pisa, Italy

10.1 Introduction

The ever increasing number of analytical techniques available for characterizations with minimum samples handling or even directly in situ has made it easier to obtain large amount of data. This aspect has consequently driven the attention toward the development of efficient data fusion methodologies to exploit the synergies of individual information provided by different techniques.

The approach based on the combination of two or more analytical techniques, e.g. LIBS and Raman spectroscopy, is generally defined as hyphenated. The two techniques can be used independently with two different instruments or with the same device. In this specific case, Raman spectroscopy will enhance the information power of LIBS by adding the molecular profile to the elemental one provided by LIBS. In this sense, the two techniques provide a complementary knowledge about the molecular structure and elemental composition of an analyzed sample, respectively, and, if measured simultaneously, they allow for a more comprehensive characterization of the system under study since, especially in the case of complex matrixes, a single analysis rarely provides a complete understanding. Optical spectroscopic techniques, particularly the laser-based spectroscopies, given the micro- or completely no-destructive feature along with high sensitivity and without sample preparation, appear as the best candidates for these purposes, making them versatile tools suitable for both in situ and ex situ analyses.

In literature, there are several examples of fields in which elemental and molecular information are needed [1], such as remote sensing of minerals [2–5], explosives or chemical agents [6, 7], and analysis of art objects [8–10]. Separate instruments are usually used, as well as hybrid units developed with a single laser and a single detector, and more recently new laser and spectrometers have been implemented in combination mainly for standoff applications [11, 12].

Chemometrics and Numerical Methods in LIBS, First Edition. Edited by Vincenzo Palleschi.
© 2023 John Wiley & Sons Ltd. Published 2023 by John Wiley & Sons Ltd.

Plenty of chemometric approaches for both calibration and classification purposes have bloomed in recent years to provide the maximum number of relevant chemical information. Calibration methods have the potential to address spectral interferences, matrix effects, and analyte signal nonlinearity. The calibration is, indeed, an empirical process, defined as the operation that determines the functional relationship between measured values and analytical quantities [13]. Classification, on the other hand, is a means to discriminate among a finite group of defined sample types. It implies that unknown samples are essentially identical to the known one, which are used as spectral training data. The limitation of the latter approach is that extrapolation beyond the known samples and training sets to a wide range of sample conditions can lead to misinterpretation of chemometric algorithms, as the "best match" may have little physical similarity to the actual unknowns.

Another constrain is the overfitting, for which model validation becomes extremely critical. Basically, common routine is related to the spectra pretreatment such as background stripping, normalization, and spectral cleaning. Approaches based on partial least squares discriminant analysis (PLS-DA) and principal component analysis (PCA) are subsequently used for sorting, noting that PCA was not originally designed as a tool for statistical discrimination. Recently, the use of algorithms based on neural networks is becoming more and more popular [14].

The aim of this chapter is to provide a brief overview of the current literature regarding the combination of chemometric algorithms and data fusion techniques involving LIBS and Raman spectroscopies, to evaluate the recent improvements in the quantitative classification of samples.

10.2 Data Fusion Background

Data fusion process allows for combining or integrating observed data, originating from two or more sensors, to produce a more specific, broad, and unified data set of the sample under study, thus achieving inferences that are not feasible using a single sensor. Recently, the benefits of data fusion strategies have been exploited in a wide range of application fields [15–18].

It is worth noting that detection limits, sensitivity, and the material properties that the instrument detects and measures vary from sensor to sensor. Combination of sensors may have advantages over a single sensor due to improved accuracy, improved precision, and reduced uncertainty.

Acquired data can be combined using a variety of approaches from raw data level to feature or decision-level fusion. Raw data can be directly combined if sensors' responses are commensurate, that is, if the sensors are measuring the same physical phenomena. On the contrary, in case of noncommensurate responses, the data must be fused at the feature level or at the decision level. Fusion at the feature level involves the extraction of representative features from the data of each sensor. Once features from observations of multiple sensors have been extracted, they are combined and concatenated into a single output, called "estimator," which acts as the input for the final decision. On the other hand, decision-level fusion involves the fusion of the decisions of each sensor.

10.2 Data Fusion Background

Generally, three levels of data fusion are encompassed (see Figure 10.1): low-level, mid-level, and high-level [19]. The first two strategies work at data level. Low-level data fusion is implemented by simply concatenating data matrices or data blocks from different data sources. The final data matrix has the same rows size as the number of samples analyzed

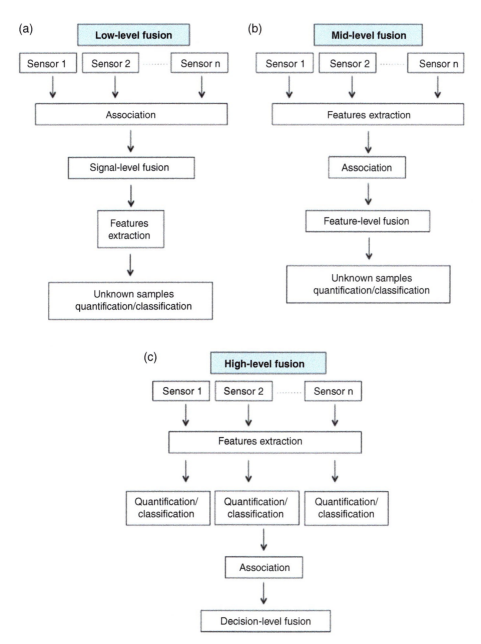

Figure 10.1 Scheme of data fusion at three different levels: (a) low-level fusion, (b) mid-level fusion, and (c) high-level fusion.

and the same columns size as the variables measured by the instruments. A single matrix is used to calculate a single classification or prediction model. Low-level data fusion, basically, considers the correlation between variables of the different data blocks. In the mid-level data fusion, first of all, the important variables from each data source are separately extracted. These informative variables are successively concatenated into a single array and used to perform classification and prediction for materials characterization. The variable selection reduces data dimensionality and therefore it is useful to treat each data block without the influence of other data set. High-level data fusion combines model outputs and the data sources at a decision level. Independent models are built for each available sensor output, and high-level data fusion combines model outputs to produce a fused response.

Although structurally more complex, sometimes an architecture based on decision-level fusion may be unsatisfactory, since the decision from each sensor is associated with the same error that the sensor gets when used alone. Moreover, it would be difficult to define a hierarchy of single decisions in the event of contradictory decisions from the two sensors.

10.3 Data Treatment

Preprocessing is generally required before chemometric modeling to increase the signal-to-noise ratio and to improve the multivariate analysis. First of all, significant differences due to baseline shifts and magnitudes of spectral signals should be removed. Since Raman and LIBS data are usually collected using different spectrometers, the dynamic range of their output spectra is not identical, meaning that the intensity values observed by each instrument span different ranges in magnitude. Due to these different responses, normalization by scaling between 0 and 1 is usually considered. It has been verified that the performance of the parameters indicating the degree of similarity is preserved, regardless of how the normalization method is realized [20]. However, such type of scaling not only ensures the distinction between new identifiers on the basis of the different frequencies and wavelengths of the spectral features but also provides an equal input of the molecular and atomic information for the final attribute, whatever be the assembling mode used. One of the most popular methods is the *unit vector* normalization. It first computes the square root of the sum of the squared values of all the measured intensities (the spectral "norm"), then each measured intensity is divided by the "norm" to obtain the normalized value, x':

$$x'_i = \frac{1}{\left|\sqrt[2]{(x_1^2 + x_2^2 + \ldots x_n^2)}\right|}$$

where x is the intensity value corresponding to either a LIBS wavelength or a Raman shift and n is the number of variables (spectral bands). This procedure allows the data from both instruments to be fused without one data set excessively dominating the other due to differences in recorded energy. Sometimes, the spectra are also smoothed by the Savitzky–Golay algorithm to remove high-frequency noise.

Nevertheless, a single LIBS or Raman spectrum could consist of hundreds to thousands of data point pairs (wavelength/wavenumber combined with intensity). To ensure that processing time is kept to a minimum and to reduce the risk of overtraining on irrelevant features, it is often useful to select a subset of the variables for input to the training algorithm. Variable selection for spectroscopy is typically approached using a priori knowledge about the chemical structure or composition of the target molecule. However, a complex environment could contribute numerous background signals from interferents, which happen to overlap with the chosen discriminating features.

A remedy to reduce the number of highly collinear features, which are a constant in spectroscopic acquisitions, is the application of PCA to obtain a lower dimensional data set. PCA is, indeed, an unsupervised data reduction technique commonly used to explore data sets of high dimensionality. It may be considered as the basic tool for data analysis that simultaneously provides a visual representation of relationships between samples and variables as well as insights into sample homogeneities and heterogeneities. PCA alone provides the information on data clustering, significant spectral regions, and outliers. Following data pretreatment and features reduction, groups in data sets are commonly discriminated with classification techniques such as linear discriminant analysis (LDA), k nearest neighbors (KNN), and PLS-DA or support vector machine (SVM). The elimination of redundant features by PCA enables to overcome a restriction of LDA, i.e. the requirements of a lower of equal number of variables and samples, and PCA-LDA has become a common approach in spectroscopic data analysis. All these procedures (data reduction and classification) can be carried out in separate data packages or in the concatenated data set to form a response.

10.4 Working with Images

The fusion process of spectral information from sensor measurements produces a new feature, which aims to more clearly identify the target. An original approach that slightly improves the low-level data fusion method was suggested by the Moros e Laserna [21], and it is based on the creation of a global picture (2D image) of the compound, which can be generated by fusing molecular and atomic responses.

After the acquisitions, the intensity values belonging to the Raman (r) and LIBS (l) spectra at each pixel ($n = 1, 2, 3, \ldots$) were arranged in two vectors, named R (Raman) and L (LIBS), respectively. For simplicity of notation, we hypothesize that the dimension of the two vectors is always the same, so the pixel-by-pixel alignment of the information required for proper data fusion is assured; anyway this does not invalidate the final result.

$$\begin{bmatrix} r_1 \\ r_2 \\ . \\ . \\ . \\ r_n \end{bmatrix} \begin{bmatrix} l_1 \\ l_2 \\ . \\ . \\ . \\ l_n \end{bmatrix}$$

As previously reported, a reduction of the information, without altering the size of the vectors, is always possible if the data therein contained have no relevant information regarding sample identification. Two preprocessing strategies, namely autoscaling (AS) and normalization (N), were used as pretreatment procedure. The first one consists of simply mean centering followed by scaling through the variance. Values for the mean and the standard deviation of each intensity vector (R and L) are calculated.

$$\bar{r} = \frac{1}{n}\sum_{i=1}^{n} r_i \qquad \bar{l} = \frac{1}{n}\sum_{i=1}^{n} l_i$$

$$S_r = \frac{1}{n-1}\sum_{i=1}^{n}(r_i - \bar{r}) \qquad S_l = \frac{1}{n-1}\sum_{i=1}^{n}(l_i - \bar{l})$$

$$r_{i(AS)} = \frac{r_i - \bar{r}}{S_r} \qquad l_{i(AS)} = \frac{l_i - \bar{l}}{S_l}$$

The normalization procedure transforms the original elements forming the intensity vectors, R and L, into new values ranging from 0 to 1. Thus, element by element, normalized data $r_{i(N)}$ and $l_{i(N)}$ are computed as follows:

$$r_{i(N)} = \frac{r_i - r_{min}}{r_{max} - r_{min}} \qquad l_{i(N)} = \frac{l_i - l_{min}}{l_{max} - l_{min}}$$

where r_{min}, l_{min} and r_{max}, l_{max} are the minimum and the maximum values across the vector, respectively. Finally, the two vectors may be combined in different way. The authors selected four different approaches: vectors concatenation, co-addition, outer sum, and product. They are described in the following paragraph. It is worth noting that the first two produced a linear vector, usually adopted as input for the low-level data fusion, whereas the last two methods may rise to a bidimensional representation.

10.4.1 Vectors Concatenation

The easiest way to proceed in data fusion is the concatenation of the spectral information, that is, the Raman and LIBS spectra from each sample are combined to give origin to a new attribute with a double size.

$$[r_1, r_2, ...r_n][l_1, l_2, ...l_n] = [r_1, r_2, ...r_n, l_1, l_2, ...l_n]$$

10.4.2 Vectors Co-addition

Taking advantage of the availability of the same number of variables for both spectra, it is possible to simply sum, pixel to pixel, the Raman and LIBS signals. Through this procedure, the new attribute preserves its vector form and the number of variables but changes its shape according to the Raman and LIBS information. It is worth noting that this process has no longer meaning in terms of spectroscopy.

$$[r_1, r_2, ...r_n] + [l_1, l_2, ...l_n] = [r_1 + l_1, r_2 + l_2, ...r_n + l_n]$$

10.4.3 Vectors Outer Sum

In this case, a fused square array ($n \times n$) is produced by summing each pixel of the transposed L vector to every pixel of the R vector.

$$\begin{bmatrix} r_1 \\ r_2 \\ \cdot \\ \cdot \\ r_n \end{bmatrix} \oplus [l_1, l_2, \ldots l_n] = \begin{bmatrix} r_1 + l_1 & r_2 + l_1 & \cdot & r_n + l_1 \\ r_1 + l_2 & r_2 + l_2 & \cdot & r_n + l_2 \\ \cdot & \cdot & \cdot & \cdot \\ \cdot & \cdot & \cdot & \cdot \\ r_1 + l_n & r_2 + l_n & \cdot & r_n + l_n \end{bmatrix}$$

10.4.4 Vectors Outer Product

Similarly to the previous approach, data are associated by multiplying the original two vectors to create a new matrix formed from all possible products between the elements of R and those of L.

$$\begin{bmatrix} r_1 \\ r_2 \\ \cdot \\ \cdot \\ r_n \end{bmatrix} \otimes [l_1, l_2, \ldots l_n] = \begin{bmatrix} r_1 l_1 & r_2 l_1 & \cdot & r_n l_1 \\ r_1 l_2 & r_2 l_2 & \cdot & r_n l_2 \\ \cdot & \cdot & \cdot & \cdot \\ \cdot & \cdot & \cdot & \cdot \\ r_1 l_n & r_2 l_n & \cdot & r_n l_n \end{bmatrix}$$

The same authors in another paper [20] proposed a slightly different and complex approach concatenating the L and R vectors to form a new matrix ($n \times 2$) and, then, data were associated by multiplying the matrix itself by its transposed producing a fused square matrix ($n \times n$), in order to produce a more complete and characteristic 2D image of the sample.

$$\begin{bmatrix} r_1 & l_1 \\ r_2 & l_2 \\ \cdot & \cdot \\ \cdot & \cdot \\ r_n & l_n \end{bmatrix} \begin{bmatrix} r_1 & r_2 & \cdots & r_n \\ l_1 & l_2 & \cdots & l_n \end{bmatrix} = \begin{bmatrix} x_{11} & x_{21} & \cdot & x_{1n} \\ x_{12} & x_{22} & \cdot & x_{2n} \\ \cdot & \cdot & \cdot & \cdot \\ \cdot & \cdot & \cdot & \cdot \\ x_{1n} & x_{2n} & \cdot & x_{nn} \end{bmatrix}$$

As observed by the same authors, when fusion is based on the product of variables, any lack of spectral features within one of the counterparts cancels the information of the other.

10.4.5 Data Analysis

Once the spectroscopic data are fused, following one of the previous approaches, the next step concerns the optimization of the classification procedures, carried out using the correlation coefficient value (r), defined as follows:

$$r = \frac{\sum_m \sum_n (A_{mn} - \overline{A})(B_{mn} - \overline{B})}{\sqrt{\left(\sum_m \sum_n (A_{mn} - \overline{A})^2\right)\left(\sum_m \sum_n (B_{mn} - \overline{B})^2\right)}}$$

where A and B are vectors or matrices representing the two targets. A high value for the correlation coefficient implies a significant similarity between the samples. As a result, the closer is the value of r to 1, the higher the similarity between the attributes under consideration and, consequently, more problematic is their identification. Generally, a threshold is chosen around 0.7–0.8 to cluster the samples. However, as r depends only on the shapes of new attributes, not on their magnitudes, an additional measure of the extent to which a pair of attributes is similarly related is also computed through the root mean square error (RMSE). The aim of this measure is to compare two attributes by providing a quantitative score that describes the degree of similarity. RMSE is calculated as follows:

$$\text{RMSE} = \sqrt{\frac{1}{MN} \sum_{m,n} (A_{m,n} - B_{m,n})^2}$$

where MN is the total number of variables in each attribute. In this case, the lower the RMSE value, the greater the similarity between the attributes from compounds being compared.

The authors reported the application of the earlier-descripted procedure in the identification of explosives by Raman and LIBS spectroscopy [20]. As shown in Figure 10.2, the use of 2D images allows a successful classification in all cases, although in favorable instances the unique information from a single sensor may be sufficient.

10.5 Applications

Although LIBS-Raman data fusion can be considered almost a recent approach, there are several applications in the literature, most of them essentially based on the "low-level" approach. One of them concerns forensic investigations [22] to examine changes in the chemical signature of oil imprints over time. LIBS signal intensities were previously normalized to the maximum intensity of each laser pulse around the CN band signal (388.29 nm) and then used to cluster the data applying the PCA; further soft independent modeling by class analogy (SIMCA) and PLS-DA were used to automatically recognize time differences in the fingerprint laser spectra. Raman spectroscopy was used basically to validate the LIBS analyses. Hoehse et al. [23] fused into a single data matrix LIBS and Raman spectra from different inks, and the number of different groups was determined through multivariate analysis, i.e. PCA, SIMCA, PLS-DA, and SVM. The spectra obtained from Raman spectroscopy and LIBS were processed for background correction. Spectral data from LIBS and Raman were evaluated separately and pretreated by the unit vector normalization. A hierarchical approach using SIMCA and PLS-DA allowed for the stepwise classification of data and separation of inks that were not identified by PCA. Finally, the reliability of the classification rules was validated through a random cross-validation

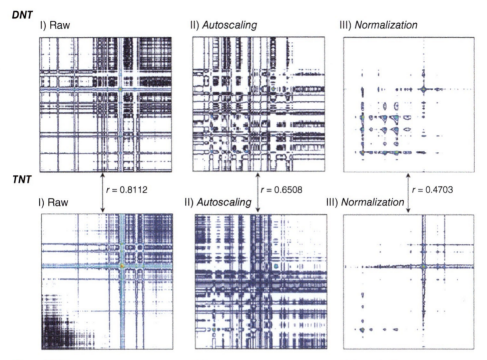

Figure 10.2 Graphical comparison of 2D images, corresponding to DNT and TNT bulk targets, generated from (I) raw, (II) autoscaled, and (III) normalized Raman and LIBS data assembled together. *Source*: From [20].

procedure. An interesting application is proposed by Prochazka et al.[24] for multivariate classification of bacteria. Five *Staphylococcus* bacterial strains and one strain of *Escherichia coli* were analyzed with LIBS and Raman, and the data were then treated separately by the unit vector normalization. To classify the bacterial strains, they used a neural network-based approach, namely a supervised version of Kohonen's self-organizing maps (SOM), whereas PCA was used to visualize the samples relationships. To demonstrate the advantage of the data fusion method, the analysis was performed at first separately on the data from each technique, and then on the merged data. The best classification accuracy was noticed for merged data. Zhao et al. [25] used LIBS and Raman, combined with Fourier transform mid-infrared (FT-IR) and chemometric techniques to quantify calcium (Ca) content in infant formula powder. Also in this case FT-IR was adopted along with Raman to reinforce the study of the molecular structures, whereas atomic absorption spectroscopy analysis was employed as a reference method to quantify Ca content in the samples. To predict Ca content, partial least squares regression (PLSR) models were developed based on LIBS, Raman, and FT-IR spectral data, respectively. Baseline correction on raw data was carried out using asymmetric least squares correction (AsLs). The optimal wavelength ranges of each spectroscopic technique used in modeling were selected on the basis of observed spectral signal intensities. PLSR models were developed using the nonlinear iterative partial least squares (NIPALS) algorithm on the calibration samples ($n = 45$) and validated using independent validation samples ($n = 6$). Leave-one-out cross-validation was carried out. To evaluate model performance, parameters such as root mean square error of cross-validation (RMSECV) and prediction (RMSEP) as well as regression coefficients of determination on cross-validation (R^2CV) and prediction (R^2P) were calculated. The authors concluded that, overall, the model developed using LIBS achieved the best prediction performance, whereas the second best-performed model was developed using the mid-level data fusion of both Raman and FT-IR spectral features.

The fields where the benefits of the combining use of LIBS and Raman are particularly noticeable are the standoff detection of explosives, as reported also in the Moros works previously [20], and the classification of minerals, particularly in the context of in situ Mars exploration.

In the first case, Waterbury et al. introduced a new proprietary LIBS enhancement technique called Townsend effect plasma spectroscopy (TEPS) integrated in a mobile instrument, TEPS-Raman Explosive Detection System (TREDS-2) [26, 27]. The data were analyzed using PLS-DA for spectral classification of the samples. Concerning the pretreatment of the spectra, background signals and noise were filtered out with sophisticated algorithms to make a clear match of a spectrum to a given material; moreover, the majority of shot to shot noise associated with TEPS were removed, as well as the long-term performance of the instrument was improved. Although Raman is an extremely useful technique for many identifying chemical or biological threats, it is challenging to get adequate signals for chemometrics analysis in reasonable amounts of time from a standoff distance. On the other hand, TEPS is extremely sensitive to many materials, though it suffers from a lack of specificity, which could lead to high false alarm rates. The results demonstrated that when taken together, these two techniques provide a powerful tool to detect hazardous materials. Botti et al. [28] used a similar approach, adopting PCA instead of PLS-DA to analyze the LIBS spectra. Different procedures were pursued, based on five ratios of the atomic line intensities, namely O/N, O/C, H/C, N/C, and O/H, or on the line intensity ratio H/N as a function of Al emission intensity, involving also C_2 and CN line intensities and taking into account their possible recombination in the plasma. The authors also included an absorption technique (LPAS; laser photoacoustic spectroscopy) and SERS (surface-enhanced Raman spectroscopy) to validate and integrate LIBS data. The results of the measurements were promising in view of the possible integration of the different techniques in a single device for trace detection. Pinkhan et al. [29] suggested the use of a genetic algorithm (GA) as a method to preselect spectral feature variables for chemometric algorithms, applied to LIBS and ultraviolet Raman spectroscopy (UVRS) data, in which the spectra consisted of approximately 10 000 and 1000 distinct spectral values, respectively. The GA-selected variables were examined using two chemometric techniques: multiclass LDA and SVM, demonstrating a slight improvement in the data discrimination when moving from a linear technique (LDA) to a more generalized nonlinear case (SVM).

For what concerns geochemical applications, Gibbons et al. [30] showed that the use of data fusion strategies improved the discrimination model and allowed correct classification of all the analyzed geological samples based on their dominant clay mineralogy. They hypothesized that the spectral concatenation offers a simple means of increasing the scientific output of these technologies. So, they fused the complete LIBS and Raman spectral responses into a new, combined data matrix, according to the low-level model. Then they employed the PCA for visualization, data reduction, and to determine the spectral features that are important for successful identification of the clay minerals. Instead, the classification was realized developing a linear discriminant model. A good deal of effort was put to the Mars exploration. Several authors studied different strategies to manage information from an integrated remote system based on a LIBS detector and a Raman spectroscopy sensor, designed to simultaneously register the elemental and molecular signatures of rocks under Martian surface conditions. Moros et al. [31] proposed a new data fusion architecture at decisions level to categorized rocks. The simultaneous LIBS-Raman approach guarantees that the atomic and molecular information gathered corresponds exactly to the same entity and to the actual conditions of the surface being inspected. This association is not secured when the sequential monitoring of the surface is performed, with a decision level justified by the potential heterogeneity of expected host rocks. The presence of impurities and veins at submillimeter and millimeter scales within the interrogated rock may, in fact, adulterate the spectral information, mainly LIBS data. The authors sketched in a flowchart the followed process (Figure 10.3). The first step involves a scrutiny of the LIBS response, with the elements being assigned by a cross-checking with a spectral database. Subsequently, the algorithm evaluates the Raman spectrum of the unknown sample against the Raman spectra from all the oxy-anions sharing the cation that are stored in the database. If no cross-checked Raman candidates meet this criterion, the algorithm starts a feedback loop. If none of these options succeeds, the rock will remain unidentified. In this way, the uncertainty on identity of an unknown sample, arising from the multielemental spectral data, may be clarified by its vibrational counterpart, and vice versa. This approach is suggested also by Rammelkamp et al. [32], although they focused their attention on the low-level approach, attaching the Raman spectrum of a sample to its LIBS counterpart without any prior analysis. In particular, they investigated pure sulfates and their mixtures as well as other Mars-relevant salts such as carbonates, chlorides, perchlorates, and sulfates in a basaltic matrix, employing PCA and PLS-DA to evaluate the data. Every LIBS spectral range was normalized separately by its total intensity (sum of all counts) and then merged to one spectrum. The Raman spectra, instead, were pretreated removing spikes (cosmic rays), subtracting the baseline, and smoothing the data using a moving average. The results indicate that quantitative information can be obtained even though mixtures did not contribute to the model. In the general procedure, best results were achieved with smoothed and by a factor of 2 weighted Raman data in the combined data set.

Manrique-Martinez et al. [33] tested the feasibility of combining several univariate and multivariate analyses with data fusion techniques on different instruments (532 and 785 nm Raman and LIBS) to evaluate the improvements in the quantitative classification of samples in binary mixtures. Regardless the interesting information achieved by the comparison of instruments with different characteristics, such as the role of the spot size in the results of the fused data sets, linked to the heterogeneity of the sample itself, they observed a great improvement introduced by the PCA + ANN (Artificial Neural Netwotk) analyses. The data fusion of Raman and LIBS did not work strictly better for the whole range in all the analyzed mixtures; however, this combination improved the classification limits based on the same data sets, with a limit between 0.5 and 3.5% depending on the sample, with a confidence interval of 95%. Finally, Konstantinidis et al. [34] proposed a new algorithm using the ChemCam preflight calibration data set and a data set from the characterization of an

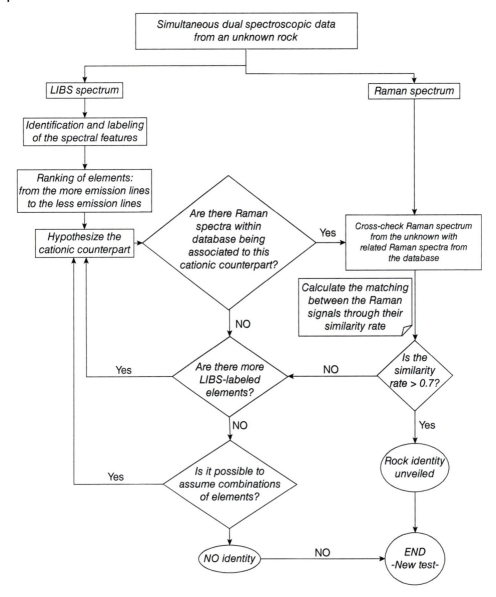

Figure 10.3 Flowchart of strategic decision making to unravel the identity of an unknown rock from its LIBS-Raman spectral responses simultaneously acquired. Source: From [31].

integrated Raman spectroscopy, LIF(Laser Induced Fluorescence) spectroscopy, and LIBS instrument (LIBS Raman Sensor – LIRS). ChemCam was the first LIBS instrument operating on another planet, and its data set was composed of 263 spectra corresponding to 69 geological standards with up to four spectra for each sample [35], while the LIRS data set consisted of 28 spectra corresponding to 28 geological standards. Konstantinidis and coworkers proposed a submodel as combination of three previous multivariate approaches

[35–37]. A dimensional reduction using independent component analysis (ICA) was performed, similar to [35]; in this way extraneous variables, which may negatively affect subsequent regression results, can be removed. The reduced spectra were ordered with respect to the L^2 distance to the origin and partitioned into clusters used to train linear mixture model (as [36]). The result is a set of *submodels*, akin to [37], which achieve very accurate results. A comparison of the results of the single three models is also given, in addition to predictions obtained by ordinary least squares regression, partial least square regression, and principal component regression (PCR). Although having very small sample size (28 samples), the mixture model was able to achieve very accurate results.

10.6 Conclusion

The studies here discussed show that the use of data fusion strategies can improve the discrimination models and allow correct classification of samples of different origin on the basis of their dominant factors. Assembling Raman and LIBS responses in a single data set provides two independent, agreeable measurements of the same physical phenomena, thereby improving the reliability of the research conclusions. In summary, the application of chemometrics to LIBS, when hyphenated to other spectroscopic techniques, brings an entire new dimension to spectral analysis as compared to a traditional univariate analysis, with potential benefits to calibration procedures and quantitative analysis, and opens the door for sophisticated classification routines. Anyway, the general advice is use chemometrics with a firm knowledge of the fundamentals of LIBS and Raman techniques and of the system under study, and not simply as a black-box methodology.

References

1 E. Ercioglu, H.M. Velioglu, and I.H. Boyaci. (2018). Chemometric evaluation of discrimination of aromatic plants by using NIRS, LIBS. *Food Analytical Methods* 11 (6): 1656–1667. https://doi.org/10.1007/s12161-018-1145-x
2 I.N. Sokolik. (2002). Remote sensing of mineral dust aerosols in the UV/visible and IR regions. In *Proceedings of SPIE - The International Society for Optical Engineering*, 265–271. https://doi.org/10.1117/12.466355
3 J.R.C. von Holdt, F.D. Eckardt, M.C. Baddock, and G.F.S. Wiggs. (2019). Assessing landscape dust emission potential using combined ground-based measurements and remote sensing data. *Journal of Geophysical Research: Earth Surface* 124, 5: 1080–1098. https://doi.org/10.1029/2018JF004713
4 R.C. Wiens, S.K. Sharma, J. Thompson, A. Misra, and P.G. Lucey. (2005). Joint analyses by laser-induced breakdown spectroscopy (LIBS) and Raman spectroscopy at stand-off distances. *Spectrochimica Acta - Part A: Molecular and Biomolecular Spectroscopy* 61, 10: 2324–2334. https://doi.org/10.1016/j.saa.2005.02.031
5 H.G.M. Edwards, S.E.J. Villar, D. Pullan, M.D. Hargreaves, B.A. Hofmann, and F. Westall. (2007). Morphological biosignatures from relict fossilised sedimentary geological specimens:

a Raman spectroscopic study. *Journal of Raman Spectroscopy* 38, 10: 1352–1361. https://doi.org/10.1002/jrs.1775

6 J.I. Steinfeld, R.W. Field, M. Gardner, M. Canagaratna, S. Yang, A. Gonzalez-Casielles, S. Witonsky, P. Bhatia, B. Gibbs, B. Wilkie, S.L. Coy, and A. Kachanov. (1999). New spectroscopic methods for environmental measurement and monitoring. In *Proceedings of SPIE - The International Society for Optical Engineering*, 28–33. Retrieved from https://www.scopus.com/inward/record.uri?eid=2-s2.0-0033313289&partnerID=40&md5=c95dc9642b6a204728373bff70fea77b

7 S.D. Christesen, A.W. Fountain, J.A. Guicheteau, T.H. Chyba, and W.F. Pearman. (2014). *Laser spectroscopy for the detection of chemical, biological and explosive threats.* https://doi.org/10.1533/9780857098733.3.393

8 M. Zuena, S. Legnaioli, B. Campanella, V. Palleschi, P. Tomasin, M.K. Tufano, F. Modugno, J. La Nasa, and L. Nodari. (2020). Landing on the moon 50 years later: a multi-analytical investigation on Superficie Lunare (1969) by Giulio Turcato. *Microchemical Journal* 157. https://doi.org/10.1016/j.microc.2020.105045

9 R. Fantoni, S. Almaviva, L. Caneve, F. Colao, M.F. De Collibus, L. De Dominicis, M. Francucci, M. Guarneri, V. Lazic, A. Palucci, G. Maddaluno, and C. Neri. (2019). In situ and remote laser diagnostics for material characterization from plasma facing components to Cultural Heritage surfaces. *Journal of Instrumentation* 14, 7. https://doi.org/10.1088/1748-0221/14/07/C07004

10 A. Botto, B. Campanella, S. Legnaioli, M. Lezzerini, G. Lorenzetti, S. Pagnotta, F. Poggialini, and V. Palleschi. (2019). Applications of laser-induced breakdown spectroscopy in cultural heritage and archaeology: A critical review. *Journal of Analytical Atomic Spectrometry* 34, 1: 81–103. https://doi.org/10.1039/c8ja00319j

11 R.D. Waterbury, A. Pal, D.K. Killinger, J. Rose, E.L. Dottery, and G. Ontai. (2008). Standoff LIBS measurements of energetic materials using a 266 nm excitation laser. In *Proceedings of SPIE – The International Society for Optical Engineering*. https://doi.org/10.1117/12.778668

12 J. Moros, J.A. Lorenzo, P. Lucena, L.M. Tobaria, and J.J. Laserna. (2010). Simultaneous Raman spectroscopy-laser-induced breakdown spectroscopy for instant standoff analysis of explosives using a mobile integrated sensor platform. *Analytical Chemistry* 82, 4: 1389–1400. https://doi.org/10.1021/ac902470v

13 D.W. Hahn and N. Omenetto. (2012). Laser-induced breakdown spectroscopy (LIBS), Part II: Review of instrumental and methodological approaches to material analysis and applications to different fields. *Applied Spectroscopy* 66, 4: 347–419. https://doi.org/10.1366/11-06574

14 B. Campanella, E. Grifoni, S. Legnaioli, G. Lorenzetti, S. Pagnotta, F. Sorrentino, and V. Palleschi. (2017). Classification of wrought aluminum alloys by ANN evaluation of LIBS spectra from aluminum scrap samples. *Spectrochimica Acta – Part B Atomic Spectroscopy* 134: 52–57. https://doi.org/10.1016/j.sab.2017.06.003

15 F. Castanedo. (2013). A review of data fusion techniques. *The Scientific World Journal* 2013: 704504. https://doi.org/10.1155/2013/704504

16 H.E. Tahir, M. Arslan, G.K. Mahunu, J. Shi, X. Zou, M A A Gasmalla, and A.A. Mariod. (2019). Data Fusion approach improves the prediction of single phenolic compounds in honey: A study of NIR and Raman spectroscopies. *eFood* 1, 2: 173–180. https://doi.org/10.2991/efood.k.191018.001

17 E. Borràs, J. Ferré, R. Boqué, M. Mestres, L. Aceña, and O. Busto. (2015). Data fusion methodologies for food and beverage authentication and quality assessment – A review. *Analytica Chimica Acta* 891: 1–14. https://doi.org/10.1016/j.aca.2015.04.042

18 M. Cristina Márquez I. López, I. Ruisánchez, and M. Pilar Callao. (2016). FT-Raman and NIR spectroscopy data fusion strategy for multivariate qualitative analysis of food fraud. *Talanta* 161: 80–86. https://doi.org/10.1016/j.talanta.2016.08.003

19 F. Desta and M. Buxton. (2018). Automation in sensing and raw material characterization – a conceptual framework. 1501–1506. https://doi.org/10.1109/IROS.2018.8593774

20 J. Moros and J.J. Laserna. (2011). New Raman–laser-induced breakdown spectroscopy identity of explosives using parametric data fusion on an integrated sensing platform. *Analytical Chemistry* 83, 16: 6275–6285.

21 J. Moros and J. Javier Laserna. (2015). Unveiling the identity of distant targets through advanced Raman-laser-induced breakdown spectroscopy data fusion strategies. *Talanta* 134: 627–639.

22 J-H Yang and J.J. Yoh. (2018). Reconstruction of chemical fingerprints from an individual's time-delayed, overlapped fingerprints via laser-induced breakdown spectrometry (LIBS) and Raman spectroscopy. *Microchemical Journal* 139: 386–393. https://doi.org/10.1016/j.microc.2018.03.027

23 M. Hoehse, A. Paul, I. Gornushkin, and U. Panne. (2012). Multivariate classification of pigments and inks using combined Raman spectroscopy and LIBS. *Analytical and Bioanalytical Chemistry* 402, 4: 1443–1450.

24 D. Prochazka, M. Mazura, O. Samek, K. Rebrošová, P. Pořízka, J. Klus, P. Prochazková, J. Novotný, K. Novotný, and J. Kaiser. (2018). Combination of laser-induced breakdown spectroscopy and Raman spectroscopy for multivariate classification of bacteria. *Spectrochimica Acta Part B: Atomic Spectroscopy* 139: 6–12. https://doi.org/10.1016/j.sab.2017.11.004

25 M. Zhao, M. Markiewicz-Keszycka, R.J. Beattie, M.P. Casado-Gavalda, X. Cama-Moncunill, C.P. O'Donnell, P.J. Cullen, and C. Sullivan. (2020). Quantification of calcium in infant formula using laser-induced breakdown spectroscopy (LIBS), Fourier transform mid-infrared (FT-IR) and Raman spectroscopy combined with chemometrics including data fusion. *Food Chemistry* 320. https://doi.org/10.1016/j.foodchem.2020.126639

26 R.D. Waterbury, A.R. Ford, J.B. Rose, and E.L. Dottery. (2009). Results of a UV TEPS-Raman energetic detection system (TREDS-2) for standoff detection. In *Chemical, Biological, Radiological, Nuclear, and Explosives (CBRNE) Sensing X*, 327–332. https://doi.org/10.1117/12.819211

27 A. Ford, R. Waterbury, J. Rose, K. Pohl, M. Eisterhold, T. Thorn, K. Lee, and E. Dottery. (2010). Extension of a standoff explosive detection system to CBRN threats. In *Chemical, Biological, Radiological, Nuclear, and Explosives (CBRNE) Sensing XI*, 275–284. https://doi.org/10.1117/12.849815

28 S. Botti, M. Carpanese, L. Cantarini, G. Giubileo, V. Lazic, S. Jovicevic, A. Palucci, and A. Puiu. (2010). Trace detection of explosive compounds by different laser-based techniques at the ENEA Laboratories. In *Chemical, Biological, Radiological, Nuclear, and Explosives (CBRNE) Sensing XI*, 193–204. https://doi.org/10.1117/12.850722

29 D.W. Pinkham, J.R. Bonick, and M.D. Woodka. (2012). Feature optimization in chemometric algorithms for explosives detection. In *Detection and Sensing of Mines, Explosive Objects, and Obscured Targets XVII*, 83571K. https://doi.org/10.1117/12.923387

30 E. Gibbons, R. Léveillé, and K. Berlo. (2020). Data fusion of laser-induced breakdown and Raman spectroscopies: Enhancing clay mineral identification. *Spectrochimica Acta Part B: Atomic Spectroscopy* 170: 105905.

31 J. Moros, M.M. ElFaham, and J. Javier Laserna. (2018). Dual-spectroscopy platform for the surveillance of mars mineralogy using a decisions fusion architecture on simultaneous LIBS-Raman data. *Analytical Chemistry* 90, 3: 2079–2087.

32 K. Rammelkamp, S. Schröder, S. Kubitza, D. Vogt, S. Frohmann. 2018. Peder Bagge Hansen, Ute Böttger, Franziska Hanke, and Heinz-Wilhelm Hübers. LIBS and Raman Data Fusion for in-situ Planetary Exploration.

33 J.A. Manrique-Martinez, G. Lopez-Reyes, A. Alvarez-Perez, T. Bozic, M. Veneranda, A. Sanz-Arranz, J. Saiz, J. Medina-Garcia, and F. Rull-Perez. (2020). Evaluation of multivariate analyses and data fusion between Raman and laser-induced breakdown spectroscopy in binary mixtures and its potential for solar system exploration. *Journal of Raman Spectroscopy* 51, 9: 1702–1717. https://doi.org/10.1002/jrs.5819

34 M. Konstantinidis, K. Cote, E.A. Lalla, G. Zhang, M.G. Daly, X. Gao, and P. Dietrich. (2019). On the application of a novel linear mixture model on laser-induced breakdown spectroscopy: Implications for Mars. *Journal of Chemometrics* 33, 10: e3174. https://doi.org/10.1002/cem.3174

35 R.C. Wiens, S. Maurice, J. Lasue, O. Forni, R.B. Anderson, S. Clegg, S. Bender, D. Blaney, B.L. Barraclough, A. Cousin, L. Deflores, D. Delapp, M.D. Dyar, C. Fabre, O. Gasnault, N. Lanza, J. Mazoyer, N. Melikechi, P-Y Meslin, H. Newsom, A. Ollila, R. Perez, R.L. Tokar, and D. Vaniman. (2013). Pre-flight calibration and initial data processing for the ChemCam laser-induced breakdown spectroscopy instrument on the Mars Science Laboratory rover. *Spectrochimica Acta Part B: Atomic Spectroscopy* 82: 1–27. https://doi.org/10.1016/j.sab.2013.02.003

36 D. Di Genova, D. Morgavi, K-U Hess, D.R. Neuville, N. Borovkov, D. Perugini, and D.B. Dingwell. (2015). Approximate chemical analysis of volcanic glasses using Raman spectroscopy. *Journal of Raman Spectroscopy* 46, 12: 1235–1244. https://doi.org/10.1002/jrs.4751

37 R.B. Anderson, S.M. Clegg, J. Frydenvang, R.C. Wiens, S. McLennan, R.V. Morris, B. Ehlmann, and M. Darby Dyar. (2017). Improved accuracy in quantitative laser-induced breakdown spectroscopy using sub-models. *Spectrochimica Acta Part B: Atomic Spectroscopy* 129: 49–57. https://doi.org/10.1016/j.sab.2016.12.002

Part IV

Quantitative Analysis

11

Univariate Linear Methods

Stefano Legnaioli, Asia Botto, Beatrice Campanella, Francesco Poggialini, Simona Raneri, and Vincenzo Palleschi

Applied and Laser Spectroscopy Laboratory, Institute of Chemistry of Organometallic Compounds, Research Area of the National Research Council, Pisa, Italy

11.1 Standards

In a univariate calibration approach, the concentration of an analyte in an unknown sample is determined by comparing a single spectral signal (in Laser-Induced Breakdown Spectroscopy [LIBS], the intensity of an emission line of the element of interest) with the one measured on one or more samples of known and different concentration. When more than one standard is used, a calibration curve is built, describing the mathematical relation between the experimental signal and the analyte concentration. The analyte concentration in the calibration samples can be obtained either from the certification given by the provider of the standards (e.g. the National Institute of Standards and Technology (NIST) in the United States or the Bundesanstalt füer Materialforschung und -prüfung (BAM) in Germany) or could be determined using other analytical techniques, such as X-ray fluorescence (XRF), inductively coupled plasma-optical emission spectroscopy (ICP-OES), inductively coupled plasma-mass spectrometry (ICP-MS), etc.

It should be noted that LIBS is mostly applied on solid samples, usually without any preliminary treatment. This is crucial to understand the peculiar differences between LIBS and other, more traditional techniques. For example, the use of diluted standard solutions is precluded. This method, which is common in conventional wet chemistry, allows to obtain many substandards at progressively lower concentration from a single solution of known concentration. Some attempts in this direction have involved geological [1] and biological materials [2], but, in general, the difficulty of obtaining a suitable number of standards for building a satisfactory calibration curve has been one of the most limiting factors in the development of LIBS as a laboratory analytical technique. It is still not too uncommon to find calibration curves built with very few standards for the quantification of given elements in unknown samples.

However, as correctly remarked in [3], having a large number of standards for a calibration curve is not enough to guarantee the trueness of the analysis. It is essential to know the parameters of the calibration curve with good precision in a range that comprises the minimum and maximum concentration of the analyte in the unknown samples. While this information is sometimes available from previous knowledge about the nature of the samples to be analyzed, in general, it is not easy to select standards of optimal composition. In applications where the unknown concentration of the analyte is expected to vary substantially, the choice of the standards for building the calibration curve can be even more challenging.

11.2 Matrix Effect

The main problem associated with the univariate calibration methods is the possible failure of the spectrochemical equivalent of the Beer's law, which states that the intensity of the signal emitted by the analyte is proportional to its concentration:

$$I = kC \qquad (11.1)$$

The definition of LIBS signal intensity is more or less agreed among the researchers working in LIBS, and it is identified as either the peak intensity of an emission line or its integrated intensity after removing the background given by the continuum emission. On the other hand, the definition of the concentration C in Eq. (11.1) is not unambiguous.

Moreover, in LIBS, there are several effects which can make Eq. (11.1) not valid. The intensity of a given emission line depends on the total number of emitters in the corresponding transition [4]. The number of atoms of the analyte emitting at a given wavelength can be considered proportional to the concentration of the same analyte only when the sampled mass remains the same in all the measurements done on the different standards used for building the calibration curve. However, this is not always the case. In some situations, the physical and chemical properties of the sample may vary, depending on the changes in concentration of the analyte. This is one of the aspects of the so-called "matrix effect," which is particularly prevalent in LIBS and may affect substantially the building of univariate calibration curves.

As discussed by Borisov et al. [5] in the framework of laser ablation ICP-MS, a typical example of the effect of the variation of the ablated mass with the composition of the standards can be observed in brass (Cu–Zn) alloys. The plot of the Cu I line intensity at 324.8 nm against the certified concentration of Cu in the samples shows an unexpected decreasing trend of the line intensity with the increase of the concentration of copper in the standards. At the same time, the calibration curve of zinc shows a nonlinear increase of the LIBS signal (334.5 nm) with the increase of Zn concentration (see Figure 11.1)

The authors explained this effect with a change of sample reflectivity at the Nd:YAG wavelength of 1064 nm. With the increase of Zn concentration, the reflectivity of the brass samples decreases, more laser energy is absorbed by the sample and a larger mass is ablated. Since brass alloys considered were Cu–Zn binary alloys, the increase in Cu concentration corresponds to a decrease in Zn concentration, causing a decrease in the ablated mass that explains the lower observed Cu I line intensity. The same consideration explains why the Zn line intensity does not increase linearly with the increase in Zn concentration.

This extreme matrix effect is typical of LIBS analysis.

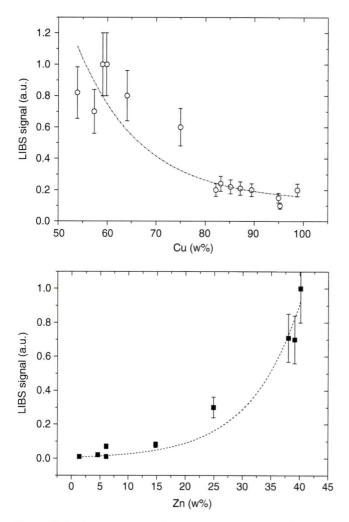

Figure 11.1 Calibration curve for Cu (left) and Zn (right). The calibration samples are Cu–Zn binary alloys.

The changes in the physical–chemical properties of the samples used for the calibration are normally negligible when working with small amounts of analyte diluted in a liquid matrix, as is the case of ICP-OES/ICP-MS applications. These changes, however, can become extremely important in LIBS analysis of solid samples, where the composition of the standards can change substantially.

11.3 Normalization

A common method used in LIBS to compensate for the variations in the ablated mass and reduce the effect of the fluctuations of the laser energy is the normalization of the measured line intensities.

This normalization can be performed using global spectral parameters (i.e. background intensity, total integrated spectral intensity) or the intensity of another line of an element whose concentration is known can be used as a reference.

For example, in the analysis of trace elements in a known matrix, the analyte signal can be normalized using the emission line of a major component of the matrix itself, whose concentration is assumed to be constant.

In the case of binary alloys, the variations in the concentration of one element directly affect the concentration of the other. In this case, it can be convenient to build a calibration curve for the ratio of the concentrations of the two elements as a function of the ratio of the intensities of the corresponding spectral lines.

In the case of the previously discussed Cu–Zn binary alloys, the calibration curve for the concentration ratio $\frac{C_{Zn}}{C_{Cu}}$, built considering the ratio of the intensities of the Zn line at 334.5 nm and the Cu line at 324.8 nm shows a good linearity (Figure 11.2). Once determined the concentration ratio on the unknown samples, the corresponding concentrations of Cu and Zn can be obtained from the relations:

$$C_{Cu} + C_{Zn} = 100\%$$
$$C_{Cu} = 100\% - C_{Cu}\frac{C_{Zn}}{C_{Cu}} = \frac{100\%}{1 + (C_{Zn}/C_{Cu})} \quad (11.2)$$

In most cases, the situation is more complicated. A good compromise is normalizing the line intensity with the integral intensity of the whole LIBS spectrum. The integral intensity is not much affected by the change in the intensity of the emission lines produced by the variation of the analyte concentration and can be a good indication of the laser energy actually released at the sample surface. This kind of normalization also partially compensates for the fluctuations of the energy of the laser from shot to shot.

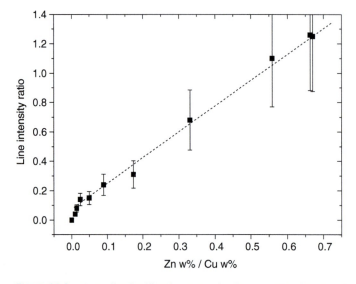

Figure 11.2 Normalized calibration curve for the concentration ratio Zn/Cu.

Another kind of normalization has been proposed, which correlates the acoustic emission associated with the ablation process with the energy released on the sample surface. However, the intensity of the sound depends on the energy of the shockwave generated during the ablation process which, in turn, depends on the energy of the laser [6]. Consequently, an estimation of the ablated mass based on the acoustic signal is not very precise.

Panne et al. [7] proposed the normalization of the spectral line intensities taking into account the changes in the plasma electron temperature that could be produced by fluctuations of the laser energy or changes in the physical–chemical characteristics of the samples. For an ideal LIBS plasma in local thermal equilibrium (LTE) [8], the line integral intensity is proportional in first approximation on the number concentration of the atoms/ions populating the upper level of the transition considered:

$$I_0 = F n^a g_k A_{ki} \frac{e^{-(E_k/k_B T)}}{U^a(T)}$$

$$a = \begin{cases} I & \text{for neutral lines} \\ II & \text{for ionized lines} \end{cases}$$

(11.3)

where A_{ki} is the transition probability, g_k is the degeneracy, and E_k is the energy of the upper level of the transition, $U(T)$ is the partition function of the species (weakly depending on the plasma electron temperature T), k_B is the Boltzmann constant, and F is a proportionality parameter correlated with the ablated mass and the light collection efficiency of the system.

The electron temperature is influenced also by the physical–chemical properties of the sample, even assuming that the ablated mass would not be influenced. The Panne approach, which is considered a precursor of the calibration-free LIBS method [9], has been recently reintroduced by Li et al. [10].

In building a calibration curve, one should also consider that the intensity of a LIBS line depends on the number concentration of the emitting species. Each spectral species of the same element shows a characteristic spectrum. The line intensities depend on the balance between the number densities of consecutive ionization stages, represented for a LIBS plasma in LTE by the Saha–Eggert relation [4]:

$$\frac{n^{II}}{n^I} = \frac{2}{n_e} \left(\frac{h^2}{2\pi m_e k_B T} \right)^{-3/2} \frac{U^{II}(T)}{U^I(T)} e^{-E_{ion}/k_B T}$$

(11.4)

where n_e is the electron number density, h is the Planck constant, m_e is the mass of the electron, k_B is the Boltzmann constant, $U(T)$ is the partition function of the corresponding specie, E_{ion} is the ionization energy of the element, and T is the electron temperature.

The ratio between the number concentration of neutral and singly ionized species of the same element, which typically are the only ones represented in LIBS plasmas, depends on the plasma temperature and electron number density. The electron number density, as the electron temperature, is a global parameter of the plasma and depends on the laser energy, as well as by the variation of sample composition and physical properties. In a recent paper [11], Castellon et al. have discussed the separate effect of the variation of plasma electron temperature and number density on the calibration curves for mercury in dental amalgams. The authors found that the most relevant effect was the variation in electron temperature between measurements of the different standard samples.

The choice of the proper normalization of the LIBS signal is important for improving the quality of the analytical results that can be obtained. The analytical software developed by Applied Spectra (United States) for their J200 LIBS instruments provides 12 different types of calibration (e.g. the integral value of the spectrum, its norm, its maximum, or with the intensity of a carbon reference line, etc.) [12]. However, having at disposal such a wide choice of possible normalization methods does not mean that one must use all of them in the attempt to find the best approach. The normalization of the spectra with the intensity of a carbon line, for example, is expected to work well for organic materials, where the matrix is mostly composed by carbon but would not work properly on samples where carbon is present in traces or when its concentration is unknown.

In the end, the choice of the normalization method should take into consideration the characteristics of the spectrum and the chemical and physical properties of the samples.

11.4 Linear vs. Nonlinear Calibration Curves

When using a univariate calibration approach, a linear calibration curve is desirable:

$$I = aC + b \tag{11.5}$$

A linear relation like that in Eq. (11.5) provides a simple analytical relation between the concentration of the analyte in unknown samples and the measurement of the signal intensity. In addition, the uncertainty of the measurement can be obtained by inverting Eq. (11.5):

$$C_{unk} = \frac{I_{unk} - b}{a} \pm \left(\frac{I_{unk} - b}{a^2}\Delta a + \frac{\Delta b}{a}\right) \tag{11.6}$$

A linear calibration curve also allows for a simple definition of the limit of detection (the minimum concentration of the analyte which can be detected) and the limit of quantification (LOQ) (the minimum concentration of the analyte which can be quantified).

Before discussing the numerical methods involved in the use of a univariate calibration approach, a few points should be clarified.

In most of the cases, the calibration curves used in LIBS are not linear. This is usually due to the phenomenon of self-absorption [13], which occurs when the photons emitted in the observed transition are re-absorbed by other atoms or ions of the same species before escaping from the plasma. The probability of re-absorption increases with the concentration of the analyte and the population of the lower level of the transition. In case of resonant lines (i.e. lines corresponding to transitions with the energy of the lower level equal or very close to zero), the self-absorption effect is typically high in LIBS and the nonlinearity of the corresponding calibration curves can be very strong. For this reason, it is advisable to avoid the use of resonant emission lines for building LIBS calibration curves, if not for the determination of elements present in traces in the samples.

However, there could be situations in which the use of resonant lines for calibration is unavoidable. In these situations, it might be useful to compensate the effects of self-absorption by considering the equation which can be derived for the propagation of radiation in an emitting and absorbing medium, which in the framework of the LTE approximation can be schematized as a plasma cylinder of length l and unit section, at a temperature T. The line integral intensity of the radiation emitted along the line of sight of the detector in a given transition can be written as [13]:

$$I = I_0 (SA)^{0.46} \tag{11.7}$$

where I_0 is given by Eq. (11.3) and SA is the self-absorption parameter, which depends on the so-called columnar density of the analyte through the optical depth at the peak of the line τ:

$$SA = \frac{1 - e^{-\tau}}{\tau} \tag{11.8}$$

with

$$\tau = \frac{1}{\Delta\lambda_0} \frac{\lambda_0^4}{8\pi^2 c} A_{ki} g_k \frac{n^a}{U^a(T)} e^{-E_i/k_B T} l \tag{11.9}$$

where λ_0 is the central wavelength of the transition, l is the optical path length viewed by the detector, c is the speed of light, and E_i is the energy of the lower level of the transition. $\Delta\lambda_0$ is the Full Width at Half-Maximum (FWHM) of the emission line for an optically thin ($\tau \ll 1$, SA $\simeq 1$) plasma. In LIBS plasma conditions, the emission line profile can be represented as a Lorentzian curve:

$$L(\lambda) = \frac{1}{\pi} \frac{(\Delta\lambda_0/2)}{(\lambda - \lambda_0)^2 + (\Delta\lambda_0/2)^2} \tag{11.10}$$

and the FWHM is typically dominated by the Stark effect [14]:

$$\Delta\lambda_0 = \omega_S n_e \tag{11.11}$$

The parameter ω_S is called Stark coefficient and depends weakly on the plasma electron temperature.

Assuming that the parameters of the plasma do not change during the analysis of the different reference samples, it can be seen from Eq. (11.8) that the calibration curve is linear with the species concentration when $\tau \ll 1$ (SA $\simeq 1$), while in the opposite situation of $\tau \gg 1$, the self-absorption parameter varies as:

$$SA \cong \frac{1}{\tau} \tag{11.12}$$

Since τ depends linearly on the species concentration (Eq. 11.9), at high concentrations the calibration changes from a linear to a square root dependence between the measured intensity and the number concentration of the emitting species:

$$I = I_0 \tau^{0.46} \cong K (n^a)^{0.5} \tag{11.13}$$

The relation (Eq. 11.13) allows to check the degree of self-absorption in the measurements. The analyte concentration at which the linear regime transforms in the square root one can be determined by studying the calibration curve in a log–log plot (curve of growth) [15]. The intersection between the linear fitting of the curve at low concentration, with slope 1, and the linear fitting of the curve at high concentration, with slope ½, gives the critical concentration of the species above which the self-absorption effects must be taken into account.

Eq. (11.7)–(11.11) also provide a way for experimentally determining the self-absorption parameter and compensating for this effect, recovering the linearity of the calibration curve even for concentrations larger than the critical one, as determined by the curve of growth method.

The key assumptions for calculating the self-absorption parameter from the LIBS measurements are considering the plasma in LTE [8] (the radiation propagation equations have been derived under this assumption) and homogeneous. Under these assumptions, the plasma temperature can be determined from the corresponding LIBS spectra using the Boltzmann or Saha–Boltzmann plot method, and the electron number density can be obtained from the measurement of the FWHM of an optically thin line, whose Stark coefficient is known. The determination of the plasma electron number density can also be obtained from the measurement of the Balmer alpha line of hydrogen (655 nm) [16]. The line is generally not self-absorbed, and its FWHM is easy to measure even with relatively low-resolution spectrometers. On the other hand, the relation between FWHM and electron number density is more complex and debated, with respect to the simple behavior described in Eq. (11.8). In spite of these difficulties, the measurement of the electron number density of the plasma through the hydrogen Balmer alpha line is currently considered as very convenient and therefore this method is widely used in LIBS analytical applications.

If the plasma electron temperature and number density are known from the LIBS spectrum, the self-absorption parameter can be determined using the relation obtained in 2005 by the Pisa group [17], which demonstrated that the ratio between the measured FWHM of an emission line $\Delta\lambda$ and its optically thin limit $\Delta\lambda_0$ (given by Eq. (11.11)) is related to the self-absorption parameter by the formula:

$$SA = \left(\frac{\Delta\lambda}{w_s n_e}\right)^{-1.85} \tag{11.14}$$

Eq. (11.14) can be thus used for compensating the effect of self-absorption in the high-concentration zone of the calibration curve, recovering a linear dependence between integrated line intensity and species concentration.

It is worth noting that a nonlinear calibration curve can also be obtained in case of optically thin lines (SA $\simeq$ 1), if the concentration of the analyte in the reference samples is not calculated correctly.

If the self-absorption effect is not considered, the LIBS signal is proportional to the number density of the emitting species. However, in most of the cases, the composition of the reference samples is given in weight %. The difference is not negligible, if the sample is composed by elements with very different atomic weight. For example, in a binary alloy composed by Cu (atomic weight 63.5 u) and Sn (atomic weight 118.7 u), an alloy composed 50% in weight of copper and 50% in weight of tin corresponds to a 65–35 alloy when the number concentration of the emitting atoms is considered.

The effect of using number concentrations or weight concentrations on the linearity of the calibration curve is shown in Figure 11.3. The results of an ideal LIBS measurement,

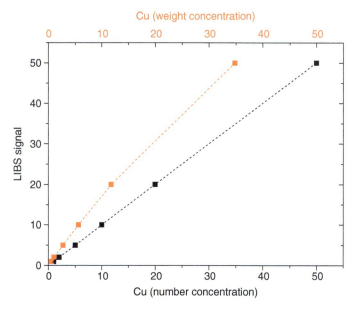

Figure 11.3 Apparent nonlinearity of the calibration curve produced by the use of a wrong scale (concentration in weight, red, against numerical concentration, black).

reported in the correct scale of number concentration, are linearly correlated with the analyte concentration, while the same results produce a nonlinear curve when plotted using the weight concentrations in the horizontal axis.

11.5 Figures of Merit of a Calibration Curve

In LIBS, there are typically several emission lines corresponding to the same species. One of the problems of building a univariate calibration curve is choosing the best line to be used for analytical purposes. In 1998, the International Union of Pure and Applied Chemistry (IUPAC) released some guidelines for calibration in analytical chemistry [18]. Assuming that a suitable number of standards (at least 10) is available, and a suitable number of independent measurements of the spectral signal is obtained (at least 10), the parameters of the best-fitting linear curve (slope a and intercept b) of the measured signals are found by minimizing the deviations of the model from the experimental data. In many spectrochemical applications, the best-fitting parameters that define the linear calibration curve are obtained through a straightforward minimization of the sum of the squared errors (least squares fitting):

$$\min_{a,b} \sum (I_i - aC_i - b)^2 \qquad (11.15)$$

As a result of this minimization, the parameters of the best-fitting calibration curve are obtained as:

$$a = \frac{\sum(C_i - \overline{C})(I_i - \overline{I})}{\sum(C_i - \overline{C})} \qquad (11.16)$$

$$b = \overline{I} - a\,\overline{C}$$

where $\overline{C}$ and $\overline{I}$ are the average values of concentration and signal, respectively, on the whole set of calibration standards.

To obtain the above results, some assumptions on the nature of the experimental data are implicitly accepted. It is assumed that the 10 or more independent measurements done on the single standards would give results described by a normal distribution, with the same standard deviation σ for all the samples (homoscedasticity). Additionally, the uncertainty on the concentrations C_i of the standards is assumed to be negligible, with respect to the uncertainty on the experimentally measured intensities I_i.

Under these premises, the uncertainty on the best-fitting curve parameters can be easily calculated as:

$$\Delta a = \sigma \sqrt{\frac{1}{\sum(C_i - \overline{C})^2}}$$

$$\Delta b = \sigma \sqrt{\left(\frac{1}{n} + \frac{\overline{C}^2}{\sum(C_i - \overline{C})^2}\right)} \qquad (11.17)$$

where n is the number of points in the calibration curve.

The parameters obtained from the minimization of Eq. (11.15) are currently used in LIBS analysis for building a linear calibration curve. LIBS, however, is different from the other spectrochemical applications, and the assumptions enumerated above for the calculation of the a and b coefficients might not be satisfied. In particular, the uncertainty on the reference samples' composition may be not negligible with respect to other experimental uncertainties, especially when the reference standards are characterized using other laboratory techniques and the experimental error in the measurement must be considered. Most importantly, the assumption of homoscedasticity of the experimental data is not satisfied in LIBS.

Modern detectors measure the light intensity in photon counts whose statistics is described by a Poisson distribution (which coincides with the normal distribution when the number of counts is large) with a standard deviation which goes as the square root of the counts (i.e. the square root of the signal intensity). An example is given in Figure 11.4.

Since the uncertainty of the measurements increases with the measured intensity, the best linear fit parameters must be determined by minimizing the weighted square differences between the experimental data and the linear model:

$$\min_{a,b} \sum \frac{(I_i - aC_i - b)^2}{\sigma_i^2} \qquad (11.18)$$

The function to be minimized is called $\chi^2 = \sum \frac{(I_i - aC_i - b)^2}{\sigma_i^2}$, and the procedure for determining the optimum a and b parameters is called method of minimum chi-square.

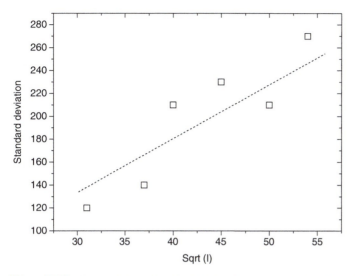

Figure 11.4 Standard deviation of a LIBS line intensity, as a function of the square root of the intensity.

With respect to Eq. (11.15), the points corresponding to high concentrations of the analyte (and higher uncertainty) are proportionally less important for the determination of the best linear fitting parameters which, in this case, can be calculated as:

$$a = \frac{\sum(1/\sigma_i^2)\sum(C_i I_i/\sigma_i^2) - \sum(C_i/\sigma_i^2)\sum(I_i/\sigma_i^2)}{\sum(1/\sigma_i^2)\sum(C_i^2/\sigma_i^2) - (\sum(C_i/\sigma_i^2))^2}$$
$$b = \frac{\sum(C_i^2/\sigma_i^2)\sum(I_i/\sigma_i^2) - \sum(C_i/\sigma_i^2)\sum(C_i I_i/\sigma_i^2)}{\sum(1/\sigma_i^2)\sum(C_i^2/\sigma_i^2) - (\sum(C_i/\sigma_i^2))^2}$$
(11.19)

The expressions in Eq. (11.19) coincide with Eq. (11.15) when $\sigma_i = \sigma$ for all the samples. Similarly, the uncertainty on the best-fit parameters can be calculated as:

$$\Delta a = \sqrt{\frac{\sum(1/\sigma_i^2)}{\sum(1/\sigma_i^2)\sum(C_i^2/\sigma_i^2) - (\sum(C_i/\sigma_i^2))^2}}$$
$$\Delta b = \sqrt{\frac{\sum(C_i^2/\sigma_i^2)}{\sum(1/\sigma_i^2)\sum(C_i^2/\sigma_i^2) - (\sum(C_i/\sigma_i^2))^2}}$$
(11.20)

A good control of the reliability of the calibration curve obtained through the minimization of the χ^2 is to check its value at the minimum. If the individual σ_i of the measurements are correctly evaluated, one would expect a value of the order of the number of degrees of freedom of the system (i.e. $n - 2$, where n is the number of reference standards used for building the calibration curve). If the value is substantially higher than $n - 2$, it is probable that the uncertainty of the measurements was underestimated. Similarly, if the value of the χ^2 at its minimum is too low, it is a sign of a probable overestimation of the measurement uncertainties.

11.5.1 Coefficient of Determination

The parameter R^2, defined as:

$$R^2 = 1 - \frac{\sum (I_i - a\,C_i - b)^2}{\sum (I_i - \bar{I})^2} \qquad (11.21)$$

is called the coefficient of determination of the linear regression. This parameter assumes values between 0 (the linear model does not represent the experimental data) and 1 (perfect coincidence between the model and the data). This parameter can also be interpreted as the square of the correlation function between the concentrations and the signals:

$$R^2 = \left(\frac{\sum (I_i - \bar{I})(C_i - \bar{\bar{C}})}{\sum (I_i - \bar{I}) \sum (C_i - \bar{\bar{C}})} \right)^2 \qquad (11.22)$$

A good linear calibration curve should be characterized by a value of R as close to 1 as possible. On the other hand, the value of R^2 is not very sensitive and it is difficult assessing the quality of a calibration curve based on small variations of this parameter.

11.5.2 Root Mean Squared Error of Calibration

The root mean squared error of calibration (RMSEC) is a measure of the capability of the linear calibration curve in representing the variability of the experimental data. It is defined as:

$$\text{RMSEC} = \sqrt{(C_i - C_i^{\text{ext}})^2} \qquad (11.23)$$

where C_i^{ext} is the value of the concentration of the analyte for the i-th sample, as estimated from the calibration curve. The RMSEC can be an estimation of the uncertainty on the results that can be obtained on unknown samples using the same calibration curve. However, in LIBS measurements, this uncertainty is likely to be overestimated at low concentrations and underestimated at high concentrations.

In this respect, is worth reminding that the best analytical results in LIBS are obtained using calibration curves that exactly bracket the expected range of concentration of the analyte in the unknown samples. If not needed, the use of reference samples characterized by a high concentration of the analyte should be avoided. This prevents possible self-absorption effects or larger absolute variability of the measured intensities. In addition, unnecessary calibration points at high concentration introduce larger uncertainties on the calibration curve parameters and a larger error in the determination of the analyte concentration in the unknown samples (see Figure 11.5).

11.5.3 Limit of Detection

The limit of detection represents the lower concentration of the analyte that can be detected using the corresponding calibration curve. In 1975, the IUPAC proposed an equation for the calculation of the Limit of Detection (LOD) [19]:

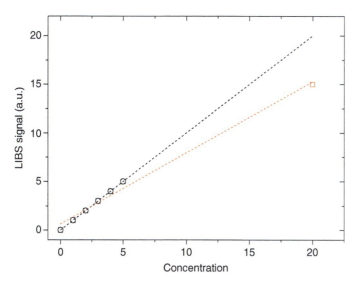

Figure 11.5 Schematic representation of the effect on the calibration curve of an isolated calibration point at high concentration. Red and black dashed lines are the best linear fits of the experimental data with and without the high concentration point.

$$\text{LOD} = \frac{3\sigma}{a} \tag{11.24}$$

where a is the slope of the calibration curve, and σ is the standard deviation of the signal measured on a blank sample, i.e. a sample in which the analyte is not present.

However, the implicit assumptions behind the derivation of Eq. (11.24) are not fulfilled in LIBS experiments.

The IUPAC definition works well with calibration curves built with many standards, usually obtained by progressive dilution of the analyte. For these calibration curves, it is safe to assume that the uncertainty on the slope and intercept is negligible with respect to other experimental uncertainties.

In LIBS, it is not unusual to find calibration curves built with four or five reference samples. Moreover, measurements on blank samples are rarely feasible.

The limits of the IUPAC definitions were discussed by Long and Winefordner in a paper published in 1983 [20]. The starting point of the Long and Winefordner criticism of the IUPAC definition for the LOD is the formula, equivalent to Eq. (11.24), for the minimum signal that can be attributed to the presence of the analyte in a sample with reasonable certainty. Under the assumption that the statistical distribution of the signal measured on the blank can be represented as a Gaussian with standard deviation σ, imposing the condition $I_{min} = 3\sigma$ means, for the properties of the Gaussian distribution, that the corresponding sample has 95% probability of being different from the blank (see Figure 11.6).

In terms of the calibration curve, $I_{min} = a\, C_{min}$ and this correspondence brings back to the IUPAC definition of LOD. Note that the intercept b of the calibration curve can be different from zero, but this is not a problem because this constant value can be subtracted from the signal by redefining it as $\bar{I} = I - b$.

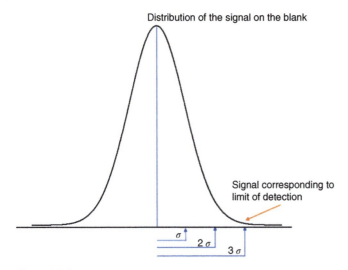

Figure 11.6 Schematic representation of the distribution of the signal on the blank, and the signal corresponding to the limit of detection (IUPAC definition).

However, redefining the signal for taking into account the intercept of the calibration curve implies that the uncertainty on the signal of the blank also depends on the uncertainty on the b parameter (Eq. (11.20)). In LIBS experiments, the uncertainty on the fitting parameters can be larger than the fluctuations of the signal on the blank (or their estimate from the fluctuations of the background, if a proper blank is not available). The authors proposed a more accurate expression for the LOD of a given element which takes into account the uncertainty on the parameters of the calibration curve, in the form:

$$\text{LOD} = \frac{3}{a}\sqrt{\sigma^2 + (\Delta b)^2 + \left(\frac{b}{a}\Delta a\right)^2} \tag{11.25}$$

Eq. (11.25) coincides with the IUPAC formula when the uncertainty on the calibration curve parameters is negligible. However, in most of the LIBS analytical applications, using Eq. (11.24) instead of Eq. (11.25) for calculating the LOD results in a severe underestimation of this parameter.

The LOD for a given analyte corresponds to the minimum concentration that can be detected in a system described by a given calibration curve. The minimum concentration of the analyte that can be quantified is typically much higher. Conventionally, this LOQ is three times the LOD. This is because the statistical distribution of the signal on the blank and the signal on the sample must be well separated for measuring the concentration of the analyte (see Figure 11.7).

In some LIBS papers, it is not uncommon to see a confusion between these two limits. Sometimes, emission lines are identified in the spectra at concentrations lower than the LOD, and/or an estimate of the concentration is given for lines barely above the LOD. Moreover, due to an incorrect use of the IUPAC definition, it is also not uncommon the evaluation of very low LODs from calibration curves but using reference samples at

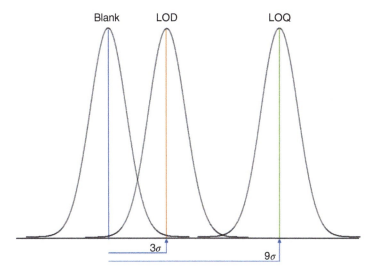

Figure 11.7 Schematic representation of the distribution of the signals corresponding to the blank, the limit of detection and the limit of quantification.

concentrations hundreds of times higher than the LOD. As a rule of thumb, considering the definition of LOD and LOQ, the minimum concentration of the reference sample used in the calibration curve should be around the LOQ, i.e. slightly higher than three times the LOD.

11.6 Inverse Calibration

Safi et al. in [21] briefly discussed the apparent incongruency between the direct relation form typically used for linear calibration (Eq. (11.5)) and the inverse relation:

$$C = f(I) \qquad (11.26)$$

which is commonly assumed when the relation between concentration and signal is not linear, and/or a multivariate approach is used.

The authors remarked that in both cases, the definition of the optimum calibration curve/surface is obtained from the minimization of an objective function given by the sum of the quadratic deviations of the predictions of the model with respect to the known values. The main difference is that in univariate linear calibration, this minimization can be performed analytically, and the analytical calculation is straightforward assuming the uncertainty on the independent variable (the concentration) is negligible with respect to the dependent variable (the LIBS signal I).

Duponchiel et al. [22] pushed forward this concept, remarking in their paper that inverse calibration is used in several spectrochemical applications not just because of the convenience of having already a relation that would give the concentration of the unknown sample from the measurement of the corresponding spectral signal, but because it can be demonstrated, from a mathematical point of view, that the predictions of an inverse linear calibration are statistically better than the ones obtained by the direct calibration approach.

When applied to LIBS, however, the use of inverse calibration does not present any practical advantage. The main reason of the failure of the inverse calibration approach in LIBS is due to the non-homoscedasticity of the LIBS calibration data. When the calibration curve is inverted, the best-fit parameters of the calibration curve are calculated in the framework of a least squares approach, being the uncertainty on the concentration of the standards typically small and approximatively the same for all the standards. In the few cases in which the inverse calibration approach could be useful (large indetermination on the reference samples composition and/or very reproducible LIBS signals), the Pisa group [23] suggested the use of a symmetrical calibration approach [24] instead, in which the errors in both the concentration of the analyte in the standards and the uncertainty on the corresponding LIBS signals are fully taken into account, in the framework of a χ^2 minimization.

The use of symmetrical approaches guarantees better performance with respect to both direct and indirect calibration and should be preferred whenever there is some doubt on the validity of the direct model. However, the direct calibration model with χ^2 minimization remains the best approach in terms of simplicity and verifiability in the large majority of LIBS applications.

11.7 Conclusion

In this chapter, we have discussed the chemometric aspects associated with the use of linear univariate calibration curves for LIBS analysis. Even though this approach is commonly used in many other spectroanalytical techniques, the peculiarities of LIBS impose a deep reconsideration of many concepts (e.g. the definition of limit of detection) which seem consolidated by decades of use. These concepts must be carefully readapted to LIBS to avoid errors in the assessment of the performance of the technique and consequent errors in the quantitative analysis.

References

1 Rifai, K., Laflamme, M., Constantin, M., Vidal, F., Sabsabi, M., Blouin, A., Bouchard, P., Fytas, K., Castello, M., Kamwa, B.N. Analysis of gold in rock samples using laser-induced breakdown spectroscopy: Matrix and heterogeneity effects. *Spectrochim. Acta – Part B At. Spectrosc.* (2017), **134**, 33–41, doi:10.1016/j.sab.2017.06.004.

2 Da Silva Gomes, M., De Carvalho, G.G.A., Santos, D., Krug, F.J. A novel strategy for preparing calibration standards for the analysis of plant materials by laser-induced breakdown spectroscopy: A case study with pellets of sugar cane leaves. *Spectrochim. Acta – Part B At. Spectrosc.* (2013), **86**, 137–141, doi:10.1016/j.sab.2013.03.009.

3 Prichard, L., Barwick, V. *Preparation of Calibration Curves A Guide to Best Practice*; Report #LGC/VAM/2003/032. Teddington, UK. LGC Ltd. (2003).

4 Tognoni, E., Palleschi, V., Corsi, M., Cristoforetti, G., Omenetto, N., Gornushkin, I., Smith, B. W., Winefordner, J.D. From sample to signal in laser-induced breakdown spectroscopy: a complex route to quantitative analysis, in: Miziolek, A.W., Palleschi, V., Schechter, I., Eds. *Laser-Induced Breakdown Spectroscopy: Fundamentals and Applications*; Cambridge University Press: Cambridge, (2006), Vol. **9780521852**; ISBN 9780511541261.

5 Borisov, O.V.V., Mao, X.L.L., Fernandez, A., Caetano, M., Russo, R.E.E. Inductively coupled plasma mass spectrometric study of non-linear calibration behavior during laser ablation of binary Cu–Zn Alloys. *Spectrochim. Acta Part B At. Spectrosc.* (1999), **54**, 1351–1365, doi:10.1016/S0584-8547(99)00083-X.

6 Campanella, B., Legnaioli, S., Pagnotta, S., Poggialini, F., Palleschi, V. Shock waves in laser-induced plasmas. *Atoms* (2019), 71–74.

7 Panne, U., Clara, M., Haisch, C., Niessner, R. Analysis of glass and glass melts during the vitrification of fly and bottom ashes by laser-induced plasma spectroscopy. Part II. Process analysis. *Spectrochim. Acta Part B At. Spectrosc.* (1998), **53**, 1969–1981, doi:10.1016/S0584-8547(98)00239-0.

8 Cristoforetti, G., De Giacomo, A., Dell'Aglio, M., Legnaioli, S., Tognoni, E., Palleschi, V., Omenetto, N. Local thermodynamic equilibrium in laser-induced breakdown spectroscopy: beyond the McWhirter criterion. *Spectrochim. Acta – Part B At. Spectrosc.* (2010), **65**, 86–95, doi:10.1016/j.sab.2009.11.005.

9 Ciucci, A., Corsi, M., Palleschi, V., Rastelli, S., Salvetti, A., Tognoni, E. New procedure for quantitative elemental analysis by laser-induced plasma spectroscopy. *Appl. Spectrosc.* (1999), **53**, 960–964, doi:10.1366/0003702991947612.

10 Li, L., Wang, Z., Yuan, T., Hou, Z., Li, Z., Ni, W. A simplified spectrum standardization method for laser-induced breakdown spectroscopy measurements. *J. Anal. At. Spectrom.* (2011), **26**, 2274–2280, doi:10.1039/c1ja10194c.

11 Castellón, E., Pacheco Martínez, P., Álvarez, J., Bredice, F., Borges, F., Muniz, M.-V., Sánchez-Aké, C., Palleschi, V., Sarmiento, R. Elemental analysis of dental amalgams by laser-induced breakdown spectroscopy technique. *Spectrochim. Acta Part B At. Spectrosc.* (2018), **149**, 229–235, doi:10.1016/j.sab.2018.05.025.

12 Castro, J.P., Pereira-Filho, E.R. Twelve different types of data normalization for the proposition of classification, univariate and multivariate regression models for the direct analyses of alloys by laser-induced breakdown spectroscopy (LIBS). *J. Anal. At. Spectrom.* (2016), **31**, doi:10.1039/c6ja00224b.

13 Rezaei, F., Cristoforetti, G., Tognoni, E., Legnaioli, S., Palleschi, V., Safi, A. A review of the current analytical approaches for evaluating, compensating and exploiting self-absorption in Laser Induced Breakdown Spectroscopy. *Spectrochim. Acta Part B At. Spectrosc.* (2020), **169**, 105878, doi:10.1016/j.sab.2020.105878.

14 Griem, H. *Spectral Line Broadening by Plasmas*. Elsevier Science, The Netherlands (1974), ISBN 9780323150941.

15 Gornushkin, I.B., Anzano, J.M., King, L.A., Smith, B.W., Omenetto, N., Winefordner, J.D. Curve of growth methodology applied to laser-induced plasma emission spectroscopy. *Spectrochim. acta, Part B At. Spectrosc.* (1999), **54**, 491–503, doi:10.1016/S0584-8547(99)00004-X.

16 Pardini, L., Legnaioli, S., Lorenzetti, G., Palleschi, V., Gaudiuso, R., De Giacomo, A., Diaz Pace, D.M., Anabitarte Garcia, F., De Holanda Cavalcanti, G., Parigger, C. On the

determination of plasma electron number density from Stark broadened hydrogen Balmer series lines in laser-induced breakdown spectroscopy experiments. *Spectrochim. Acta – Part B At. Spectrosc.* (2013), **88**, 98–103, doi:10.1016/j.sab.2013.05.030.

17 El Sherbini, A.M., El Sherbini, T.M., Hegazy, H., Cristoforetti, G., Legnaioli, S., Palleschi, V., Pardini, L., Salvetti, A., Tognoni, E. Evaluation of self-absorption coefficients of aluminum emission lines in laser-induced breakdown spectroscopy measurements. *Spectrochim. Acta Part B At. Spectrosc.* (2005), **60**, 1573–1579, doi:10.1016/j.sab.2005.10.011.

18 Danzer, K., Currie, L.A. Guidelines for calibration in analytical chemistry. Part I. Fundamentals and single component calibration (IUPAC Recommendations 1998). *Pure Appl. Chem.* (1998), **70**, 993–1014, doi:10.1351/pac199870040993.

19 IUPAC. Nomenclature, symbols, units and their usage in spectrochemical analysis—II. data interpretation Analytical chemistry division. *Spectrochim. Acta Part B At. Spectrosc.* (1978), **33**, 241–245, doi:10.1016/0584-8547(78)80044-5.

20 Long, G.L., Winefordner, J.D. Limit of detection. A closer look at the IUPAC definition. *Anal. Chem.* (1983), **55**, 712A–724A.

21 Safi, A., Campanella, B., Grifoni, E., Legnaioli, S., Lorenzetti, G., Pagnotta, S., Poggialini, F., Ripoll-Seguer, L., Hidalgo, M., Palleschi, V. Multivariate calibration in Laser-Induced Breakdown Spectroscopy quantitative analysis: The dangers of a 'black box' approach and how to avoid them. *Spectrochim. Acta Part B At. Spectrosc.* (2018), **144**, 46–54.

22 Duponchel, L., Bousquet, B., Pelascini, F., Motto-Ros, V. Should we prefer inverse models in quantitative LIBS analysis? *J. Anal. At. Spectrom.* (2020), **35**, 794–803.

23 Poggialini, F., Campanella, B., Legnaioli, S., Raneri, S., Palleschi, V. About the use of inverse calibration in laser-induced breakdown spectroscopy quantitative analysis. *Spectrochim. Acta – Part B At. Spectrosc.* (2020), **170**, doi:10.1016/j.sab.2020.105917.

24 Wien, T. Linear Regression with Errors in X and Y. Available online: https://www.mathworks.com/matlabcentral/fileexchange/26586-linear-regression-with-errors-in-x-and-y (accessed on Apr 10, 2020).

12

Partial Least Squares

Zongyu Hou[1,2], Weiran Song[1,2], and Zhe Wang[1,2]

[1] State Key Lab of Power Systems, International Joint Laboratory on LowCarbon Clean Energy Innovation, Department of Energy and Power Engineering, Tsinghua University, Beijing, China
[2] Shanxi Research Institute for Clean Energy, Tsinghua University, Taiyuan, China

12.1 Overview

Laser-induced breakdown spectroscopy (LIBS) data contain several tens of thousands of variables due to the wide detection range and high spectral resolution. High dimensionality coupled with small sample size poses a serious challenge to the accuracy and robustness of ordinary least squares regression. Moreover, a common characteristic for spectral data is collinearity (also called multicollinearity), which refers to the situation where two or more variables in a statistical model are linearly related [1]. High collinearity produces unstable estimates of the regression coefficients, and this ill-conditioned regression model practically fits training data well but generalizes poorly on test data [2].

Partial least squares (PLS) regression is a standard chemometrics method that has been applied in LIBS. It was one of the most popular chemometrics methods not only for LIBS but also for other spectrum analysis. By stably estimating regression coefficients from low-dimensional projection, it can efficiently deal with the small sample size, high dimensionality, and high collinearity problems in LIBS data. PLS relies on the basic assumption that the investigated system or process is driven by a set of underlying latent variables (LVs, also called latent vectors, score vectors, or components). It extracts LVs by projecting both input data X and response Y onto a subspace such that the pairwise covariance between the LVs of X and Y is maximized as follows:

$$\max \mathbf{w}^T \mathbf{X}^T \mathbf{Y} \mathbf{c},$$
$$s.t. \mathbf{w}^T \mathbf{w} = \mathbf{c}^T \mathbf{c} = 1,$$

where $\mathbf{w}$ and $\mathbf{c}$ are weight vectors of X and Y, respectively. It yields the following eigen problem:

$$\mathbf{w}^T \mathbf{X}^T \mathbf{Y} \mathbf{c} = \lambda \mathbf{w},$$

which follows a framework of LV multivariate regression [3]. To ensure the mutual orthogonality of the LVs, this procedure is iteratively carried out by using deflation scheme [4] which subtracts from **X** and **Y** the information explained by their rank-one approximations based on score vectors. When response contain multi-categorical information, PLS can be used to classify data via dummy matrix coding, namely, PLS-DA [5]. Nonlinear iterative partial least squares (NIPALS) [6] and SIMPLS [7] are two frequently used PLS regression algorithms. Since standard PLS is a linear model, some variants of PLS were also developed to improve its capability of nonlinear regression.

12.2 Partial Least Squares Regression Algorithms

In this section, PLS and its several variants were introduced. NIPALS and SIMPLS belong to common PLS but with different algorithms. Kernel PLS, locally weighted PLS, and dominant factor PLS were variants of common PLS.

12.2.1 Nonlinear Iterative PLS

The first algorithm used in PLS regression was NIPALS, which can be summarized as follows:

a) Calculate the quantities **w** (PLS weights for **X**), **t** (PLS scores for **X**), **p** (PLS loadings for **X**), and q (PLS loading for **y**) as follows:

$$\mathbf{w} = \mathbf{X}^T \mathbf{y}$$

$$\mathbf{w} = \mathbf{w}/\|\mathbf{w}\|$$

$$\mathbf{t} = \mathbf{X}\mathbf{w}$$

$$\mathbf{p} = \mathbf{X}\mathbf{t}/(\mathbf{t}^T \mathbf{t})$$

$$q = \mathbf{y}^T \mathbf{t}/(\mathbf{t}^T \mathbf{t})$$

b) Deflate **X** and **y** by subtracting the computed LVs:

$$\mathbf{X} = \mathbf{X} - \mathbf{t}\mathbf{p}^T$$

$$\mathbf{y} = \mathbf{y} - \mathbf{t}q$$

c) Go to step a to compute the next LVs until reaching the required number of LVs.
d) Store **w**, **t**, **p**, and q in **W**, **T**, **P**, and **q**, respectively.
e) Calculate the regression coefficients as:

$$\mathbf{b} = \mathbf{W}^T (\mathbf{P}\mathbf{W}^T)^{-1} \mathbf{q}$$

12.2.2 SIMPLS Algorithm

When the number of LVs is not high, SIMPLS provides similar performance with less computational cost compared to NIPALS [8]. The SIMPLS algorithm is summarized as follows:

a) Calculate **s** as:

$$\mathbf{s} = \mathbf{X}^T \mathbf{y}$$

b) Calculate the quantities **w**, **t**, **p**, and q as follows:

$$\mathbf{r} = \mathbf{s}$$
$$\mathbf{t} = \mathbf{X}\mathbf{w}$$
$$\mathbf{t} = \mathbf{t}/\|\mathbf{t}\|$$
$$\mathbf{w} = \mathbf{w}/\|\mathbf{w}\|$$
$$\mathbf{p} = \mathbf{X}^T \mathbf{t}$$
$$q = \mathbf{y}^T \mathbf{t}$$

c) Store **w**, **t**, **p**, and q in **W**, **T**, **P**, and **q**, respectively.
d) Update **s** as:

$$\mathbf{s} = \mathbf{s} - \mathbf{P}(\mathbf{P}^T\mathbf{P})^{-1}\mathbf{P}^T\mathbf{s}$$

e) Go to step b to calculate the next LV until reaching the required number of LVs.
f) Calculate the regression vector as:

$$\mathbf{b} = \mathbf{R}\mathbf{q}$$

12.2.3 Kernel Partial Least Squares

Classical linear regression algorithms often degrade in performance under nonlinear conditions [9]. To tackle this issue, kernel approaches map the input data into Hilbert feature space where the nonlinear relationship among original variables becomes linear.

Kernel partial least squares (KPLS) maps the original data **X** into Hilbert feature space $F(\phi: \mathbf{R}^M \rightarrow \mathbf{F})$ as:

$$\mathbf{K}_{ij} = K(x_i, x_j) = (\phi(x_i), \phi(x_j)) = \phi(x_i)^T \phi(x_j)$$

and then constructs a regression or classification model. The mapping procedure is performed by kernel function which calculates the similarity between two sample vectors. According to the Covers theorem, the nonlinear relationship among variables in the original data space becomes linear in the feature space after such mapping [10]. The Gaussian kernel was usually used due to its efficiency which is defined as:

$$\mathbf{K}_{ij} = e^{-\frac{\|x_i - x_j\|^2}{2\sigma^2}},$$

where σ is the Gaussian width of the kernel function. KPLS can effectively capture the nonlinearity and improve the prediction performance. However, it is not possible to directly see the contribution of each variable with respect to the prediction model [11]. Furthermore, kernel methods are prone to overfitting if the data set has limited numbers of samples [12].

12.2.4 Locally Weighted Partial Least Squares

Locally weighted PLS (LW-PLS) is a Just-In-Time (JIT) method that constructs a local regression model based on the similarity between a given query and training samples [13]. It reduces the degree of nonlinearity by respectively enlarging and lessening the influence of neighboring and remote samples toward a PLS model. The nth sample ($n = 1, 2, ..., N$) of input and output variables is expressed as:

$$\mathbf{x}_n = [x_{n1}, x_{n2}, ..., x_{nM}]^T$$

$$\mathbf{y}_n = [y_{n1}, y_{n2}, ..., y_{nL}]^T$$

where M and L denote the number of input and output variables, respectively. Let $\mathbf{X} \in \mathbf{R}^{N \times M}$ and $\mathbf{Y} \in \mathbf{R}^{N \times L}$ be the input and output variable matrices whose nth row are $\mathbf{x}_n^T$ and $\mathbf{y}_n^T$. To predict the output of a given query $\mathbf{x}_q$, the similarity ω_n between $\mathbf{x}_q$ and $\mathbf{x}_n$ is calculated, and then a local PLS model is built by weighting samples with a similarity matrix $\mathbf{\Omega} \in \mathbf{R}^{N \times N}$ defined by:

$$\mathbf{\Omega} = \text{diag}(\omega_1, \omega_2, ..., \omega_N)$$

where diag ($\cdot$) represents a diagonal matrix and ω_n is defined on the basis of the Euclidean distance as follows:

$$\omega_n = \exp\left(-\frac{\varphi d_n}{\sigma_d}\right)$$

$$d_n = \sqrt{(\mathbf{x}_n - \mathbf{x}_q)^T (\mathbf{x}_n - \mathbf{x}_q)}$$

where φ is a localization parameter and σ_d is a standard deviation of d_n. The predicted output $\hat{\mathbf{y}}_q$ is calculated through the following procedure.

a) Set K to the desired number of LVs and initialize $k = 1$.
b) Calculate the similarity matrix $\mathbf{\Omega}$.
c) Calculate $\mathbf{X}_k$, $\mathbf{Y}_k$, and $\mathbf{x}_{q,k}$:

$$\mathbf{X}_k = \mathbf{X} - \mathbf{1}_N[\bar{x}_1, \bar{x}_2, ..., \bar{x}_M]$$

$$\mathbf{Y}_k = \mathbf{Y} - \mathbf{1}_N[\bar{y}_1, \bar{y}_2, ..., \bar{y}_L]$$

$$\mathbf{x}_{q,k} = \mathbf{x}_q - [\bar{x}_1, \bar{x}_2, ..., \bar{x}_M]^T$$

$$\bar{x}_m = \frac{\sum_{n=1}^{N} \omega_n x_{nm}}{\sum_{n=1}^{N} \omega_n}$$

$$\bar{y}_l = \frac{\sum_{n=1}^{N} \omega_n y_{nl}}{\sum_{n=1}^{N} \omega_n}$$

where $\mathbf{1}_N \in \mathbf{R}^N$ is a vector of ones.

d) Derive the kth LV of $\mathbf{X}$:

$$\mathbf{t}_k = \mathbf{X}_k \mathbf{w}_k$$

where $\mathbf{w}_k$ is the eigenvector of $\mathbf{X}_k^T \mathbf{\Omega} \mathbf{Y}_k \mathbf{Y}_k^T \mathbf{\Omega} \mathbf{X}_k$, which corresponds to the maximum eigenvalue.

e) Derive the kth loading vector of $\mathbf{X}$ and the kth regression coefficient vector:

$$\mathbf{p}_k = \frac{\mathbf{X}_k^T \mathbf{\Omega} \mathbf{t}_k}{\mathbf{t}_k^T \mathbf{\Omega} \mathbf{t}_k}$$

$$q_k = \frac{\mathbf{Y}_k^T \mathbf{\Omega} \mathbf{t}_k}{\mathbf{t}_k^T \mathbf{\Omega} \mathbf{t}_k}$$

f) Derive the kth LV of $\mathbf{x}_q$:

$$t_{q,k} = \mathbf{x}_{q,k}^T \mathbf{w}_k$$

g) If $k = K$, finish the output estimate:

$$\hat{\mathbf{y}}_q = [\bar{y}_1, \bar{y}_2, ..., \bar{y}_L]^T + \sum_{k=1}^{K} t_{q,k} q_k$$

Otherwise, set

$$\mathbf{X}_{k+1} = \mathbf{X}_k - \mathbf{t}_k \mathbf{p}_k^T$$

$$\mathbf{Y}_{k+1} = \mathbf{Y}_k - \mathbf{t}_k \mathbf{q}_k^T$$

$$\mathbf{x}_{q,k+1} = \mathbf{x}_{q,k} - t_{q,k} \mathbf{p}_k$$

h) Set $k = k + 1$ and go to step d.

LW-PLS effectively reduces global nonlinearity and outperforms the conventional PLS in developing soft sensors [14, 15].

12.2.5 Dominant Factor-based Partial Least Squares

Although PLS is a popular multivariate quantification model, it ignores the physical background of LIBS measurement and is mostly dependent on statistical correlation or curve fitting, which may lead to noise overfitting and eventually ruin the measurement accuracy for samples outside the matrix of the calibration sample set. Dominant factor-based PLS combines the advantages of univariate model and PLS method and avoids the overuse of the unrelated noise in the spectrum for PLS application [16–18]. It uses a conventional univariate model or other physical background-based models to calibrate the major concentration information (called the dominant factor) and applies PLS to compensate for the residual errors in the dominant factor. The dominant factor-based PLS model can be expressed as follows:

$$C = C_{\text{dominant}} + \sum_{j=1}^{n} b_j I_j + b_0,$$

where C indicates the concentration of the analyzed element, b_j and b_0 are coefficients determined by PLS regression for residual error correction, and I_j includes all the lines that appear in the spectrum except the dominant factor. The dominant factor PLS model enabled PLS to treat nonlinearity and self-absorption in LIBS data, with the advantage of higher accuracy and robustness over wide samples.

12.3 Partial Least Squares Discriminant Analysis

PLS-DA is an adaption of multivariate PLS regression for classification purpose, which transforms the category information into numerical responses using dummy matrix coding [5]. The category information is initiated as a null matrix **Y** in which rows and columns are equal to the number of instances N and categories c, respectively. If the values of a category c_i integrally increase from 1, the corresponding element in c_i-th column of **Y** is set to 1. Suppose $\mathbf{1}_k$ is a k-by-1 vector of all ones and, $\mathbf{0}_k$ is a k-by-1 vector of all zeros. After arranging the sequence of category information, **Y** can be reordered as:

$$\mathbf{Y} = \begin{pmatrix} \mathbf{1}_{n_1} & \mathbf{0}_{n_1} & \cdots & \mathbf{0}_{n_1} \\ \mathbf{0}_{n_2} & \mathbf{1}_{n_2} & \cdots & \mathbf{0}_{n_2} \\ \vdots & \vdots & \ddots & \vdots \\ \mathbf{0}_{n_c} & \mathbf{0}_{n_c} & \cdots & \mathbf{1}_{n_c} \end{pmatrix}_{N \times c}$$

or

$$\mathbf{Y} = \begin{pmatrix} \mathbf{1}_{n_1} & \mathbf{0}_{n_1} & \cdots & \mathbf{0}_{n_1} \\ \mathbf{0}_{n_2} & \mathbf{1}_{n_2} & \cdots & \mathbf{0}_{n_2} \\ \vdots & \vdots & \ddots & \mathbf{1}_{n_c-1} \\ \mathbf{0}_{n_c} & \mathbf{0}_{n_c} & \cdots & \mathbf{0}_{n_c} \end{pmatrix}_{N \times (c-1)}$$

where $\sum_{i=1}^{c} n_i = N$. Then the category $\hat{\mathbf{y}}_q$ of a query vector $\mathbf{x}_q$ is predicted as:

$$\hat{\mathbf{y}}_q = \mathbf{x}_q \cdot \mathbf{b}_{\text{PLS}}.$$

For binary classification, a threshold of 0.5 is commonly applied for category decision. For multiclass problem, the category decision follows the maximum assignment criterion [19, 20], where the column of the maximum value in vector $\hat{\mathbf{y}}_q$ returned is the predicted category.

12.4 Results of Partial Least Squares in LIBS

As one of the most widely used chemometrics methods, PLS has been applied in many different fields of LIBS application. In this section, we summarize the results of PLS application in LIBS within several main fields.

12.4.1 Coal Analysis

Coal is one of the most abundant primary energy sources in the world. LIBS combined with PLS showed great potential for real-time coal analysis, which is thirstily needed for the coal industry. Coal is an inhomogeneous material consisting of complex organic and inorganic molecules and containing nearly all the elements in the earth. Thus, LIBS coal analysis suffers from strong matrix effects and interferences, which make the univariate model and calibration-free model cannot achieve satisfactory results. Chemometrics methods such as PLS were effective tools to analyze the elements, calorific value, ash content, and volatile content, etc., of coal using LIBS.

Yao et al. [21] applied PLS to measure the ash content with standard ash values in the range 7.74–51.02 wt.%. Using 13 samples for calibration and four samples for validation, the correlation coefficient was 0.97 and the relative standard deviation (RSD) of repeated measurements was in the range 6.53–7.71%. Li et al. [22] used 44 coal samples to test the optimization of PLS model with different preprocessing methods for calorific heat value measurement. Better results were obtained relative to a standard offline method using PLS model with 11-point smoothing combined with the second-order derivation. The R_c (correlation coefficient of calibration set), R_v (correlation coefficient of validation set), RMSECV, and RMSEP showed 0.9909, 0.9972, 0.47 MJ/kg, and 0.276 MJ/kg, respectively, using 33 samples for calibration and 11 for validation.

Li et al. [23] measured the ash content in 58 coal slices. The data were processed using PLS–PCA combined model to show R^2, RMSEP, and ARE of 0.98, 0.77, and 4.029%, respectively (40 samples for calibration and 18 for validation). Yuan et al. [24] proposed a PLS-wavelet transform hybrid model to determine the C-content in 24 bituminous coal samples. Calibration and prediction results showed: R^2 of 0.98, RMSEP of 1.94% (seven samples for validation), and ARE of 1.67% using the hybrid model.

Yuan et al. [25] determined the caloric heat value, ash content, volatile content, and carbon content using 53 collected coal samples on air-dried basis. The dominant factor PLS model showed a remarkable improvement over the conventional PLS model. For example, the R^2 of heat value measurement improved from 0.94 to 0.97, while RMSEP, RMSE, and ARE decreased from 1.63 to 1.33 MJ/kg, 1.05 to 0.82 MJ/kg, and 3.55 to 2.71%, respectively, using 40 samples for calibration and 13 for validation.

The best results insofar were achieved by Hou et al.[26] using a hybrid model which includes three main processes: spectral line standardization, identification based on self-adaptive database, and dominant factor PLS model to measure the heat value, C, H, ash, and volatile contents in 77 standard bituminous coal samples. The average measurement errors for carbon, hydrogen, volatiles, ash, and heat values are 0.42, 0.05, 0.07, 0.17%, and 0.07 MJ/kg, respectively. The measurement errors were lower than the values stated by the national standard of China for chemical methods [27]. Table 12.1 summarizes the coal analysis results using LIBS-PLS. Sheta et al. reviewed the LIBS application on coal analysis [33], which provided more complete information for reference.

Table 12.1 Summary of coal analysis using LIBS-PLS.

Element	Experimental	Data processing	Results	Ref.
C	266 nm, 4 ns, 18 mJ/pulse, 10 Hz, delay time 200 ns, gate width 1.1 ms	MLR for model, PLS for residual correction (38 samples for calibration and 6 for validation)	$R^2 = 0.99$, RMSECV < 0.13%, RMSEP < 2.46%	[28]
C	532 nm, 120 mJ, 1 Hz, delay time 2 μs, gate width 1 ms	Nonlinearized multivariate dominant factor PLS (19 samples for calibration, 14 for prediction), 4 principle components	$R^2 = 0.933$, RMSEP = 3.77%, RMSEC = 3.28%	[29]
C	532 nm, 120 mJ, 1 Hz, delay time 2 μs, gate width 1 ms	PLS model based on dominant factor (19 samples for calibration, 14 for prediction)	$R^2 = 0.999$, RMSEP = 4.47%, RMSEC = 2.92%	[18]
C, H, ash, volatile, heat value	1064 nm, 90 mJ/pulse, delay time 0.5 μs, gate width 1 ms; pellets	Spectrum standardization, identification, and dominant factor-based PLS (69 samples for calibration and 8 for validation)	AAE = 0.42% (C), 0.05% (H), 0.07% (ash < 15%), 0.17% (15% < ash < 30%), 0.23% (ash > 30%), 0.03% (volatile < 20%), 0.11% (20% < volatile < 40%), 0.07 MJ/kg (heat value)	[26]
C	532 nm, 5 ns, 120 mJ/pulse, delay time 2 μs, gate width 1 ms; pellets	Dominant factor-based PLS (16 samples for calibration and 8 for validation)	$R^2 = 0.9849$, ARE = 1.82%	[30]
C, H	532 nm, 5 ns, 70 mJ/pulse, delay time 1.5 μs, gate width 1 ms; pellets	SVM and PLS (40 samples for calibration and 13 for validation)	$R^2 = 0.97$, ARE = 3.93%	[31]
C	532 nm, 5 ns, 120 mJ/pulse, 1 Hz, delay time 2 μs, gate width 1 ms, pellets	Pretreatment with environmental denoising and background noise reduction, 17 samples for calibration, and 7 for validation	RMSEP = 1.94%, ARE = 1.67%	[24]
Moisture, ash, volatile, heat value	532 nm, 5 ns, 70 mJ/pulse, 1 Hz, delay time 1.5 μs, gate width 1 ms, pellets	Dominant factor-based PLS (40 samples for calibration and 13 for validation)	$R^2 = 0.97$ (moisture), 0.93 (ash content), 0.97 (volatile matter content), 0.97 (heat value); RESEP = 0.87% (moisture), 3.49% (ash content), 1.41% (volatile matter content), 1.33 MJ/Kg (heat value); ARE = 26.2% (moisture), 12% (ash content), 5.47% (volatile matter content), 2.71% (heat value)	[25]
Ash content, volatile matter, heat value	1064 nm, 6 ns, 100 mJ/pulse, 8 Hz, delay time 1.58 μs, pellets	PLS with SVM classification, 101 coal samples	$R^2 = 0.9945$ (ash content), 0.9888 (volatile content), 0.9906 (heat value); RMSEP = 0.9065 (ash content), 0.7719 (volatile content), 0.4093 (heat value)	[32]

12.4.2 Metal Analysis

Metal is one of the most important materials in industry and living. LIBS is a promising fast analysis method in the whole process of metal industry, such as production, quality testing, waste recovery, etc. PLS is an effective chemometrics method for quantification, especially for complicated spectrum, such as samples containing Fe and Ni.

Afgan et al. [34] applied dominant factor-based PLS regression with spectral standardization for elements measurement in steel, with a handheld micro-LIBS system. Average absolute measurement errors of Si, Cr, Mn, and Ni achieved 0.019, 0.039, 0.013, and 0.001%, respectively, which was 2–5 times better than the common PLS method. The average RSD for these elements was less than 5%.

Sun et al. [35] analyzed the Si, Mn, Cr, Ni, and V elements in molten steel, which can be used to optimize the smelting process of steel. They compared the performance of univariate model and PLS model, and results showed that the RMSEP achieved 0.065 (Si), 0.051 (Mn), 0.195 (Cr), 0.168 (Ni), and 0.038 (V) using PLS, with a significant improvement than univariate model.

Hernandez et al.[36] applied LIBS and PLS for quantification of noble metals in Au–Ag–Cu alloys, commonly encountered in jewelry pieces. The relative error achieved ~2%. They believed that the interferences due to unknown components present in the samples, a common drawback in the analysis of commercial samples such as jewels, can be at negligible levels using the PLS algorithm. Table 12.2 summarizes the metal analysis results using LIBS-PLS.

12.4.3 Rocks, Soils, and Minerals Analysis

Rocks, soils, and minerals analysis are important for many different fields, such as mineral exploration, environmental protection and space exploration. The most famous LIBS application, ChemCam, loaded with Curiosity Rover for Mars exploration, was aimed to detect the mineral contents on the surface of Mars.

Ollila et al.[51] analyzed the trace element geochemistry (Li, Ba, Sr, and Rb) using Curiosity's ChemCam and got early results for Gale crater from Bradbury Landing Site to Rock nest. They applied PLS model with other pretreatment methods such as denoising, continuum removing, and normalization and finally achieved RMSEP to 55 ppm, 648 ppm, and 35 ppm for Li, Ba, and Ru, respectively. Dyar et al.[52] simulated the ambient gas of Mars and applied LIBS-PLS for rock analysis. With 70 samples for calibration and 30 samples for validation, the RMSEP reached 6, 3, 0.2, 2, 3, 0.75, 0.4, 0.45, 0.05% for Si, Al, Ti, Fe, Mg, Ca, Na, K, and P, respectively.

Except for natural raw rocks and minerals, PLS was also applied for manmade materials similar to minerals such as slags and concretes. Zhang et al.[53] applied combination of PLS, SVM, and GA algorithms for elements measurement in slags. With 16 samples for calibration and 4 for validation, the RMSE achieves 0.1356 for TiO_2 analysis. They confirmed that the LIBS technique coupled with SVM and PLS methods is a promising approach to achieve the online analysis and process control of slag field. Zhang et al.[54] applied LIBS-PLS for concrete measurement. The RMSE of chlorine (Cl) prediction reached 0.4233, which can be used to detect the quality of concrete. The result of rocks, soils, and mineral analysis were summarized in Table 12.3. LIBS-PLS-DA was also applied for mineral classification or discrimination, with the results also summarized in Table 12.3.

Table 12.2 Summary of metal analysis using LIBS-PLS.

Sample	Element	Experimental	Data processing	Results	Ref.
Steel	Cr, Ni, Mn, Si	Handheld micro-LIBS system	Spectrum standardization and dominant factor-based PLS	Absolute errors 0.019, 0.039, 0.013, and 0.001% for Si, Cr, Mn, and Ni, respectively	[34]
Steel	Cr, Ni, Mn, Si, Co, Mo	532 nm, 4 ns, 30 mJ, 10 Hz, gate delay 1.5 μs	PLS	SEP = 0.28 (Cr), 0.17 (Ni), 0.05 (Mn), 0.051 (Si), 0.15 (Mo), 0.12 (Co)	[37]
Steel	Cr, Ni	532 nm, 6 ns, 90–95 mJ, 1 Hz, delay time 1.1–1.3 μs, gate width 20 μs	PLS	Cr: accuracy 2.5–3%, RSD 3.5–5% Ni: accuracy 5–8%, RSD 7–9%	[38]
Steel	Cr, Ni, Mn	1064 nm, 7 ns, 95 mJ, 10 Hz, delay time 2 μs, gate with 0.5 μs	Univariate (integrated peak normalized by total light), Univariate (integrated peak normalized by an internal standard iron peak), PLS	SEP = 0.740 (Cr), 0.331 (Ni), 0.099 (Mn); RSD = 6.02 (Cr), 4.60 (Ni), 9.00 (Mn)	[39]
Steel	Cr	1064 nm, 200 mJ/pulse, 10 Hz, delay time 4000 ns, gate width 1.05 ms; pellets	PLS2 (100 data for calibration and 50 for validation)	For calibration curve, R^2 = 0.8864, For PLS, R^2 = 0.9953	[40]
Steel	Si, Mn, Cr, Ni and Cu	1064 nm, 10 ns, 80 mJ/pulse, 20 Hz, delay time 1.5 μs, gate width 2 ms	PLS (10 samples for calibration and 4 for validation)	R^2 = 0.873, RESE = 1.760	[41]
Low-alloy steel	Mn	532 nm, 16–18 ns, 470 mJ/pulse, delay time 1 μs, gate width 8 μs	(8 samples for calibration and 6 for validation)	RSD < 10%, 83% < recover rate < 121%	[42]
Stainless steel	Cr, Ni, Ti, Nb	532 nm, 3–5 ns, 12 mJ/pulse, 15 Hz, delay time 10 μs, gate width 10 μs	PLS		[43]
Low alloy steel	Mn, Cr, Ni, Ti	1064 nm, 7 ns, 0.27 mJ/pulse, 10 Hz, gate width 10 ms; pellets	PLS (163 samples for calibration and for validation)	R^2 = 0.9849, RMSE = 0.065%	[44]

Sample	Elements	Laser parameters	Method	Results	Ref.
Molten steel	Si, Mn, Cr, Ni, V	1064 nm, 6–8 ns, 20 and 10 Hz	Univariate model and PLS	RMSEP = 0.065 (Si), 0.051 (Mn), 0.195 (Cr), 0.168 (Ni), 0.038 (V); MSD = 0.047 (Si), 0.025 (Mn), 0.095 (Cr), 0.111 (Ni), 0.014 (V)	[35]
Molten steel	Mn, Si	1064 nm, 150 mJ/pulse, 10 Hz, delay time 2 µs, gate width 1 ms; liquid	PLS (200 samples for calibration and 100 for validation)	$R^2 = 0.996$	[45]
Aluminum alloy	Fe, Mg, Mn, Ni, Si, Zn, Cu	1064 nm, 4–15 µJ, 1–10 kHz	PLS	RMSE = 0.38% (Cu), 0.06–0.18% (other elements)	[46]
Brass	Cu	532 nm, 5 ns, 90 mJ/pulse, delay time 2.25 µs, gate width 2 ms; pellets	PLS (20 samples for calibration and 9 for validation)	$R^2 = 0.992$, RMSEP = 1.30%	[47]
Brass	Zn, Sn, Pb, Fe	10 ns, 50 mJ/pulse, 1 Hz, delay time 1000 ns, gate width 2.1 ms	Partial least squares correction (60 average LIBS spectra, 54 from standards, and 6 from unknown sample)	$R^2 = 0.99$ (Zn), 0.97 (Sn), 0.96 (Fe), 0.74 (Pb)	[48]
Au-Ag-Cu alloy	Au, Ag	532 nm, 5 ns, 0.32~1.73 mJ, 1 Hz	PLS (9 samples for Au calibration, 7 samples for Ag calibration; 5 samples for Au and Ag prediction, respectively), 3–4 latent variables	$R^2 = 0.999$, REP = 0.92~1.12% (Au), $R^2 = 0.993–0.996$, REP = 0.64–0.82 (Ag)	[49]
Jewelry alloy	Au, Ag	532 nm, 6 ns, 10 Hz, delay time 300 ns, gate width 3000 ns; pellets	PLS-DA (entire laser-induced emission spectra from 266 to 340 nm, total 990 experiment points, five replicates were considered in all cases)	Relative error ~2%. Earring I: mean predicted value 60.6%, SD = 2.3%; Earring II: mean predicted value 62.3%, SD = 1.2%; Ring: mean predicted value 58.1%, SD = 4.7%	[36]
Alloys	Ag, Cu, Ni, Zn, In	532 nm, 1.7 mJ/pulse, 1 Hz; solutions	Dark current subtraction and PLS (25 samples for calibration and 7 for validation)	SEP (standard error of prediction) = 3.0535% (Ag), 2.9566% (Cu), 2.6841% (Ni), 1.3303% (Zn), 2.8869% (In)	[50]

Table 12.3 Summary of rocks, soils, and mineral analysis using LIBS-PLS.

Sample	Element	Experimental	Data processing	Results	Ref.
Minerals of Mars	Li, Ba, Sr, Rb	1067 nm, 5 ns, 9.5/10/14 mJ, 3 Hz, pellets	Remove non-LIBS signal, wavelet denoise, remove continuum, normalization, univariate model, and PLS model	RMSEP = 55 ppm (Li), 648 ppm (Ba), 35 ppm (Ru)	[51]
Rock slab sample	Multi elements, Si, Ti, Al etc.	1064 nm, 17 mJ/pulse, 10 Hz, pressed powder	PCA, PLS, multilayer perceptron artificial neural networks (MLP ANNs) and cascade correlation (CC) ANNs (90 rock slabs, corresponding powder, and 22 geostandards were split into training, validation, and test sets)	RMSE = 3.07 wt% (SiO_2), 0.87 wt% (TiO_2), 2.36 wt% (Al_2O_3), 2.20 wt% (Fe_2O_3), 0.08 wt% (MnO), 1.74 wt% (MgO), 1.14 wt% (CaO), 0.85 wt% (Na_2O), 0.81 wt% (K_2O)	[55]
Rocks	Na, K, Ti, P, Mg, Si, Al, Fe, Ca	1064 nm, 17 mJ/pulse, focusing length 9 m, 7 Torr CO_2 pressure; pellet	PLS (70 samples for calibration and 30 for validation)	MSEP = 12.55, RMSEP = 8.97; For PLS-1, MSEP = 13.07, RMSEP = 9.07	[56]
Rocks	Si, Al, Fe, Mn, Mg, Ca Na, K,	1064 nm, 17±1 mJ/pulse, 10 ns, focusing length 9 m, 7 Torr CO_2 pressure; pellet	PLS	For PLS-2 model, R^2 = 0.999 (Si), 0.997 (Al), 0.997 (Fe), 0.945 (Mn), 0.999 (Mg), 0.994 (Ca), 0.982 (Na), 0.996 (K); for validation, R^2 = 0.997 (Si), 0.981 (Al), 0.992 (Fe), 0.918 (Mn), 0.997 (Mg), 0.987 (Ca), 0.966 (Na), 0.988 (K)	[57]
Oily soil sample	Cu, Zn, Cr, Ni	1064 nm, 8 ns, 1 Hz, 150 mJ/pulse, delay time 3000 ns, gate width 1 ms; pellet	PLS	R^2 = 0.99 RMSE = 0.01	[58]
Rocks	Si, Al, Ti, Fe, Mg, Ca, Na, K, P	7 Torr CO_2, 1064 nm, 10 ns, 17 mJ/pulse, 10 Hz, non-gate	PLS (70 samples for calibration and 30 for validation)	RMSEP = 6, 3, 0.2, 2, 3, 0.75, 0.4, 0.45, 0.05%, R^2 = 0.87, 0.76, 0.86, 0.78, 0.92, 0.94, 0.66, 0.75, 0.85	[52]
Sulfur minerals	S	7 Torr CO_2, 1064 nm, 10 ns, 17 mJ/pulse, 10 Hz, non-gate	PLS (12 samples, 7 spectra for calibration, and 7 for validation)	RMSEP = 2~37.85%	[59]

Sample	Elements	Laser parameters	Method	Results	Ref
Soils	Al, Ba, Ca, etc.	532 nm, 5 ns, 25 mJ, 10 Hz, delay time 2 μs, gate width 10 s	PLS (one-third samples for calibration and the left for prediction)	R^2 = 0.557–0.904, RMSEP = 20% (Ti, Al), 40% (Ba, Ca)	[60]
Rocks	Fe, Mg, Si, Ti, Al, Mn, Ca, NaH, K, O	9.3 mbar CO_2, Nd:KGW laser 1067 nm, 14 mJ/pulse, 3 Hz	PLS (163305 spectra in total from 707 targets, 93196, 16910, and 53199 spectra for calibration, validation, and prediction, respectively), 15 principle components	R^2 > 0.99, RMSEP = 3%	[61]
Glass and slags	Al, Ba, Ca, Fe, Mg, Na, Si	1064 nm, 5 ns, 1 Hz, delay time 50 μs, gate width 5 ms	PLS (19 pellets for calibration, 6 steelmaking slag RM pellets for validation, and 7 glasses and the steel slag samples for prediction)	RPD = 1.84 (Al), 1.43 (Ba), 2.04 (Ca), 1.63 (Fe), 2.23 (Mg), 1.19 (Na), 1.92 (Si)	[62]
Iron ore	Acidity	400 nm, 8 ns, 30 mJ/pulse, 3 Hz, delay time 5 μs, gate width 5 μs pellets	50 samples for calibration and 10 for validation; 6 standards and 54 mixed samples	R^2 = 0.9903, RMSE = 0.0048, ARE = 3.65%	[63]
Iron ore	TFe, SiO_2, Al_2O_3, CaO, MgO	532 nm, 6 ns, 3 mJ/pulse, 10 Hz, delay time 5 μs, gate width 5 μs; pellets	Sparse partial least squares for variable selection and regression, LS-SVM for correction of the residual error (24 samples for calibration and 8 for validation; 6 standards and 24 mixed samples)	RMSEP = 0.6242% (TFe), 0.3569% (SiO_2), 0.0456% (Al_2O_3), 0.0962% (MgO), 0.2157% (CaO)	[64]
Soil	Pb	1064 nm, 8 ns, 110 mJ/pulse, 2 Hz, delay time 1.4 μs; pellets	Smoothing and mean-centering (30 samples for calibration and 10 for validation)	ARE = 7.43%	[65]
Copper mine	Cu	1064 nm, 10 ns, 0.01 mJ/pulse, 8000 Hz; pellets	PLS based on double genetic algorithm (8 samples for calibration and 3 for validation)	R^2 = 0.9964, RMSECV = 0.2631%	[66]
Mineral	Ca, Cu, Ag	532 nm, 1.73 mJ/pulse, 1 Hz; tablets	PLS (8 samples for calibration and 3 for validation)	R^2 = 0.998, REP = 0.21%	[67]
Rocks	Si, Ca, Mg, Fe, Al	1064 nm, 6 ns, 36 mJ, 1 Hz, delay time 1.5 μs; pellets	SVR and PLS	ForPLS, R^2 = 0.9966 (Si), 0.9976 (Ca), 0.9844 (Mg), 0.9854 (Fe), 0.9905 (Al); RMSEP (wt%) = 1.1788 (Si), 1.3456 (Ca), 0.4169 (Mg), 0.3946 (Fe), 0.1320 (Al); RSD = 16.09% (Si), 12.75% (Ca), 4.97% (Mg), 1.87% (Fe), 1.42% (Al)	[68]

(Continued)

Table 12.3 (Continued)

Sample	Element	Experimental	Data processing	Results	Ref.
Glasses	Th, Fe, Cr, Ni, Al, F, Sr	532 nm, 6 ns, 440 mJ, 1 Hz, delay time 25 μs; solid glass	PLS	RSD: 5.3% (Th), 5.1% (Fe), 7.1% (Al), 8.1% (F), 10.4% (Sr), 7.7% (Ni), 6.4% (Cr)	[69]
Soils	Cr	355 nm, 7 ns, 3 mJ, delay time 150 ns, gate width 2 μs	PLS and ANN	REP = 9.9%, RSD = 9.8%	[70]
Slags	Ca, Si, Al, Mg, Fe, Mn, Ti	1064 nm, 10 ns, 80 mJ/pulse, 5 Hz, delay time 1.5 μs, gate width 2 ms, pellets	PLS (16 samples for calibration and 4 for validation)	$R^2 = 0.9848$, RESE = 0.1356 (TiO_2)	[53]
Concrete	Cl	1064 nm, 6 ns, 70 mJ/pulse, 10 Hz. Delay time 14 μs and gate width 4 μs for CaCl radical emission, block	PLS (21 samples for validation)	$R^2 = 0.9344$, RESE = 0.4233 wt%	[54]
Soil	Mn	532 nm, 8 ns, 40 mJ/pulse, 3 Hz, delay time 4 μs, gate width 10 μs, pellets	GA-PLS (29 samples for calibration and 14 for validation)	$R^2 = 0.9989$, RMSEP = 0.0167 %, MPE = 5.20 %	[71]
Garnet	Classification	1064 nm, 9 ns, 70 mJ/pulse, delay time 1500 ns, gate width1 ms, pieces	PLS-DA (157 samples of different position from 92 worldwide and 25 LIBS spectra per garnet sample)	Predicted value 85–94.7% with 5–26 variables; classification performance for individual garnet types ranging from 21.4–85.7%	[72]
Rocks	Classification	1064 nm, 500 ns, 25 mJ, delay time 200 ns,	PCA, SIMCA and PLS-DA	Correct rate: 85.9% (PLS-DA)	[73]
Geomaterials	Classification	1064 nm, 7 ns, 320 mJ/pulse, double pulse, interpulse delay 1 μs, delay time 2 μs, gate width 50 μs	Normalization and mean-centering (19 samples, 20 spectra for calibration, and 5 for validation; 11 classes)	100% correct classification, 1.1% misclassified, 0% unidentified	[74]

12.4.4 Organics Analysis

Fast analysis of organics is an important requirement for a wide variety of fields, such as chemical engineering, explosive detection, pharmaceuticals industry, foods, etc. Lucia et al.[75] designed a remote LIBS system and applied PLS-DA to identify the explosive. For focusing distance of 30 m, the correct classification was 94%; for focusing distance of 50 m, the correct classification was 80%. Wang et al. [76] applied PLS-DA to discriminate the explosives from plastics and other materials. Excellent separation between explosives and nonexplosives was achieved with 100% accuracy for calibration set and test set.

Liu et al. [77] applied PLS-DA to classify different kinds of plastics. The input spectral range was optimized and the accuracy was 93.93%. Yang et al.[78] applied PLS and spectral smoothing to detect the Cd content in oranges. With 30 samples for calibration and 10 for validation, the RMSEP reached 15.1 ppm. Braga et al. [79] applied PLS to measure the B, Cu, Fe, Mg, and Zn content in plant materials. Results showed that the RMSEP reached 5.6, 11.8, 113, 49, and 3.7 ppm for B, Cu, Fe, Mg, and Zn, respectively. The results of organics analysis using LIBS-PLS were summarized in Table 12.4, with diverse samples such as leaves, meat, milk powder, oil, seeds, etc.

12.5 Conclusion

PLS is an effective tool to deal with the small sample size, high dimensionality, and high collinearity problems in LIBS data and has been widely applied in different applications such as coal analysis, metal analysis, minerals analysis, and organics analysis. PLS usually can get a better calibration result than univariate model. However, PLS also has two shortages. Firstly, it ignores the physical background of LIBS measurement and is mostly dependent on statistical correlation or curve fitting, so noise over fitting is easy to occur especially when there are not enough calibration samples. Secondly, PLS is a linear correlation method and cannot process the nonlinear correlations due to self-absorption and matrix effect. Thus, the physical background-based model needs to be integrated with PLS to improve its robustness, and the nonlinear PLS such as kernel PLS should be developed and widely applied to enhance its nonlinear capability. We believe that PLS will improve its performance based on these two directions and will play a more important role in chemometrics for LIBS analysis.

Table 12.4 Summary of organics analysis using LIBS-PLS.

Sample	Element	Experimental	Data processing	Results	Ref.
Explosive	Classification	1064 nm, 5 ns, 10 Hz, 335 mJ/pulse, delay time 2000 ns, gate width 0.1 ms; powder	PLS-DA	For focusing distance 30 m, correct classification 0.9403, false positive 0.016; for focusing distance 50 m, correct classification 0.8, false positive 0.078	[75]
Explosive residues	Classification	1064 nm, 9 ns, 200 mJ/pulse, gate width 1 ms; 1064 nm, 5 ns, 250 mJ/pulse, double pulse, interpulse delay 3 µs, delay time 2 µs, gate width 50 µs; powder spread on substrates	Line intensities and line ratios, 9 classes (19 samples for calibration and 9 for validation; 2 classes)	TP rate = 72.8%, FP rate = 2.2%	[80]
Explosive residues	Classification	1064 nm, 5 ns, 275 mJ/pulse, double pulse, interpulse delay 3 µs, delay time 2 µs, gate width 100 µs; powder spread on glass slides/dissolved powder on an aluminum substrate	Line intensities and line ratios; user-specified threshold (7 samples for calibration and 3 for validation; 6 classes)	TP rate = 100%, FP rate = 0	[81]
Chemical and biological threats	Classification	1064 nm, 5 ns, 275 mJ/pulse, double pulse, interpulse delay 3 µs, delay time 2 µs, gate width 100 µs; powder spread on glass slides/dissolved powder on an aluminum substrate	Line intensities and line ratios (25 samples for calibration and 6 for validation; 6 classes)	TP rate = 100%, FP rate = 0	[82]
Pharmaceutical samples	Classification	532 nm, 7 ns, 10 Hz, 25 mJ/pulse, delay time 1000 ns, gate width 500 ns; pellet	PLS-DA	Average correct classification: 0.9403	[83]
Plastics	Classification	1064 nm, 5 ns, 10 Hz, delay time 2 µs, gate width 6 µs	PLS-DA (40 samples for calibration, 11 for prediction)	Classification accuracy: 91–100% (Cd), 82–91% (Cr), 73–91% (Pb)	[84]
Plastics	Classification	1064 nm, 8 ns, 90 mJ/pulse, delay time 1.0 µs, gate width 1.05 ms; pieces	PLS-DA based on Spectral Windows (13 samples for calibration and 7 for validation)	Accuracy: 93.93%, efficiency: 0.0021 s	[77]
Plastics	Classification	1064 nm, 8 ns, 60 mJ/pulse, delay time 2.0 µs, gate width 2 ms	PLS-DA based on Variable Importance (180 samples for calibration and 20 for validation)	Accuracy: 99.55%, efficiency: 0.096 ms	[85]

Sample	Analyte	Laser parameters	Method	Results	Ref.
Polymer	Classification	266 nm, 5 ns, 10 mJ/pulse, 20 Hz, delay time 200 ns, gate width 1 μs	23 samples for calibration and 8 for validation; 2 classes	100% correct classification	[86]
Oranges	Cd	1064 nm, 8 ns, 130 mJ/pulse, delay time 1.3 μs	First derivative, moving window partial least squares (30 samples for calibration and 10 for validation)	$R_v = 0.9953$, RMSEP $= 15.10 \times 10^{-6}$, ARE $= 7.43\%$	[78]
Leaves	P, K, Ca, Mg, Mn, Fe, Zn, B	1064 nm, 4.5 μs, 110 mJ/pulse, 10 Hz, delay time 2.0 μs, pellets	Background correlation, classical least squares regression, PLS (26 samples for calibration and 15 for validation)	LOD $= 0.02$ (P), 0.21 (K), 0.08 (Ca), 0.03 (Mg), 1.3 (Fe), 1.2 (Zn), 0.5 (B)	[87]
Orange	Cd	1064 nm, 8 ns, 200 mJ/pulse, gate width 2 μs, dried naturally	PLS (39 samples for calibration and 13 for validation)	$R^2 = 0.9806$, relative error $= 10.94\%$	[88]
Plant materials	B, Cu, Fe, Mg, Zn	532 nm, 12 ns, 140 mJ/pulse, 10 Hz, delay time 900 ns, gate width 1100 ns; pellets	PLS regression (15 calibration samples, the numbers of latent variables were 3, 2, 2, 2, and 2 for B, Cu, Fe, Mn, and Zn, respectively)	For univariate regression, RMSEP (mg/kg) = 11.6 (B), 3.2 (Cu), 39 (Fe), 55 (Mg), 2.5 (Zn); For PLS, RMSEP (mg/kg) = 5.6 (B), 11.8 (Cu), 113 (Fe), 49 (Mg), 3.7 (Zn)	[79]
Plants	Mg, Ca, Na	1064 nm, 7 ns, 80–140 mJ/pulse, 10 Hz; pellets	PLS (100 samples for calibration and 100 for validation)	Accuracy $> 98\%$	[89]
Seeds	Classification	532 nm, 5 ns, 80 mJ/pulse, 20 Hz, delay time 500 ns, gate width 2 ms	PLS-DA (255 samples for calibration and 81 for validation)	Accuracy: 97%	[90]
Bone	Classification	1064 nm, 4 ns, 100 mJ/pulse, 1 Hz, delay time 2 μs, gate width 100 ms	PLS-DA (50 samples for calibration and 50 for validation)	Accuracy: 58%	[91]
Tooth	Classification	1064 nm, 6 ns, 90 mJ/pulse, 2 Hz, delay time 1 μs, gate width 20 μs	PLS-DA (252 spectra from 4 samples for calibration, 22 spectra from 2 samples for prediction), 20 latent variables	Classification accuracy: 100%	[92]

(Continued)

Table 12.4 (Continued)

Sample	Element	Experimental	Data processing	Results	Ref.
Milk powder, sweet, and acid whey powder	Adulteration	532 nm, 18 mJ/pulse, 1 Hz, delay time 2000 ns, gate width 3000 ns; pellets	PCA-PLS (96 object for calibration and 48 for validation)	For adulteration with sweet whey, R^2 for calibration 0.999, R^2 for validation 0.981, RSD = 13.93%, REP = 22.64%, LOD = 1.5%; For adulteration with acid whey, R^2 for calibration 0.996, R^2 for validation 0.985, RSD = 6.37%, REP = 16.57%, LOD = 0.55%	[93]
Meat	Zn, Mg, Ca, Na, K	1064 nm, 38 mJ/pulse, 4 Hz, delay time 300 ns, gate width 1.05 ms; pellets	PLS (6 samples for calibration and 6 for validation)	For pork adulterated beef, RMSEC = 2.7, R^2 = 0.99; for chicken adulterated beef, RMSEC = 1.18 R^2 = 0.999	[94]
Glucose	B	1064 nm, 6 ns, 40 mJ/pulse, 10 Hz, delay time 2 μs, gate width 2 μs, pellets	Genetic algorithm and partial least squares regression combination model (9 samples for calibration and 4 for validation)	R^2 = 0.988, RMSEC = 0.2187 wt%, RMSEP = 0.8667 wt%, MPE = 10.9685%	[95]
Honey	Ca, Mg, Na	1064 nm, 6 ns, 180 mJ, 10 Hz; solution	PLS	R^2 = 0.923 (Ca), 0.950 (Mg), 0.909 (Na), RSD = 1.55~7.40% (Ca), 1.95~9.22% (Mg), 3.42–10.24% (Na)	[96]
Gelatin	Edible gelatin adulteration	1064 nm, 8 ns, 100 mJ/pulse, 1 Hz, delay time 0 μs, gate width 1 ms, pellets	PLS support vector machine method (2 samples for calibration and 11 for validation)	R^2 = 0.9790, RESEP = 5.69%	[97]
Motor oil	O, N, H, CN, C_2	1064 nm, 90 mJ/pulse, 5 ns, minimum gate time <2 ns, gate width 1–4 μs; focusing length 30 m	PLS	Predicted value: 0.91 (Chevron sample), 1.099 (Mobil sample), 1.163 (Shell sample), 0.970 (Valvoline)	[98]

References

1 R.L. Somorjai, B. Dolenko, R.J.B. Baumgartner, Class prediction and discovery using gene microarray and proteomics mass spectroscopy data: Curses, caveats, cautions, *Bioinformatics*, **19** (2003), 1484–1491.
2 C.H. Mason, W.D. Perreault, Collinearity, power, and interpretation of multiple regression analysis, *Journal of Marketing Research*, **28** (1991), 268.
3 Z. He, H. Zhou, J. Wang, S. Zhai, A unified framework for contrast research of the latent variable multivariate regression methods, *Chemometrics and Intelligent Laboratory Systems*, **143** (2015), 136–145.
4 A. Höskuldsson, PLS regression methods, *Journal of Chemometrics*, **2** (1988), 211–228.
5 M. Barker, W. Rayens, Partial least squares for discrimination, *Journal of Chemometrics*, **17** (2003), 166–173.
6 H. Wold, Nonlinear iterative partial least squares (NIPALS) modelling: Some current developments, *Multivariant Analysis*, **1973**: pp. 383–407 1973. https://doi.org/10.1016/B978-0-12-426653-7.50032-6.
7 S.D.J.C. Jong, I.L. Systems, SIMPLS: An alternative approach to partial least squares regression, *Chemometrics and Intelligent Laboratory Systems*, **18** (1993), 251–263.
8 J.P.A. Martins, R.F. Teofilo, M.M.C. Ferreira, Computational performance and cross-validation error precision of five PLS algorithms using designed and real data sets, *Journal of Chemometrics*, **24** (2010), 320–332.
9 U. Thissen, M. Pepers, B. Üstün, W.J. Melssen, L.M.C. Buydens, Comparing support vector machines to PLS for spectral regression applications, *Chemometrics and Intelligent Laboratory Systems*, **73** (2004), 169–179.
10 R. Rosipal, L.J. Trejo, Kernel partial least squares regression in reproducing Kernel Hilbert space, *Journal of Machine Learning Research*, **2** (2002), 97–123.
11 G.J. Postma, P.W.T. Krooshof, L.M.C. Buydens, Opening the kernel of kernel partial least squares and support vector machines, *Analytica Chimica Acta*, **705** (2011), 123–134.
12 F. Despagne, D.L. Massart, P.J.A.C. Chabot, Development of a robust calibration model for nonlinear in-line process data, *Analytical Chemistry*, **72** (2000), 1657–1665.
13 S. Kim, M. Kano, H. Nakagawa, S. Hasebe, Estimation of active pharmaceutical ingredients content using locally weighted partial least squares and statistical wavelength selection, *International Journal of Pharmaceutics*, **421** (2011), 269–274.
14 T. Uchimaru, M.J.I.P. Kano, Sparse sample regression based just-in-time modeling (SSR-JIT): Beyond locally weighted approach, *IFAC-PapersOnLine*, **49** (2016), 502–507.
15 X. Zhang, M. Kano, Y.J.C. Li, Locally weighted kernel partial least squares regression based on sparse nonlinear features for virtual sensing of nonlinear time-varying processes, *Computers & Chemical Engineering*, **104** (2017), 164–171.
16 Z. Wang, J. Feng, L. Li, W. Ni, Z. Li, A multivariate model based on dominant factor for laser-induced breakdown spectroscopy measurements, *Journal of Analytical Atomic Spectrometry*, **26** (2011), 2289–2299.
17 Z. Wang, J. Feng, L. Li, W. Ni, Z. Li, A non-linearized PLS model based on multivariate dominant factor for laser-induced breakdown spectroscopy measurements, *Journal of Analytical Atomic Spectrometry*, **26** (2011), 2175–2182.

18 J. Feng, Z. Wang, L. West, Z. Li, W.D. Ni, A PLS model based on dominant factor for coal analysis using laser-induced breakdown spectroscopy, *Analytical and Bioanalytical Chemistry*, **400** (2011), 3261–3271.

19 D. Ballabio, V.J.A.M. Consonni, Classification tools in chemistry. Part 1: linear models. *PLS-DA*, **5** (2013), 3790–3798.

20 N.F. Pérez, J. Ferré, R. Boqué, Calculation of the reliability of classification in discriminant partial least-squares binary classification, *Chemometrics and Intelligent Laboratory Systems*, **95** (2009), 122–128.

21 S. Yao, J. Lu, M. Dong, K. Chen, J. Li, J. Li, Extracting coal ash content from laser-induced breakdown spectroscopy (LIBS) spectra by multivariate analysis, *Applied Spectroscopy*, **65** (2011), 1197–1201.

22 W. Li, J. Lu, M. Dong, S. Lu, J. Yu, S. Li, J. Huang, J. Liu, Quantitative analysis of calorific value of coal based on spectral preprocessing by laser-induced breakdown spectroscopy (LIBS), *Energy & Fuels*, **32** (2018), 24–32.

23 A. Li, S. Guo, N. Wazir, K. Chai, L. Liang, M. Zhang, Y. Hao, P.F. Nan, R.B. Liu, Accuracy enhancement of laser induced breakdown spectra using permittivity and size optimized plasma confinement rings, *Optics Express*, **25** (2017), 27559–27569.

24 T.B. Yuan, Z. Wang, Z. Li, W.D. Ni, J.M. Liu, A partial least squares and wavelet-transform hybrid model to analyze carbon content in coal using laser-induced breakdown spectroscopy, *Analytica Chimica Acta*, **807** (2014), 29–35.

25 T. Yuan, Z. Wang, S.-L. Lui, Y. Fu, Z. Li, J. Liu, W. Ni, Coal property analysis using laser-induced breakdown spectroscopy, *Journal of Analytical Atomic Spectrometry*, **28** (2013), 1045–1053.

26 Z. Hou, Z. Wang, T. Yuan, J. Liu, Z. Li, W. Ni, A hybrid quantification model and its application for coal analysis using laser induced breakdown spectroscopy, *Journal of Analytical Atomic Spectrometry*, **31** (2016), 722–736.

27 L.-B. Guo, D. Zhang, L.-X. Sun, S.-C. Yao, L. Zhang, Z.-Z. Wang, Q.-Q. Wang, H.-B. Ding, Y. Lu, Z.-Y. Hou, Z. Wang, Development in the application of laser-induced breakdown spectroscopy in recent years: A review, *Frontiers of Physics*, **16** (2021), 22500.

28 M. Dong, L. Wei, J. Lu, W. Li, S. Lu, S. Li, C. Liu, J.H. Yoo, A comparative model combining carbon atomic and molecular emissions based on partial least squares and support vector regression correction for carbon analysis in coal using LIBS, *Journal of Analytical Atomic Spectrometry*, **34** (2019), 480–488.

29 J. Feng, Z. Wang, L. Li, Z. Li, W. Ni, A nonlinearized multivariate dominant factor-based partial least squares (PLS) model for coal analysis by using laser-induced breakdown spectroscopy, *Applied Spectroscopy*, **67** (2013), 291–300.

30 X. Li, Z. Wang, Y. Fu, Z. Li, W. Ni, A model combining spectrum standardization and dominant factor based partial least square method for carbon analysis in coal using laser-induced breakdown spectroscopy, *Spectrochimica Acta Part B-Atomic Spectroscopy*, **99** (2014), 82–86.

31 X. Li, Y. Yang, G. Li, B. Chen, W. Hu, Accuracy improvement of quantitative analysis of calorific value of coal by combining support vector machine and partial least square methods in laser-induced breakdown spectroscopy, *Plasma Science and Technology*, **22** (2020), 074014.

32 W. Zhang, Z. Zhuo, P. Lu, J. Tang, H. Tang, J. Lu, T. Xing, Y. Wang, LIBS analysis of the ash content, volatile matter, and calorific value in coal by partial least squares regression based on ash classification, *Journal of Analytical Atomic Spectrometry*, **35** (2020), 1621–1631.

33 S. Sheta, M.S. Afgan, Z.Y. Hou, S.C. Yao, L. Zhang, Z. Li, Z. Wang, Coal analysis by laser-induced breakdown spectroscopy: A tutorial review, *Journal of Analytical Atomic Spectrometry*, **34** (2019), 1047–1082.

34 M.S. Afgan, Z.Y. Hou, Z. Wang, Quantitative analysis of common elements in steel using a handheld mu-LIBS instrument, *Journal of Analytical Atomic Spectrometry*, **32** (2017), 1905–1915.

35 L. Sun, H. Yu, Z. Cong, Y. Xin, Y. Li, L. Qi, In situ analysis of steel melt by double-pulse laser-induced breakdown spectroscopy with a Cassegrain telescope, *Spectrochimica Acta Part B-Atomic Spectroscopy*, **112** (2015), 40–48.

36 J. Amador-Hernandez, L.E. Garcia-Ayuso, J.M. Fernandez-Romero, M.D.L. de Castro, Partial least squares regression for problem solving in precious metal analysis by laser induced breakdown spectrometry, *Journal of Analytical Atomic Spectrometry*, **15** (2000), 587–593.

37 A. Sarkar, V. Karld, S.K. Aggarwal, G.S. Maurya, R. Kumar, A.K. Rai, X. Mao, R.E. Russo, Evaluation of the prediction precision capability of partial least squares regression approach for analysis of high alloy steel by laser induced breakdown spectroscopy, *Spectrochimica Acta Part B-Atomic Spectroscopy*, **108** (2015), 8–14.

38 M. Singh, V. Karki, A. Sarkar, Optimization of conditions for determination of Cr and Ni in steel by the method of laser-induced breakdown spectroscopy with the use of partial least squares regression, *Journal of Applied Spectroscopy*, **83** (2016), 497–503.

39 C.B. Stipe, B.D. Hensley, J.L. Boersema, S.G. Buckley, Laser-induced breakdown spectroscopy of steel: A comparison of univariate and multivariate calibration methods, *Applied Spectroscopy*, **64** (2010), 154–160.

40 Z.-b. Cong, L.-x. Sun, Y. Xin, Y. Li, L.-f. Qi, Z.-j. Yang, Quantitative analysis of alloy steel based on laser induced breakdown spectroscopy with partial least squares method, *Spectroscopy and Spectral Analysis*, **34** (2014), 542–547.

41 T. Zhang, L. Liang, K. Wang, H. Tang, X. Yang, Y. Duan, H. Li, A novel approach for the quantitative analysis of multiple elements in steel based on laser-induced breakdown spectroscopy (LIBS) and random forest regression (RFR), *Journal of Analytical Atomic Spectrometry*, **29** (2014), 2323–2329.

42 S. Kashiwakura, K. Wagatsuma, Optimization of partial-least-square regression for determination of manganese in low-alloy steel by single-shot laser-induced breakdown spectroscopy, *ISIJ International*, **58** (2018), 1705–1710.

43 S. Kashiwakura, K. Wagatsuma, Selection of atomic emission lines on the mutual identification of austenitic stainless steels with a combination of laser-induced breakdown spectroscopy (LIBS) and partial-least-square regression (PLSR), *ISIJ International*, **60** (2020), 1245–1253.

44 H. Kim, S.-H. Nam, S.-H. Han, S. Jung, Y. Lee, Laser-induced breakdown spectroscopy analysis of alloying elements in steel: Partial least squares modeling based on the low-resolution spectra and their first derivatives, *Optics and Laser Technology*, **112** (2019), 117–125.

45 C. Ma, J. Cui, Quantitative analysis of composition in molten steel by LIBS based on improved partial least squares, *Laser Technology*, **40** (2016), 876–881.

46 A. Freedman, F.J. Iannarilli, J.C. Wormhoudt, Aluminum alloy analysis using microchip-laser induced breakdown spectroscopy, *Spectrochimica Acta Part B-Atomic Spectroscopy*, **60** (2005), 1076–1082.

47 X. Li, Z. Wang, S.-L. Lui, Y. Fu, Z. Li, J. Liu, W. Ni, A partial least squares based spectrum normalization method for uncertainty reduction for laser-induced breakdown spectroscopy measurements, *Spectrochimica Acta Part B-Atomic Spectroscopy*, **88** (2013), 180–185.

48 J.M. Andrade, G. Cristoforetti, S. Legnaioli, G. Lorenzetti, V. Palleschi, A.A. Shaltout, Classical univariate calibration and partial least squares for quantitative analysis of brass samples by laser-induced breakdown spectroscopy, *Spectrochimica Acta Part B-Atomic Spectroscopy*, **65** (2010), 658–663.

49 L.E. Garcia-Ayuso, J. Amador-Hernandez, J.M. Fernandez-Romero, M.D.L. de Castro, Characterization of jewellery products by laser-induced breakdown spectroscopy, *Analytica Chimica Acta*, **457** (2002), 247–256.

50 A. Jurado-Lopez, M.D.L. de Castro, Rank correlation of laser-induced breakdown spectroscopic data for the identification of alloys used in jewelry manufacture, *Spectrochimica Acta Part B-Atomic Spectroscopy*, **58** (2003), 1291–1299.

51 A.M. Ollila, H.E. Newsom, B. Clark, III, R.C. Wiens, A. Cousin, J.G. Blank, N. Mangold, V. Sautter, S. Maurice, S.M. Clegg, O. Gasnault, O. Forni, R. Tokar, E. Lewin, M.D. Dyar, J. Lasue, R. Anderson, S.M. McLennan, J. Bridges, D. Vaniman, N. Lanza, C. Fabre, N. Melikechi, G.M. Perrett, J.L. Campbell, P.L. King, B. Barraclough, D. Delapp, S. Johnstone, P.-Y. Meslin, A. Rosen-Gooding, J. Williams, The MSL Science Team, Trace element geochemistry (Li, Ba, Sr, and Rb) using Curiosity's ChemCam: Early results for Gale crater from Bradbury Landing Site to Rocknest, *Journal of Geophysical Research, Planets*, **119** (2014), 255–285.

52 M.D. Dyar, M.L. Carmosino, E.A. Breves, M.V. Ozanne, S.M. Clegg, R.C. Wiens, Comparison of partial least squares and lasso regression techniques as applied to laser-induced breakdown spectroscopy of geological samples, *Spectrochimica Acta Part B-Atomic Spectroscopy*, **70** (2012), 51–67.

53 T. Zhang, S. Wu, J. Dong, J. Wei, K. Wang, H. Tang, X. Yang, H. Li, Quantitative and classification analysis of slag samples by laser induced breakdown spectroscopy (LIBS) coupled with support vector machine (SVM) and partial least square (PLS) methods, *Journal of Analytical Atomic Spectrometry*, **30** (2015), 368–374.

54 W. Zhang, R. Zhou, P. Yang, K. Liu, J. Yan, P. Gao, Z. Tang, X. Li, Y. Lu, X. Zeng, Determination of chlorine with radical emission using laser-induced breakdown spectroscopy coupled with partial least square regression, *Talanta*, **198** (2019), 93–96.

55 R.B. Anderson, R.V. Morris, S.M. Clegg, J.F. Bell, III, R.C. Wiens, S.D. Humphries, S.A. Mertzman, T.G. Graff, R. McInroy, The influence of multivariate analysis methods and target grain size on the accuracy of remote quantitative chemical analysis of rocks using laser induced breakdown spectroscopy, *Icarus*, **215** (2011), 608–627.

56 T.F. Boucher, M.V. Ozanne, M.L. Carmosino, M.D. Dyar, S. Mahadevan, E.A. Breves, K.H. Lepore, S.M. Clegg, A study of machine learning regression methods for major elemental analysis of rocks using laser-induced breakdown spectroscopy, *Spectrochimica Acta Part B-Atomic Spectroscopy*, **107** (2015), 1–10.

57 S.M. Clegg, E. Sklute, M.D. Dyar, J.E. Barefield, R.C. Wiens, Multivariate analysis of remote laser-induced breakdown spectroscopy spectra using partial least squares, principal component analysis, and related techniques, *Spectrochimica Acta Part B-Atomic Spectroscopy*, **64** (2009), 79–88.

58 Y. Ding, G. Xia, H. Ji, X. Xiong, Accurate quantitative determination of heavy metals in oily soil by laser induced breakdown spectroscopy (LIBS) combined with interval partial least squares (IPLS), *Analytical Methods*, **11** (2019), 3657–3664.

59 M.D. Dyar, J.M. Tucker, S. Humphries, S.M. Clegg, R.C. Wiens, M.D. Lane, Strategies for mars remote laser-induced breakdown spectroscopy analysis of sulfur in geological samples, *Spectrochimica Acta Part B-Atomic Spectroscopy*, **66** (2011), 39–56.

60 M.E. Essington, G.V. Melnichenko, M.A. Stewart, R.A. Hull, Soil metals analysis using laser-induced breakdown spectroscopy (LIBS), *Soil Science Society of America Journal*, **73** (2009), 1469–1478.

61 E. Ewusi-Annan, D.M. Delapp, R.C. Wiens, N. Melikechi, Automatic preprocessing of laser-induced breakdown spectra using partial least squares regression and feed-forward artificial neural network: Applications to Earth and Mars data, *Spectrochimica Acta Part B-Atomic Spectroscopy*, **171** (2020), 105930.

62 L. Gomez-Nubia, J. Aramendia, S. Fdez-Ortiz de Vallejuelo, J. Manuel Madariaga, Analytical methodology to elemental quantification of weathered terrestrial analogues to meteorites using a portable laser-induced breakdown spectroscopy (LIBS) instrument and partial least squares (PLS) as multivariate calibration technique, *Microchemical Journal*, **137** (2018), 392–401.

63 Z.Q. Hao, C.M. Li, M. Shen, X.Y. Yang, K.H. Li, L.B. Guo, X.Y. Li, Y.F. Lu, X.Y. Zeng, Acidity measurement of iron ore powders using laser-induced breakdown spectroscopy with partial least squares regression, *Optics Express*, **23** (2015), 7795–7801.

64 Y.M. Guo, L.B. Guo, Z.Q. Hao, Y. Tang, S.X. Ma, Q.D. Zeng, S.S. Tang, X.Y. Li, Y.F. Lu, X.Y. Zeng, Accuracy improvement of iron ore analysis using laser-induced breakdown spectroscopy with a hybrid sparse partial least squares and least-squares support vector machine model, *Journal of Analytical Atomic Spectrometry*, **33** (2018), 1330–1335.

65 H. Hu, L. Huang, J. Tu, X. Xu, M. Yao, T. Chen, M. Liu, P. Yang, C. Wang, Determination of Pb in soil by laser induced breakdown spectroscopy coupled with interval partial least squares, *Applied Laser*, **35** (2015), 104–109.

66 H. Li, M. Huang, H. Xu, High accuracy determination of copper in copper concentrate with double genetic algorithm and partial least square in laser-induced breakdown spectroscopy, *Optics Express*, **28** (2020), 2142–2155.

67 J.L. Luque-Garcia, R. Soto-Ayala, M.D.L. de Castro, Determination of the major elements in homogeneous and heterogeneous samples by tandem laser-induced breakdown spectroscopy-partial least square regression, *Microchemical Journal*, **73** (2002), 355–362.

68 Q. Shi, G. Niu, Q. Lin, T. Xu, F. Li, Y. Duan, Quantitative analysis of sedimentary rocks using laser-induced breakdown spectroscopy: comparison of support vector regression and partial least squares regression chemometric methods, *Journal of Analytical Atomic Spectrometry*, **30** (2015), 2384–2393.

69 M. Singh, V. Karki, R.K. Mishra, A. Kumar, C.P. Kaushik, X. Mao, R.E. Russo, A. Sarkar, Analytical spectral dependent partial least squares regression: a study of nuclear waste glass from thorium based fuel using LIBS, *Journal of Analytical Atomic Spectrometry*, **30** (2015), 2507–2515.

70 J.B. Sirven, B. Bousquet, L. Canioni, L. Sarger, Laser-induced breakdown spectroscopy of composite samples: Comparison of advanced chemometrics methods, *Analytical Chemistry*, **78** (2006), 1462–1469.

71 X.-H. Zou, Z.-Q. Hao, R.-X. Yi, L.-B. Guo, M. Shen, X.-Y. Li, Z.-M. Wang, X.-Y. Zeng, Y.-F. Lu, Quantitative analysis of soil by laser-induced breakdown spectroscopy using genetic algorithm-partial least squares, *Chinese Journal of Analytical Chemistry*, **43** (2015), 181–186.

72 D.C. Alvey, K. Morton, R.S. Harmon, J.L. Gottfried, J.J. Remus, L.M. Collins, M.A. Wise, Laser-induced breakdown spectroscopy-based geochemical fingerprinting for the rapid analysis and discrimination of minerals: The example of garnet, *Applied Optics*, **49** (2010), C168–C180.

73 J.-B. Sirven, B. Salle, P. Mauchien, J.-L. Lacour, S. Maurice, G. Manhes, Feasibility study of rock identification at the surface of Mars by remote laser-induced breakdown spectroscopy and three chemometric methods, *Journal of Analytical Atomic Spectrometry*, **22** (2007), 1471–1480.

74 J.L. Gottfried, R.S. Harmon, F.C. De Lucia, Jr., A.W. Miziolek, Multivariate analysis of laser-induced breakdown spectroscopy chemical signatures for geomaterial classification, *Spectrochimica Acta Part B-Atomic Spectroscopy*, **64** (2009), 1009–1019.

75 F.C. De Lucia, Jr., J.L. Gottfried, C.A. Munson, A.W. Miziolek, Multivariate analysis of standoff laser-induced breakdown spectroscopy spectra for classification of explosive-containing residues, *Applied Optics*, **47** (2008), G112–G121.

76 Wang, Q.Q., Liu, K., Zhao, H. et al. Detection of explosives with laser-induced breakdown spectroscopy. *Frontier in Physics* **7**, 701–707 (2012), https://doi.org/10.1007/s11467-012-0272-x.

77 K. Liu, D. Tian, X. Deng, H. Wang, G. Yang, Rapid classification of plastic bottles by laser-induced breakdown spectroscopy (LIBS) coupled with partial least squares discrimination analysis based on spectral windows (SW-PLS-DA), *Journal of Analytical Atomic Spectrometry*, **34** (2019), 1665–1671.

78 H. Yang, L. Huang, M. Liu, T. Chen, G. Rao, C. Wang, M. Yao, Detection of cadmium in navel orange by laser induced breakdown spectroscopy combined with moving window partial least square, *Laser & Optoelectronics Progress*, **54** (2017), 083002-083001-083002-083008.

79 J.W. Batista Braga, L.C. Trevizan, L.C. Nunes, I.A. Rufini, D. Santos, Jr., F.J. Krug, Comparison of univariate and multivariate calibration for the determination of micronutrients in pellets of plant materials by laser induced breakdown spectrometry, *Spectrochimica Acta Part B-Atomic Spectroscopy*, **65** (2010), 66–74.

80 J.L. Gottfried, F.C. De Lucia, Jr., A.W. Miziolek, Discrimination of explosive residues on organic and inorganic substrates using laser-induced breakdown spectroscopy, *Journal of Analytical Atomic Spectrometry*, **24** (2009), 288–296.

81 J.L. Gottfried, F.C. De Lucia, Jr., C.A. Munson, A.W. Miziolek, Strategies for residue explosives detection using laser-induced breakdown spectroscopy, *Journal of Analytical Atomic Spectrometry*, **23** (2008), 205–216.

82 J.L. Gottfried, F.C. De Lucia, Jr., C.A. Munson, A.W. Miziolek, Standoff detection of chemical and biological threats using laser-induced breakdown spectroscopy, *Applied Spectroscopy*, **62** (2008), 353–363.

83 N.C. Dingari, I. Barman, A.K. Myakalwar, S.P. Tewari, M.K. Gundawar, Incorporation of support vector machines in the LIBS toolbox for sensitive and robust classification amidst unexpected sample and system variability, *Analytical Chemistry*, **84** (2012), 2686–2694.

84 Q. Godoi, F.O. Leme, L.C. Trevizan, E.R. Pereira-Filho, I.A. Rufini, D. Santos, Jr., F.J. Krug, Laser-induced breakdown spectroscopy and chemometrics for classification of toys relying on toxic elements, *Spectrochimica Acta Part B-Atomic Spectroscopy*, **66** (2011), 138–143.

85 K. Liu, D. Tian, H. Wang, G. Yang, Rapid classification of plastics by laser-induced breakdown spectroscopy (LIBS) coupled with partial least squares discrimination analysis based on variable importance (VI-PLS-DA), *Analytical Methods*, **11** (2019), 1174–1179.

86 S. Gregoire, M. Boudinet, F. Pelascini, F. Surma, V. Detalle, Y. Holl, Laser-induced breakdown spectroscopy for polymer identification, *Analytical and Bioanalytical Chemistry*, **400** (2011), 3331–3340.

87 L.C. Nunes, J.W.B. Braga, L.C. Trevizan, P.F. de Souza, G.G. Arantes de Carvalho, D. Santos Junior, R.J. Poppi, F.J. Krug, Optimization and validation of a LIBS method for the determination of macro and micronutrients in sugar cane leaves, *Journal of Analytical Atomic Spectrometry*, **25** (2010), 1453–1460.

88 X. Zhang, M. Yao, M. Liu, Quantitative analysis of cadmium in navel orange by laser-induced breakdown spectroscopy combined with partial least squares, *Acta Physica Sinica*, **62** (2013), 044211-044211-044211-044216.

89 G. Kim, J. Kwak, J. Choi, K. Park, Detection of nutrient elements and contamination by pesticides in spinach and rice samples using laser-induced breakdown spectroscopy (LIBS), *Journal of Agricultural and Food Chemistry*, **60** (2012), 718–724.

90 M.R. Martelli, F. Brygo, A. Sadoudi, P. Delaporte, C. Barron, Laser-induced breakdown spectroscopy and chemometrics: A novel potential method to analyze wheat grains, *Journal of Agricultural and Food Chemistry*, **58** (2010), 7126–7134.

91 S. Moncayo, S. Manzoor, F. Navarro-Villoslada, J.O. Caceres, Evaluation of supervised chemometric methods for sample classification by laser induced breakdown spectroscopy, *Chemometrics and Intelligent Laboratory Systems*, **146** (2015), 354–364.

92 M. Gazmeh, M. Bahreini, S.H. Tavassoli, Discrimination of healthy and carious teeth using laser-induced breakdown spectroscopy and partial least square discriminant analysis, *Applied Optics*, **54** (2015), 123–131.

93 G. Bilge, B. Sezer, K.E. Eseller, H. Berberoglu, A. Topcu, I.H. Boyaci, Determination of whey adulteration in milk powder by using laser induced breakdown spectroscopy, *Food Chemistry*, **212** (2016), 183–188.

94 G. Bilge, H.M. Velioglu, B. Sezer, K.E. Eseller, I.H. Boyaci, Identification of meat species by using laser-induced breakdown spectroscopy, *Meat Science*, **119** (2016), 118–122.

95 Z. Zhu, J. Li, Y. Guo, X. Cheng, Y. Tang, L. Guo, X. Li, Y. Lu, X. Zeng, Accuracy improvement of boron by molecular emission with a genetic algorithm and partial least squares regression model in laser-induced breakdown spectroscopy, *Journal of Analytical Atomic Spectrometry*, **33** (2018), 205–209.

96 K.W. Se, S.K. Ghoshal, R.A. Wahab, Laser-induced breakdown spectroscopy unified partial least squares regression: An easy and speedy strategy for predicting Ca, Mg and Na content in honey, *Measurement*, **136** (2019), 1–10.

97 H. Zhang, S. Wang, D. Li, Y. Zhang, J. Hu, L. Wang, Edible gelatin diagnosis using laser-induced breakdown spectroscopy and partial least square assisted support vector machine, *Sensors*, **19** (2019), 4225.

98 A.A. Bol'shakov, J.H. Yoo, C. Liu, J.R. Plumer, R.E. Russo, Laser-induced breakdown spectroscopy in industrial and security applications, *Applied Optics*, **49** (2010), C132–C142.

13

Nonlinear Methods

Francesco Poggialini, Asia Botto, Beatrice Campanella, Stefano Legnaioli, Simona Raneri, and Vincenzo Palleschi

Applied and Laser Spectroscopy Laboratory, Institute of Chemistry of Organometallic Compounds, Research Area of the National Research Council, Pisa, Italy

Laser-induced breakdown spectroscopy (LIBS) spectra are characterized by a redundancy of information that makes the use of multivariate methods for building calibration surfaces particularly suited. In the previous chapter, the case of linear calibration surfaces in all the coordinates considered was discussed. However, many effects in LIBS may produce a nonlinear dependence of the signal from the concentration of the elements in the sample. In the past decades, several multivariate nonlinear chemometric methods have been proposed and successfully tested for LIBS analytical applications. In this chapter, we will present a general overview and some examples on the application of these techniques.

13.1 Introduction

The general relation between LIBS signals (line intensities) and concentrations in a sample has the form:

$$C_i = f_i(I_i) \tag{13.1}$$

where the functions f_i can assume arbitrary form and are, in general, nonlinear.

As discussed in the section on linear methods (Chapter 4.1), in some limit cases a relation may exist between the concentration of a single element and several emission lines (of the same element, or of other elements in the sample). In these cases, a single nonlinear function f can be considered:

$$C = f(I_i) \tag{13.2}$$

The preliminary measurement of LIBS signals on samples of known concentration has the purpose of determining the form of the functions f_i. The measurement of the intensities of the same lines on unknown samples could be successively exploited for calculating the

---

Chemometrics and Numerical Methods in LIBS, First Edition. Edited by Vincenzo Palleschi.
© 2023 John Wiley & Sons Ltd. Published 2023 by John Wiley & Sons Ltd.

concentration(s) of the analyte(s) in the plasma, and then in the sample, assuming that the measurement conditions would be exactly the same during the calibration and measurement phases.

In this respect, the multivariate nonlinear methods suffer from the same problems previously discussed for the linear methods, i.e. the increase in the number of variables in the calculation, which would force the use of a large number of standards to cover all the possible variations of the parameters and avoid the risk of the so-called overfitting.

In an overfitting case, the parameters of the nonlinear multivariate functions to be determined at the calibration stage are comparable in number, or even larger, than the number of experimental data from the calibration samples. In this case, the nonlinear best fitting strategies would provide parameters that can reproduce perfectly the data obtained from the calibration samples, but the functions determined in this way would fail dramatically when applied to unknown samples [1]. This is the reason why the nonlinear multivariate techniques most used in LIBS (as well as in other spectrochemical methods) always apply an internal validation to find the best functional relation between concentrations and intensities relatively unaffected by the overfitting problem.

However, the proper validation of the calibration process can be only obtained through a process of external validation, i.e. the analysis of reference samples of known composition that were not used for building the multidimensional calibration surface in Eq. (13.1). Although this validation procedure could be difficult to apply in LIBS, given the problem of collecting enough standards for the calibration process, it is a necessary step for performing meaningful quantitative LIBS analysis and should not be bypassed in any case.

13.2 Multivariate Nonlinear Algorithms

13.2.1 Artificial Neural Networks

13.2.1.1 Conventional Artificial Neural Networks

Artificial neural networks (ANNs) or simply neural networks (NNs) are nowadays the most used algorithm for nonlinear multivariate quantitative analysis of materials by LIBS. Similar to linear methods, ANNs are often used for classification. This topic has been discussed in Section 3.3 of this book.

In the framework of LIBS quantitative analysis, ANNs are often applied in the feed-forward multilayer configuration [2]. The optimization of the ANN parameters is obtained through a backpropagation process in which the results of the ANN (C_{pred}) are compared with the known composition of the reference samples (C_{std}) for calculating an objective function to be minimized (for example, the root mean squared error of calibration – RMSEC):

$$\text{RMSEC} = \sqrt{\frac{\sum (C_{pred} - C_{std})^2}{n}} \quad (13.3)$$

Through an iterative process, the parameters of the ANN are gradually adjusted until they reach the global minimum of the objective function. Basically, the trueness of the predictions of the network is used as feedback to refine the same parameters, until further changes in the ANN weights and biases do not produce a further decrease of the RMSEC.

The structure of a feed-forward multilayer NN is a sequence of layers, in which all the elements (neurons) of a layer receive as input a linear combination of the elements of the previous layer and then apply to this number a function (linear or nonlinear). This layer, in turn, acts as input for the next layer, which is not directly accessible to the operator and for this reason is called "hidden layer." The neurons of a given layer are only connected with the neurons of the previous and the next layer, while there is no connection between neurons belonging to the same layer. From a mathematical point of view, there is no limit to the number of neurons in each layer and on the number of hidden layers. The only obvious constraint is that the first layer (input layer) should have a number of neurons equal to the number of inputs and the last layer (output layer) should have a number of neurons equal to the desired outputs (see Figure 13.1).

Practically, if we would like to measure by LIBS the concentration of three elements in a sample, and the calibration (in ANN jargon, the "training" of the network) is done considering the intensity of 5 spectral lines in the spectra, the number of inputs would be 5 and the number of outputs would be 3.

The parameters to be determined are the weights w_{ji} and a_{kz} and the biases b_i and z_h.

A few examples would clarify how the algorithm of the ANN works. Let us consider a linear univariate model, where the concentration of a given analyte would be obtained from the intensity of a single LIBS line in the form:

$$C = wI + b \tag{13.4}$$

The ANN that could be built for obtaining the w and b parameters from a set of samples of known concentration has a single neuron in the input layer (the measured line intensity), one in the output layer (the corresponding concentration) and just one neuron in the hidden layer (Figure 13.2).

The hidden neuron applies to the input the identity function $f(x) = x$, i.e. it passes the input value to the output neuron, multiplied by the factor w and added to the "bias" b (in this case, the a and z parameters can be taken as 1 and 0, respectively, without any loss of generality).

The parameters w and b can be determined by iteratively minimizing the RMSEC cost function or by minimizing the sum of the squares of the prediction errors, which is the same as performing a least square minimization as described in Section 4.1.

Once the weight and the bias have been determined (the model requires the optimization of just these two parameters), the concentration of the analyte in an unknown sample, on which we measure a line intensity I_{unk}, can be immediately calculated from Eq. (13.4).

A slightly more complicated situation corresponds to the case where the analyte concentration can be related to a LIBS line through a nonlinear correspondence:

$$C = f(I) \tag{13.5}$$

If we assume the knowledge of the f function, it would be straightforward to substitute the identity function, which is appropriate when the dependence between concentration and intensity is linear, with this same function (Figure 13.3).

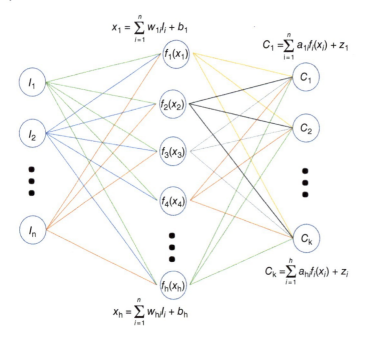

Figure 13.1 Schematic representation of an artificial neural network with n inputs, k outputs, and h neurons in the hidden layer.

Figure 13.2 An artificial neural network for the determination of a univariate linear model (Eq. 13.4).

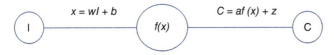

Figure 13.3 An artificial neural network for the determination of a univariate nonlinear model (Eq. 13.5).

However, in LIBS as in many other techniques, the functional dependence of the concentration on the experimentally determined line intensity is rarely known. Depending on the range of concentrations analyzed, we can observe either a linear dependence or an exponential variation (increasing or decreasing). This is the situation in which the ANN algorithm works better. This approach does not require the function $f(x)$, applied

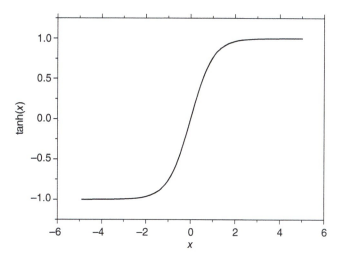

Figure 13.4 The sigmoidal function defined by Eq. (13.6).

by the hidden neuron, to be identical to the function in Eq. (13.5). There are many possible choices for the transfer function that could be used in building an ANN. In typical LIBS applications, where we expect a more or less "regular" dependence of the concentration on the signal and vice versa, a good choice is represented by the sigmoidal function (Figure 13.4):

$$f(x) = \tanh(x) = \frac{e^x - e^{-x}}{e^x + e^{-x}} \tag{13.6}$$

The function defined by Eq. (13.6) contains in the same curve a positive exponential ($x \ll -1$), a linear dependence ($-1 < x < 1$), and a negative exponential ($x \gg 1$), with all the possible transitions between a regime and the other.

In the last two examples, a univariate case is considered, for simplicity. A further step toward a deeper understanding of the ANN algorithm is the simplified multivariate situation in which the concentration of a single analyte is linearly correlated to the intensities of several spectral lines. We have seen in Chapter 12 that this case can be modeled using a partial least squares (PLS) approach. In the framework of ANN formalism to determine the optimum parameters of the network, we would build the ANN represented in Figure (13.5).

During the training phase, the optimum weights w_i and the bias b will be determined in order to minimize the RMSEC. Even though the method is different, the parameters are optimized by the ANN as in the PLS algorithm, to minimize the same objective function.

The general case of a nonlinear multivariate dependence between concentrations and intensities is more complicated, but the idea should be clear, from the previous example. Let us consider the equivalent of the ANN described in Figure 13.5, but representing a possible nonlinear dependence between the inputs (I_i) and the output C. In this case, the number of neurons in the hidden layer cannot be determined a priori. However, if we are describing a system where the dependence is not too far from linearity, it is reasonable

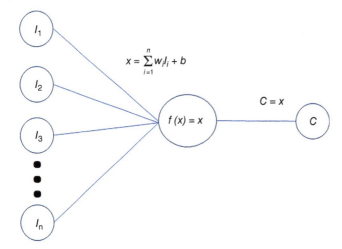

Figure 13.5 An artificial neural network for the determination of a multivariate linear model.

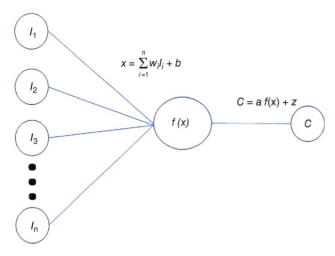

Figure 13.6 An artificial neural network for the determination of a multivariate nonlinear model with a single output.

to think that a single neuron in the hidden layer would be enough for the analysis (see Figure 13.6)

While the ANN algorithm is able to manage a large number of hidden neurons, it should be clear that the number of free parameters to be optimized diverges quickly with the increase of the number of inputs and the number of hidden neurons.

A number of papers dealing with the use of ANNs does not consider the danger of having too many free parameters, compared to the number of different cases (spectra) used for the training of the network [1]. One of the telltale signs of this problem, which is technically called "overfitting," is the exceptionally low values of the errors or calibration (RMSEC). Sometimes this result is mistakenly presented as a demonstration of the goodness of the

ANN for analytical purpose. In the presence of overfitting, the predictive capabilities of the ANN are irremediably degraded.

Although the optimization algorithms used by most of the commercial software attempt to reduce the possible effects of overfitting by performing a sort of internal validation, the only meaningful parameter that would tell the real predictive capability of the algorithm is the root mean squared error of prediction (RMSEP) (Eq. 13.3):

$$\text{RMSEP} = \sqrt{\frac{\sum (C_{\text{pred}} - C_{\text{std}})^2}{n}} \qquad (13.7)$$

The RMSEP refers to the errors in the predictions of the ANN, when applied to the analysis of external standards of know composition, that have not been used for the training of the ANN. The analytical performance of the Artificial Neural Network must be evaluated on samples of known composition that have not been used for training the ANN (external validation).

For the application of the ANN technique in LIBS experiments, the major difficulty is finding a number of samples sufficient for the calibration to avoid overfitting. However, there are several methods that can be used for performing a good external calibration, the most used is called "Leave-One-Out" (LOO). In the LOO approach, all the samples except one are used for the ANN training. Then, its analytical performances are tested on the sample not used for the training, and finally the samples are rotated, in order to have one of them not used for the calibration, and therefore ready for the external validation.

The advantages of this method are evident: a few samples are enough for the external evaluation of the performances of the network. On the other hand, the RMSEP defined in Eq. (13.7) must be calculated considering the outputs of different networks and is not so obvious that these networks would have similar performances. Moreover, since the ANN can be used only for predicting concentrations between the minimum and maximum used for the training (extrapolation of the prediction outside these limits would not produce meaningful results), at least two of the reference samples (the ones corresponding to the minimum and maximum concentration of the analytes to be determined) could not be used for external validation in the LOO procedure. In LIBS, it is not unusual to have only four or five samples available for the validation. Nevertheless, it is essential to perform an external validation of the ANN before using it for analytical purpose. Particular care should be given to values of the RMSEC that appear too good, and it should be remembered that this parameter is only used for the optimization of the ANN parameters. The RMSEC does not give information on the predictive analytical performances of the network. For that, RMSEP should be considered instead.

To reduce the risk of overfitting, the number of inputs and neurons in the hidden layer must be kept to the minimum. In ref. [3] a hierarchical method is presented for choosing the LIBS spectral features, which better reproduce the composition of the standards. The method selects first the single spectral feature, which gives the best analytical results, used as input in an ANN, then the second that, together with the already selected one, will give the best analytical result in a two-inputs ANN. It was found that good analytical performances were obtained selecting only five spectral features, among over 3500 spectral points acquired.

In recent years, several papers have been devoted to developing genetic algorithms for feature selection [4–7]. In ref. [8] a feature selection method is proposed, based on the analysis of the variance of the LIBS signal on all the spectra acquired on the reference samples. Only the spectral points showing a large variability between one sample and the other (the most dependent on the change in composition of the samples) were retained and used as input of the ANN. In all the cases an experienced LIBS operator would determine the optimum set of spectral features much faster and more efficiently than any sophisticated computer algorithm. However, if the use of automatic feature selection algorithms is deemed appropriate, the result must always be checked to ensure that the spectral points selected would correspond to a meaningful physical feature.

The optimization on the number of neurons in the hidden layer can be performed experimentally, by studying the effect on the RMSEP. However, some heuristic considerations can guide a reasonable choice for this number. As in the case of a single output ANN, we expect to need a hidden neuron for each nonlinearity in the relation between inputs and outputs. A good starting point for the optimization is thus putting the number of hidden neurons equal to the number of outputs, and then testing values around that number. Commercial analytical software is able to perform this optimization automatically, between a minimum and maximum number of neurons given by the operator.

The training of the ANN (i.e. the determination of the weights w_{ji} and a_{kz} and the biases b_i and z_h, in the general case) can be long and complicated, because the predictions of the ANN must be validated on a subset of spectra randomly chosen, and the whole training process repeated several times to avoid the error backpropagation for the minimization of the RMSEC, typically a Levenberg–Marquardt type steepest-descent algorithm [9], would remain stuck on local minima of the RMSEC curve.

However, once the network is trained, the evaluation of the composition of an unknown sample from the measured line intensities is extremely fast, using the processing scheme depicted in Figure (13.1). The ANN is thus particularly suited for online monitoring of industrial processes [10–12] and environmental [13–16], biological, and medical applications [17, 18].

As a final remark on this topic, we would like to stress that the parameters of the ANN, once defined the optimum number of inputs and neurons in the hidden layer, are optimized through the minimization of the RMSEC (Eq. (13.3)). One of the strengths of LIBS is the possibility of determining, in a single measurement, both the major (%) and the minor (ppm) elements in the sample. However, in the minimization of the mean squared error, the errors in the minor elements would be negligible with respect to the errors in the major elements. Therefore, single ANN would not be able to provide a meaningful estimate of both the major and minor elements in the sample. This problem has suggested, in several applications, the use of two or more ANNs, to be applied in sequence, for determining separately the composition of the samples both for major and minor elements [8, 19]. In the case described in ref. [8], this strategy has allowed to improve the analytical performances of a commercial handheld LIBS instrument, largely outperforming the factory-installed software [20].

13.2.1.2 Convolutional Neural Networks

Convolutional neural networks (CNNs) are the core of so-called deep learning methods. The potential advantages in the use of these methods in LIBS would be a complete automatic and unsupervised feature selection, combined with the robustness and analytical performances of the conventional ANN.

A CNN is an ANN with many hidden layers, which are able to individuate and select characteristic features at different scales. The number of possible inputs (in LIBS, several thousands of spectral points) is transformed to a set of meaningful data, which are then linked to the concentrations to be determined with the same training procedure used in conventional ANN.

Since the CNN is principally used for image analysis (mainly object recognition and face detection), it has become customary in LIBS papers dealing with CNN, the transformation of the LIBS spectra in gray scale images (for example, a spectrum with 4096 spectral points is converted into a square image of 64×64 pixels, whose intensity would correspond to the spectral intensities of the original vector (Figure 13.7)).

Many predesigned CNNs are available and freely usable, for example in the Matlab® environment. However, the transformation of the spectrum in an image may compress the dynamics of the data to 8 bits (the intensity of grayscale images is represented by an integer comprised between 0 and 255) and is not justified by any reason, if not the possibility of using an existing CNN without adapting it to one-dimensional (1D) spectra. Only in a few works the CNN is built maintaining the 1D original spectrum as input [21, 22].

However, even in the few situations in which the CNNs are used correctly, their usefulness in LIBS analysis is very limited. We have already discussed the danger of using automatic feature selection methods in LIBS. The kind of materials that are typically analyzed with this technique are relatively limited, and experienced LIBS operators in their life would probably deal with no more than 30/40 elements of the periodic table. The criteria for using or not using a given elemental line for analytical purposes are rather simple (good signal, limited self-absorption, minimal interference with other lines, etc.). A CNN can be useful for identifying a feature (a face, road signs, an animal, etc.) in a given image, when this feature can appear in different positions, sizes, and rotations. The spectral lines in a LIBS spectrum are always in the same position, and they have similar shapes. Therefore, the advantages of using a CNN for feature selection in LIBS is quite limited if compared with the complexity of the algorithms involved, if not in very particular situations. One of these could be the analysis of LIBS spectra acquired on complex materials with low-resolution spectrometers (for example, steel analysis with handheld LIBS instrumentation [8]); in that

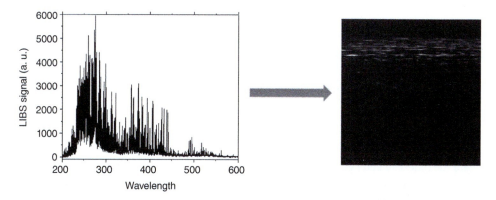

Figure 13.7 Conversion of a one-dimensional spectrum into a 2D image.

case, the human experience could be challenged by the interference and superposition of the characteristic spectral features of the elements of interest, and an automatic feature-selection approach based on CCN could have some justification. It should be considered, however, that the CNN algorithm requires, by its nature, a training involving several thousands of LIBS spectra, acquired from different samples of known composition. In usual LIBS applications, it is very difficult to meet these requirements, which are necessary for avoiding the risk of overfitting; for overcoming this problem, several methods of data augmentation have been proposed. Képeš et al. [23] recently proposed the bootstrapping method, a simple trick that can be used for increasing the number of LIBS spectra acquired on a sample. The idea is averaging the n spectra in groups of k (with $k \ll n$), to obtain many independent realizations of the averaged spectra; a positive side effect of bootstrapping is the smoothing out of the effect of possible outliers in the LIBS spectral set. The number of independent samples can be increased, on the other hand, by combining with different "weights" the spectra acquired on two or more samples and associating the new spectra to the weighted average of the composition of the samples. The synthetic linear combination of the spectra is only a rough approximation of the real spectra (nonlinear effects such as self-absorption are not correctly reproduced in the combined spectra); however, for the analysis of samples whose composition can vary in a relatively narrow range of concentrations, the synthetic spectra would not be very far from the real ones. The reliability of these data augmentation methods should be, in any case, always tested through external validation (on real spectra).

13.2.2 Other Nonlinear Multivariate Approaches

The ANN algorithm is very powerful and can provide excellent results, when used with the proper attention. However, the risk of using the ANNs as "black boxes," without checking the reliability of the results obtained, is high. As remarked in [1], the black box in fact is perfectly transparent, if we are interested in the analysis of its parameters. Nevertheless, on some occasions it might be useful adopting other methods, either faster or more easily controllable.

13.2.2.1 The Franzini–Leoni Method

The Franzini–Leoni method [24] is a variation of the Lucas–Tooth and Pine algorithm [25] widely applied in X-ray fluorescence for quantitative analysis of geological materials. The method was adapted to LIBS, for the first time, for the compositional analysis of silicate rocks [24], although the method is not limited to geological applications. The assumption is that the concentrations of the minerals in the rock/elements in the sample could be related to the intensity of the corresponding emission lines in the LIBS spectrum in the form:

$$C_i = A_i I_i \left(1 + \sum W_{ij} I_j\right) + B_i \tag{13.8}$$

where the indexes I and j go from 1 to the number of minerals/elements to be analyzed. In the case discussed in ref. [24], nine oxides where considered (MnO, Fe_2O_3, SiO_2, K_2O, TiO_2, CaO, MgO, Al_2O_3, Na_2O) and the emission lines of the corresponding elements where measured on 19 samples of silicate rocks of certified composition. Twenty-five spectra were acquired per samples, obtaining a total of 475 spectra for the determination of the A_i, W_i, and B_i coefficients. The total number of coefficients to be determined is $N = \frac{n(n-1)}{2} + 1$, where n is the number of elements considered.

When compared with different alternative strategies (univariate calibration curves, multivariate linear approach (PLS), and ANN), the Franzini–Leoni method demonstrates an excellent predictive capability, comparable to the one of an optimized ANN (number of hidden neurons equal to the number of the elements to be analyzed).

The main interest of this method is in the possibility of being treated in the framework of a multivariate linear approach, using as inputs the products $I_i I_j$ (with i and j ranging from 1 to n).

Since the PLS results are easier to control, with respect to a typical ANN, this approach can be considered as a valid alternative; the effect of the nonlinear components in the determination of the concentrations can be easily estimated, given the parameters A_i, W_i, and B_i obtained from the linear algorithm.

13.2.2.2 The Kalman Filter Approach

The Kalman filter approach uses a procedure that is common in sensor engineering for finding the optimal results deriving from two or more estimates of the same physical quantity, obtained with different instruments.

The classical example of application of the Kalman filter is the determination of the instantaneous position of a vehicle, starting from the data obtained by the Global Positioning System (GPS) and the accelerometer. The Kalman filter method considers the different accuracies of the two measurements and determines the result that is more compatible with the information provided by the two instruments.

The Kalman filter method has been adapted to LIBS by Palleschi and Palleschi [26] in 2018. The authors considered that, in analytical chemistry, the quantity to be determined can be the concentration of a given element, while the "sensors" to determine this concentration can be the intensities of two emission lines of the same element, with the corresponding calibration curves described by the two (possibly not linear) functions $f_1(I_1)$ and $f_2(I_2)$.

In this framework, the best estimate of the concentration of the analyte would be given by the formula:

$$C = f_1(I_1) + K(I_2 - f_2^{-1}(f_1(I_1))) \tag{13.9}$$

where f_2^{-1} is the inverse of the f_2 function and K is the Kalman filter, which represents a sort of correction of the estimate of the result of the first sensor in view of the results of the second sensor. The amount of the correction to be applied depends, among the other things, on the standard deviations z_1 and z_2 of the measured line intensities. In fact, the value of K is determined by minimizing the uncertainty on the result. If the uncertainty on the measurement of a sensor is much larger than the one obtained from the other sensor, the Kalman filter will ignore the measurement in favor of the less uncertain estimate. When the uncertainties are comparable, the best compromise is found between the two values obtained by the sensors.

In case of a linear dependence between the concentration and the line intensities (for the sake of simplicity, we will assume zero intercept of the calibration curves):

$$C_1 = \frac{I_1}{a_1}$$
$$C_2 = \frac{I_2}{a_2} \quad (13.10)$$

It can be easily demonstrated that, under the assumption of Eq. (13.10), the optimum concentration C is given by:

$$C = \frac{I_1}{a_1} + K\left(I_2 - a_2\frac{I_1}{a_1}\right). \quad (13.11)$$

With

$$K = \frac{1}{a_2}\frac{a_2^2 z_1^2}{a_2^2 z_1^2 + z_2^2}. \quad (13.12)$$

The resulting uncertainty on C is given by:

$$\sigma_C = \sqrt{\frac{z_1^2 z_2^2}{a_2^2 z_1^2 + z_2^2}}. \quad (13.13)$$

It can be demonstrated that the evaluation of the concentration of the analyte and the associated uncertainty do not depend on the order in which the lines are considered. The Kalman filter method can be extended to the analysis of the results of more sensors (more spectral lines, in case of LIBS analysis), providing the best evaluation (i.e. minimum uncertainty) of the resulting concentration.

In the general case, when the calibration curves are not linear, the slopes of the calibration curves can be substituted by the local linearization of the curves at the measured intensities:

$$\frac{1}{a_1} = \frac{d}{dI}f_1(I)\bigg|_{I_1}$$
$$\frac{1}{a_2} = \frac{d}{dI}f_2(I)\bigg|_{I_2} \quad (13.14)$$

In this case, the algorithm is called extended Kalman filter method. From a mathematical point of view, there is no guarantee that the result is the one minimizing the variance of the result, as in the linear case. However, in typical LIBS conditions extreme nonlinear effects are not expected. The simplicity of the approach might suggest the use of this method to find the optimum concentration when one or more lines emitted by the analyte show self-absorption effects, leading to a nonlinearity of the corresponding calibration curve.

13.2.2.3 Calibration-Free Methods

The calibration-free LIBS method (CF-LIBS) was introduced in 1999 by Palleschi and colleagues [27], following the general idea that the complex link between line intensities and elemental concentrations in the proper experimental conditions could be described by the basic equations describing the optical emission of a plasma in local thermal equilibrium [28], i.e. a plasma in which all the processes are at equilibrium at the same temperature T (the velocity distribution of atoms and electrons described by the Maxwell equation, the energy distribution among the atomic and ionic energy levels described by the Boltzmann equation, the ionization equilibrium given by the Saha–Eggert equation) [29]. This approximation is valid when the characteristic time for reaching the equilibrium through electron-atoms collisions is much shorter than the time of the radiative processes, which are not at the equilibrium.

The CF-LIBS method has been applied to the analysis of many different materials [30–34]. The application of the method requires the measurement, from the LIBS spectrum, of the plasma electron temperature and the electron number density, which appear in the Boltzmann and Saha–Eggert equations. There are many methods for determining these parameters, mostly based on the construction of a Boltzmann or Saha–Boltzmann plot [35] and the determination of the Stark broadening [36] of suitable reference lines [37, 38]. However, these methods require a numerical fitting of several emission lines, a procedure that often takes some time and is not easily applicable when a large number of LIBS spectra must be processed (in elemental mapping applications, for example, millions of LIBS spectra can be acquired to form a single elemental image).

To make a quantitative analysis feasible, while retaining the advantages of the calibration-free approach, two different strategies have been developed.

The first was a classification of the LIBS spectra, using the method of self-organizing maps (SOM) proposed by Palleschi and coworkers in 2015 [39], followed by CF-LIBS analysis of the prototype spectra representing the different clusters [40] (either the average of the spectra in the cluster, or the LIBS spectrum representing the centroid of the cluster, see Section 3.1 for more details on the SOM method).

The second method, also proposed by Palleschi and coworkers in 2015 [41], exploits the possibility of determining the electron temperature using the spectral information of the lines that normally would be used for the determination of the electron temperature from the Boltzmann plot (plus the hydrogen Balmer alpha line for the determination of the electron number density) using an ANN approach, demonstrated in [42].

According to this method, the same spectral information used for the determination of the electron temperature and number density are used for building an ANN, whose outputs are the elemental concentrations and the electron number density and temperature, determined by conventional CF-LIBS on a limited subset of the spectra.

Once properly trained, the above ANN can be used for a very fast analysis of many unknown spectra, providing for each of them the corresponding concentrations, the electron temperature, and number density. The optimization of the network done in [41] revealed that the optimum number of hidden neurons was equal to the number of outputs (number of elements whose concentration must be determined plus 2, corresponding to the two outputs of electron temperature and number density).

13.3 Conclusion

There are many different methods for the analysis of materials by LIBS exploiting multivariate nonlinear models. In this chapter, we have discussed the most important and recent techniques that should be sufficient to deal with most analytical problems. In all cases, it is crucial in avoiding the problems of overfitting and always performs an external validation of the predictions of the algorithm used. As a general remark, as discussed throughout all this book, it should be always considered that the increase in complexity of the analytical treatment brings the risk of introducing overfitting artifacts or, in any case, would make more difficult the control and physical/chemical interpretation of the results obtained. For this reason, the analyst should also verify if a simpler alternative is effective, starting from a univariate calibration approach (which allows an immediate check of the limits of detection and trueness of the results obtained) and then progressively moving to multivariate linear and finally to multivariate nonlinear algorithms, if needed.

References

1 Safi, A., Campanella, B., Grifoni, E., Legnaioli, S., Lorenzetti, G., Pagnotta, S., Poggialini, F., Ripoll-Seguer, L., Hidalgo, M., Palleschi, V. Multivariate calibration in Laser-Induced Breakdown Spectroscopy quantitative analysis: The dangers of a 'black box' approach and how to avoid them. *Spectrochim. Acta - Part B At. Spectrosc.* (2018), **144**, 46–54, https://doi.org/10.1016/j.sab.2018.03.007.
2 D'Andrea, E., Pagnotta, S., Grifoni, E., Lorenzetti, G., Legnaioli, S., Palleschi, V., Lazzerini, B. An artificial neural network approach to laser-induced breakdown spectroscopy quantitative analysis. *Spectrochim. Acta - Part B At. Spectrosc.* (2014), **99**, https://doi.org/10.1016/j.sab.2014.06.012.
3 D'Andrea, E., Lazzerini, B., Palleschi, V., Pagnotta, S. Determining the composition of bronze alloys by means of high-dimensional feature selection and Artificial Neural Networks. In Proceedings of the Conference Record - IEEE Instrumentation and Measurement Technology Conference; 2015; Vol. 2015-July.
4 Kumar Myakalwar, A., Spegazzini, N., Zhang, C., Kumar Anubham, S., Dasari, R.R., Barman, I., Kumar Gundawar, M. Less is more: Avoiding the LIBS dimensionality curse through judicious feature selection for explosive detection. *Sci. Rep.* (2015), **5**, 1–10, https://doi.org/10.1038/srep13169.
5 Nunes, L.C., da Silva, G.A., Trevizan, L.C., Santos Júnior, D., Poppi, R.J., Krug, F.J. Simultaneous optimization by neuro-genetic approach for analysis of plant materials by laser induced breakdown spectroscopy. *Spectrochim. Acta - Part B At. Spectrosc.* (2009), **64**, 565–572, https://doi.org/10.1016/j.sab.2009.05.002.
6 Shen, Q., Zhou, W., Li, K. Quantative elemental analysis using laser induced breakdown spectroscopy and neuro-genetic approach. *Zhongguo Jiguang/Chinese J. Lasers* (2011), **38**, https://doi.org/10.3788/CJL201138.0308044.
7 Shen, Q., Zhou, W., Li, K. Quantitative analysis of Ni, Zr and Ba in soil by combing neuro-genetic approach and laser induced breakdown spectroscopy. In Proceedings of the Infrared, Millimeter Wave, and Terahertz Technologies; SPIE, 2010; Vol. 7854, p. 78543Q.
8 Poggialini, F., Campanella, B., Legnaioli, S., Pagnotta, S., Raneri, S., Palleschi, V. Improvement of the performances of a commercial hand-held laser-induced breakdown spectroscopy instrument for steel analysis using multiple artificial neural networks. *Rev. Sci. Instrum.* (2020), **91**, 073111, https://doi.org/10.1063/5.0012669.
9 Rumelhart, D.E., Hinton, G.E., Williams, R.J. Learning representations by back-propagating errors. *Nature* (1986), **323**, 533–536, https://doi.org/10.1038/323533a0.

10 Legnaioli, S., Campanella, B., Poggialini, F., Pagnotta, S., Harith, M.A., Abdel-Salam, Z.A., Palleschi, V. Industrial applications of laser-induced breakdown spectroscopy: A review. *Anal. Methods* (2020), **12**, 1014–1029, https://doi.org/10.1039/c9ay02728a.

11 Legnaioli, S., Campanella, B., Pagnotta, S., Poggialini, F., Palleschi, V. Determination of Ash Content of coal by Laser-Induced Breakdown Spectroscopy. *Spectrochim. Acta - Part B At. Spectrosc.* (2019), **155**, 123–126, https://doi.org/10.1016/j.sab.2019.03.012.

12 Lorenzetti, G., Legnaioli, S., Grifoni, E., Pagnotta, S., Palleschi, V. Laser-based continuous monitoring and resolution of steel grades in sequence casting machines. *Spectrochim. Acta - Part B At. Spectrosc.* (2015), **112**, 1–5, https://doi.org/10.1016/j.sab.2015.07.006.

13 Sirven, J.-B., Bousquet, B., Canioni, L., Sarger, L., Tellier, S., Potin-Gautier, M., Hecho, I. Le Qualitative and quantitative investigation of chromium-polluted soils by laser-induced breakdown spectroscopy combined with neural networks analysis. *Anal. Bioanal. Chem.* (2006), **385**, 256–262, https://doi.org/10.1007/s00216-006-0322-8.

14 El Haddad, J., Villot-Kadri, M., Ismaël, A., Gallou, G., Michel, K., Bruyère, D., Laperche, V., Canioni, L., Bousquet, B. Artificial neural network for on-site quantitative analysis of soils using laser induced breakdown spectroscopy. *Spectrochim. Acta Part B At. Spectrosc.* (2013), **79–80**, 51–57, https://doi.org/10.1016/j.sab.2012.11.007.

15 El Haddad, J., Bruyère, D., Ismaël, A., Gallou, G., Laperche, V., Michel, K., Canioni, L., Bousquet, B. Application of a series of artificial neural networks to on-site quantitative analysis of lead into real soil samples by laser induced breakdown spectroscopy. *Spectrochim. Acta Part B At. Spectrosc.* (2014), **97**, 57–64, https://doi.org/10.1016/j.sab.2014.04.014.

16 Zhang, X., Zhang, F., Kung, H., Shi, P., Yushanjiang, A., Zhu, S. Estimation of the Fe and Cu contents of the surface water in the ebinur lake basin based on LIBS and a machine learning algorithm. *Int. J. Environ. Res. Public Health* (2018), **15**, 2390, https://doi.org/10.3390/ijerph15112390.

17 Moncayo, S., Manzoor, S., Rosales, J.D.D., Anzano, J., Caceres, J.O.O. Qualitative and quantitative analysis of milk for the detection of adulteration by laser induced breakdown spectroscopy (LIBS). *Food Chem.* (2017), **232**, 322–328, https://doi.org/10.1016/j.foodchem.2017.04.017.

18 Sezer, B., Bilge, G., Boyaci, I.H. Capabilities and limitations of LIBS in food analysis. *TrAC Trends Anal. Chem.* (2017), **97**, 345–353, https://doi.org/10.1016/j.trac.2017.10.003.

19 D'Andrea, E., Lazzerini, B., Palleschi, V. Combining Multiple Neural Networks to Predict Bronze Alloy Elemental Composition. In Proceedings of the Advances In Neural Networks: Computational Intelligence For ICT; Bassis, S and Esposito, A and Morabito, FC and Pasero, E, Ed.; Springer International Publishing AG: Gewerbestrasse 11, Cham, Ch-6330, Switzerland, 2016; Vol. 54, pp. 345–352.

20 Afgan, M.S., Hou, Z., Wang, Z. Quantitative analysis of common elements in steel using a handheld μ-LIBS instrument. *J. Anal. At. Spectrom.* (2017), **32**, 1905–1915.

21 Cao, X., Zhang, L., Wu, Z., Ling, Z., Li, J., Guo, K. Quantitative analysis modeling for the ChemCam spectral data based on laser-induced breakdown spectroscopy using convolutional neural network. *Plasma Sci. Technol.* (2020), **22**, 10, https://doi.org/10.1088/2058-6272/aba5f6.

22 Li, L.N., Liu, X.F., Xu, W.M., Wang, J.Y., Shu, R. A laser-induced breakdown spectroscopy multi-component quantitative analytical method based on a deep convolutional neural network. *Spectrochim. Acta - Part B At. Spectrosc.* (2020), **169**, https://doi.org/10.1016/j.sab.2020.105850.

23 Képeš, E., Pořízka, P., Kaiser, J. On the application of bootstrapping to laser-induced breakdown spectroscopy data. *J. Anal. At. Spectrom.* (2019), **34**, 2411–2419, https://doi.org/10.1039/C9JA00304E.

24 Pagnotta, S., Lezzerini, M., Campanella, B., Legnaioli, S., Poggialini, F., Palleschi, V. A new approach to non-linear multivariate calibration in laser-induced breakdown spectroscopy analysis of silicate rocks. *Spectrochim. Acta - Part B At. Spectrosc.* (2020), **166**, https://doi.org/10.1016/j.sab.2020.105804.

25 Rousseau, R.M. Fundamental algorithm between concentration and intensity in XRF analysis 1—theory. *X-Ray Spectrum.* (1984), **13**, 115–120, https://doi.org/10.1002/xrs.1300130306.

26 Palleschi, A., Palleschi, V. An extended Kalman filter approach to non-linear multivariate analysis of laser-induced breakdown spectroscopy spectra. *Spectrochim. Acta - Part B At. Spectrosc.* (2018), **149**, 271–275, https://doi.org/10.1016/j.sab.2018.09.003.

27 Ciucci, A., Corsi, M., Palleschi, V., Rastelli, S., Salvetti, A., Tognoni, E. New procedure for quantitative elemental analysis by laser-induced plasma spectroscopy. *Appl. Spectrosc.* (1999), **53**, 960–964, https://doi.org/10.1366/0003702991947612.

28 Cristoforetti, G., De Giacomo, A., Dell'Aglio, M., Legnaioli, S., Tognoni, E., Palleschi, V., Omenetto, N. Local thermodynamic equilibrium in laser-induced breakdown spectroscopy: beyond the McWhirter criterion. *Spectrochim. Acta - Part B At. Spectrosc.* (2010), **65**, 86–95, https://doi.org/10.1016/j.sab.2009.11.005.

29 Tognoni, E., Palleschi, V., Corsi, M., Cristoforetti, G., Omenetto, N., Gornushkin, I., Smith, B.W., Winefordner, J.D. From sample to signal in laser-induced breakdown spectroscopy: A complex route to quantitative analysis, in:; *Laser-Induced Breakdown Spectroscopy: Fundamentals and Applications* Miziolek, A.W., Palleschi, V., Schechter, I., Eds.; Cambridge University Press: Cambridge, (2006), Vol. **9780521852**; ISBN 9780511541261.

30 Corsi, M., Cristoforetti, G., Giuffrida, M., Hidalgo, M., Legnaioli, S., Palleschi, V., Salvetti, A., Tognoni, E., Vallebona, C. Analysis of biological tissues by laser induced breakdown spectroscopy technique. In Proceedings of the ICALEO 2003 - 22nd International Congress on Applications of Laser and Electro-Optics, Congress Proceedings; 2003.

31 Praher, B., Palleschi, V., Viskup, R., Heitz, J., Pedarnig, J.D. Calibration free laser-induced breakdown spectroscopy of oxide materials. *Spectrochim. Acta - Part B At. Spectrosc.* (2010), **65**, 671–679, https://doi.org/10.1016/j.sab.2010.03.010.

32 Borgia, I., Burgio, L.M.F., Corsi, M., Fantoni, R., Palleschi, V., Salvetti, A., Squarcialupi, M.C., Tognoni, E. Self-calibrated quantitative elemental analysis by laser-induced plasma spectroscopy: Application to pigment analysis. *J. Cult. Herit.* (2000), **1**, https://doi.org/10.1016/S1296-2074(00)00174-6.

33 Senesi, G.S., Tempesta, G., Manzari, P., Agrosì, G. An innovative approach to meteorite analysis by laser-induced breakdown spectroscopy. *Geostand. Geoanalytical Res.* (2016), **40**, https://doi.org/10.1111/ggr.12126.

34 Gerhard, C., Hermann, J., Mercadier, L., Loewenthal, L., Axente, E., Luculescu, C.R.R., Sarnet, T., Sentis, M., Viöl, W. Quantitative analyses of glass via laser-induced breakdown spectroscopy in argon. *Spectrochim. Acta - Part B At. Spectrosc.* (2014), **101**, 32–45, https://doi.org/10.1016/j.sab.2014.07.014.

35 Yalçin, Ş., Crosley, D.R.R., Smith, G.P.P., Faris, G.W.W. Influence of ambient conditions on the laser air spark. *Appl. Phys. B Lasers Opt.* (1999), **68**, 121–130, https://doi.org/10.1007/s003400050596.

36 Griem, H.R. Stark broadening. *Adv. At. Mol. Phys.* (1976), **11**, 331–359, https://doi.org/10.1016/S0065-2199(08)60033-0.
37 Senesi, G.S.S., Benedetti, P.A.A., Cristoforetti, G., Legnaioli, S., Palleschi, V. Hydrogen Balmer alpha line behavior in laser-induced breakdown spectroscopy depth scans of Au, Cu, Mn, Pb targets in air. *Spectrochim. Acta - Part B At. Spectrosc.* (2010), **65**, 557–564, https://doi.org/10.1016/j.sab.2010.05.010.
38 Pardini, L., Legnaioli, S., Lorenzetti, G., Palleschi, V., Gaudiuso, R., De Giacomo, A., Diaz Pace, D.M., Anabitarte Garcia, F., De Holanda Cavalcanti, G., Parigger, C. On the determination of plasma electron number density from Stark broadened hydrogen Balmer series lines in Laser-Induced Breakdown Spectroscopy experiments. *Spectrochim. Acta - Part B At. Spectrosc.* (2013), **88**, 98–103, https://doi.org/10.1016/j.sab.2013.05.030.
39 Pagnotta, S., Grifoni, E., Legnaioli, S., Lezzerini, M., Lorenzetti, G., Palleschi, V. Comparison of brass alloys composition by laser-induced breakdown spectroscopy and self-organizing maps. *Spectrochim. Acta - Part B At. Spectrosc.* (2015), **103–104**, 70–75, https://doi.org/10.1016/j.sab.2014.11.008.
40 Pagnotta, S., Lezzerini, M., Campanella, B., Gallello, G., Grifoni, E., Legnaioli, S., Lorenzetti, G., Poggialini, F., Raneri, S., Safi, A., et al. Fast quantitative elemental mapping of highly inhomogeneous materials by micro-Laser-Induced Breakdown Spectroscopy. *Spectrochim. Acta - Part B At. Spectrosc.* (2018), **146**, 9–15, https://doi.org/10.1016/j.sab.2018.04.018.
41 D'Andrea, E., Pagnotta, S., Grifoni, E., Legnaioli, S., Lorenzetti, G., Palleschi, V., Lazzerini, B., D'Andrea, E., Pagnotta, S., Grifoni, E., et al. A hybrid calibration-free/artificial neural networks approach to the quantitative analysis of LIBS spectra. *Appl. Phys. B-LASERS Opt.* (2015), **118**, 353–360, https://doi.org/10.1007/s00340-014-5990-z.
42 Borges, F.O., Cavalcanti, G.H., Gomes, G.C., Palleschi, V., Mello, A. A fast method for the calculation of electron number density and temperature in laser-induced breakdown spectroscopy plasmas using artificial neural networks. *Appl. Phys. B Lasers Opt.* (2014), **117**, https://doi.org/10.1007/s00340-014-5852-8.

14

Laser Ablation-based Techniques – Data Fusion

Jhanis Gonzalez

Applied Spectra, Inc., West Sacramento, CA, USA
Lawrence Berkeley National Laboratory, Berkeley, CA, USA

14.1 Introduction

In the analyses of highly complex heterogeneous samples, such as rocks, biological tissues, environmental, and food-related samples, a single analytical technique might not have the sensitivity, resolution, or selectivity to determine its multi-elemental composition. Similarly, a single analytical technique, typically, is insufficient to specify the molecular, elemental, and isotopic compositions of the sample. In most cases, a combination of analytical techniques is needed to obtain a complete picture of the sample composition and other defining characteristics. Another advantage that arises from using multiple analytical techniques and instruments is that this gives the ability to verify the results through direct comparison of quantification results between techniques or through validation of pattern trends in the acquired data.

This approach involving multiple analytical techniques and instrument capabilities allows for a more comprehensive sample composition representation. As shown by Duce et al. [1], multiple analytical approaches were used for the characterization of carbon-based black pigments. The authors combined information from thermogravimetry, thermogravimetry coupled with Fourier transform infrared (FTIR) spectroscopy, and differential scanning calorimetry. They also used spectroscopic techniques such as laser-induced breakdown spectroscopy (LIBS), FTIR spectroscopy, Raman spectroscopy, and X-ray powder diffraction. This approach allowed them to identify the different constituents (both organic and inorganic) of the pigments and their relative contents, and, for the first time, they reported quantitative differences based on the type of pigment (bone, vine, and lamp) and manufacturer. Bazin et al. presented another example [2] of the effectiveness of multiple analytical and instrument methods. In this case, involving the study of crystal formation in kidneys, they proposed a combination of classical observations by field-emission scanning electron microscopy combined with energy-dispersive spectroscopy, Raman measurements, and FTIR spectroscopy. This set of diagnostic tools helped clinicians gather information regarding the nature and the spatial distribution, at the subcellular scale, of the crystal morphology and different chemical phases present in kidney biopsies, and, as a result, were able to obtain more precise diagnoses.

The use of multiple complementary techniques, as detailed in the previous two examples, shows extremely comprehensive studies where each utilized technique provides a piece of information that could complete the sample characterization puzzle. These approaches, however, have a couple of significant consequences: first, increased amount of data to be processed, where each technique's dataset must be reduced and analyzed independently; and second, if data fusion is applied, there will likely be significant data preprocessing and the need for complex scaling factors due to the considerable differences between the nature of the measurements. Factors to be considered include the type of signal, amount of material analyzed per measurement, and spatial (lateral and depth) resolution of the measurement.

14.2 Data Fusion of Multiple Analytical Techniques

Data fusion is defined as a method that combines datasets from different sources (techniques or sensors) to produce a single dataset for building analytical models. The case for using data fusion stems from the idea that one particular technique or sensor might not be able to thoroughly characterize a sample. In contrast, combined data from multiple techniques or sensors can offer a more complete representation of the sample composition. There are a variety of data fusion techniques that can be generally categorized as *low-level*, *mid-level*, and *high level fusion*, based on the type of information used for the fusion, Figure 14.1.

14.2.1 Low-level Fusion

The most basic data fusion level and involves a simple concatenation of the raw signals from the different sensors that are then processed as a single signal. This approach requires very

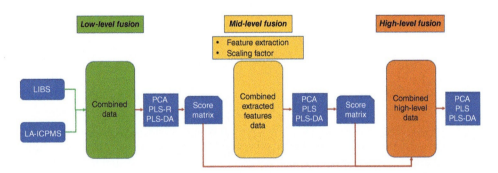

Figure 14.1 Schema for data fusion of laser ablation-based techniques.

little input from the user since it is based on the raw signals' concatenation without any preprocessing. However, the same characteristic that makes it the simplest of the fusion categories is also the one that often has unpredictable impacts on the outcome. By concatenating the signals without any pretreatment, irrelevant information such as noise or redundant data is often included, which can negatively impact the results, particularly analysis precision.

Likely, as a result of this unpredictability, we were unable to find many successful implementations of low-level fusion for conventional analytical techniques. One recent example of its success was demonstrated by Rammelkamp et al. [3], when they demonstrated the applicability of this approach with the fusion of LIBS and Raman data in the context of Mars exploration. They showed that the low-level fusion of these datasets improved the identification of sulfates and salts compared to the individual datasets. However, they recognized that the multivariate analyses performance is affected by the weighting of the individual spectra in the combined one; thus, exploring higher data fusion levels is needed.

14.2.2 Mid-level Fusion

Consists of extracting features from the original datasets, and then combining and processing them together. This data fusion category requires more user involvement, particularly in selecting the features for the fusion. *Mid-level fusion* is particularly relevant when a few features extracted from the original dataset are able to provide all the necessary information to describe the sample of interest. The complexity for analyzing spectra arises from the fact that it cannot be easily modeled using a few features without proper in-depth analysis. However, "feature extraction" can be carried out, in this case by extracting information for LIBS in the form of integrated optical emission signals and from integrated mass signals for laser ablation inductively coupled plasma mass spectrometry (LA-ICP-MS). The main difference between the *mid-level fusion* and the *low-level fusion* is that in the latter, the starting point is from the extracted features instead of the raw spectrum. However, once the features are extracted, the same algorithms used for *low-level fusion* could be used.

An example of *mid-level fusion* was presented by Jacobson et al. [4]. They combined the data from liquid chromatography/mass spectrometry (LC/MS) and proton nuclear magnetic resonance (^{1}H-NMR) to analyze rat urine to obtain metabolic fingerprints. They tested three methods: concatenated data (this is a straightforward method, but it might require appropriate scaling to prevent one dataset from being dominant), full hierarchical modeling (this is a way of modeling data on two levels, the first level is performed on separate blocks of the datasets, and a higher level is performed on the scores from the lower-level analysis), and batch modeling (this is one type of hierarchical modeling that also operates at two levels and is frequently used to extract time-sensitive information). In this example, data fusion from concatenation and full hierarchical modeling generally improved the classification results. They determined in this case that block scaling to equal sums of standard deviations between the two data blocks was the most successful scaling method for the concatenated data case.

14.2.3 High-level Fusion

Performed on the classification outputs of all the individual techniques. The classification output is called the "identity declaration." In high-level fusion, the identity declarations of each sensor are combined to give the final identity declaration. This method can be applied to all types of analytical measurements since it combines class assignments and not analytical signals. For a *high-level fusion*, Bayesian hierarchical modeling is proposed as an alternative to block scaling. The scores from modeling each technique (lower-level multivariate analysis (MVA)) could then be used for further data analysis (higher-level MVA). In addition to modeling data from different origins, hierarchical modeling is a variable reduction tool that enhances interpretation.

An example of *high-level fusion* was presented by Roussel et al. [5]. They proposed a high-level fusion method based on the Bayesian inference to combine various sensors' outputs to classify musts of white grapes according to their variety. The sensors used in this study were aroma sensors, FTIR, and ultraviolet (UV) spectrometers. They demonstrated that an effective fusion method lead to a significant improvement in the grape variety discrimination with a misclassification error of 4.7%. In contrast, the best individual sensor (FTIR) gave a misclassification error twice as high as 9.6%.

14.3 Data Fusion of Laser Ablation-Based Techniques

14.3.1 Introduction

Following this discussion of data fusion from different techniques, we explore implementing these data fusion approaches, with a particular focus on when multiple techniques are utilized for the simultaneous analysis of solid samples. Fittingly the focus is on two laser ablation-based techniques triggered by the same event, to take advantage of simultaneous collection. The fact that these two techniques originated from the same event allows for more aggressive assumptions when planning data fusion, particularly regarding the measurement of spatial resolution and amount of material analyzed.

The first one of these two techniques employed simultaneously is known as laser-induced breakdown spectroscopy (LIBS). LIBS is a versatile technique for direct elemental analysis of samples (liquid, gas, or solid) and is based on optical emission spectroscopy measurements expressed as atomic and ionic emission lines. The second is laser ablation inductively coupled plasma mass spectrometry (LA-ICP-MS). LA-ICP-MS is a powerful analytical mass spectrometry technique that enables highly sensitive elemental and isotopic analysis to be performed directly on solid samples. Measurements are typically based on the mass-to-charge ratio or time-of-flight of ions, depending on the type of mass spectrometer utilized.

These two techniques yield complementary information. Individually LIBS and LA-ICP-MS possess several distinctive characteristics. LIBS has been recognized for its unique advantages of fast, in situ, multi-elemental analysis from H to Pu. Coupling laser ablation with ICP-MS-based instruments provides additional isotopic information and enhanced sensitivity when compared to LIBS alone. Analysis taking advantage of these two techniques simultaneously can be advantageous, as every laser pulse for ablation provides both the optical plasma for emission spectroscopy and the particles for ICP mass spectrometry.

The development of tandem laser ablation-based systems that combine LIBS and LA-ICP-MS has provided unique techniques and methodologies for analyzing solid samples. One new capability is having the adaptability and flexibility that multiple sensors give researchers to analyze heterogeneous samples (Figure 14.2). The use of multiple independent but simultaneous signal detection sensors helps ensure that an element of interest's sensitivity needs are met. Tandem laser ablation-based approaches have been readily available since 2014, when the first commercial tandem laser ablation-based systems were introduced by Applied Spectra, Inc. However, Fernandez et al. [6] introduced this method in 1995, when they studied the correlation between spectral atomic emission intensity from laser-induced plasmas (LIPs) and inductively coupled plasma optical emission spectroscopy (ICP-OES). This study showed that their signals exhibited an excellent correlation for various materials and laser powers. They noted that the observed correlation was significant for two reasons. First, the LIP's remarkable correlation demonstrates that changes in ICP emission intensity appear to be primarily due to changes in the laser–material interaction and not from sample transport. Second, the correlation was of significant importance because it created the possibility of using the LIP signal as an internal standard to improve the ICP-OES analysis.

In 2006, Latkoczy et al. [7] proposed a data fusion between laser ablation-based techniques. They developed a compact LIBS/LA-ICP-MS system to analyze the distribution of major and trace elements in a magnesium-based alloy sample. In this study, they reported

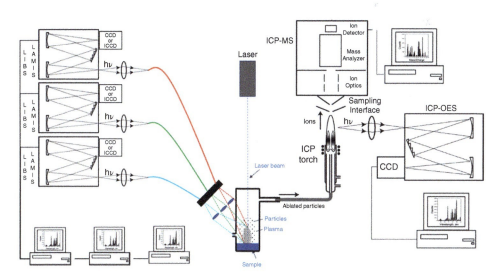

Figure 14.2 Layout of a multisensor (tandem) LIBS^n and LA-ICP(-MS or -OES) system.

a successful application of a calibration approach where they used signal intensities from LA-ICP-MS for internal standardization of the LIBS data. In the conclusion of this study, they declared that one of the significant limitations of this approach is that LIBS can only detect major elements due to the low sensitivity of the configuration used. Even though they provided some possible ways to overcome this limitation, they also concluded that implementing simultaneous LIBS–LA-ICP-MS is not straightforward and could present tremendous challenges.

Fortunately, most, if not all, of the concerns expressed in Latkoczy's study have been overcome. This has been systematically demonstrated in several studies since the introduction of the first commercial tandem system. Bonta et al. [8] showed the impact of the complementary nature of collecting LIBS and LA-ICP-MS data simultaneously by highlighting its benefits for analyzing biological samples. They demonstrated that LA-ICP-MS is ideally suited for laterally resolved analyses of trace elements in tissues (e.g. Fe, Zn, Cu). At the same time, LIBS is optimal for mapping some of the major (C, H, O) and minor elements (Na, K, Ca, Mg) in these biological tissues. Moreover, LIBS also allowed them to include the analysis of elements that are significantly affected by high background signals (e.g. K), or polyatomic interferences (e.g. Mg) often impact ICP-MS detection.

Several other studies also focused on LIBS's simultaneous use with LA-ICP-OES and LA-ICP-MS [8–20]. In Dong et al. [19] LIBS and LA-ICP-MS were used to simultaneously measure minor elements (C, Si, Ca, Al, Mg) and trace elements (V, Ba, Pb, U) in coal samples. They also demonstrated that a correlation exists between Time of Flight-Mass Spectrometry (TOF-MS) and total minor emission from LIBS. Subedi et al. [18] demonstrated the potential of both LIBS and LA-ICP-MS for the characterization of printing inks (toners, inkjets, intaglio, and offset.). Oropeza et al. [12] combined LIBS and LA-ICP-OES for quantitative analyses of S, Ni, and V by using external calibration with a carbon line as an internal standard for studying asphaltene samples. As shown earlier, these examples focus on using this tandem approach for the simultaneous data collection while still processing and reporting the data separately.

Next, the focus shifts to two examples where our scientific teams performed mid-level data fusion of laser ablation-based techniques. The first example was published in 2016 by Lee et al. [15]. Lee et al. performed simultaneous LIBS and LA-ICP-MS analysis for the classification of edible salt samples. In the second example, in 2020, Dong et al. [9] showed that the classification analysis model based on the combined ICP-MS (TOF) and LIBS data was superior to using TOF or LIBS data separately.

14.3.2 Classification of Edible Salts

Salt is a natural ionic-compound mixture with a NaCl matrix and is used universally as a food seasoning. Salts are classified based on their source, where sea salts are produced by evaporation of seawater and rock salts extracted from underground. Although the main component of unrefined salts is NaCl, the chemical composition exhibits a broad mixture of minor elements. In the case of sea salts, the chemical composition is not purely that of seawater. The concentrations of the minor metallic mineral elements, K, Mg, and Ca, dissolved in seawater, show considerable variation with the density of brine water extracted from the reservoir into the evaporation and crystallization areas of saltpans. For rock salts, their chemical composition is originally based on that of the ancient seawater. However, the inclusion of rock particles, interaction with underground freshwater, and chemical reactions underground all impact the chemical composition of rock salts. Analysts could use all these chemical composition variations in unrefined salts as reliable fingerprints for classifying them according to their geographical origin or discriminating a particular salt from others.

Edible sea salts are recognized for their rich mineral composition, typically containing K, Mg, and Ca, with concentrations ranging from several thousand ppm to a few %. K, Ca, and Mg play essential roles in the classification of salts. The ICP-MS, however, suffers from significant issues analyzing these elements (and other light-mass elements masses <56), predominantly related to interferences caused by the overlap of abundant molecular species present in ICP. Despite this, ICP-MS offers tremendous sensitivity for mid- and heavy elements (mass >56), where molecular interferences are scarcer than the light part of the mass spectrum. LIBS addresses this issue, since it can efficiently detect light elements such as Li, Al, Ti, Si, and Fe, if their concentrations in these salts are at the ppm level and above.

To investigate the effect the environment (natural or contamination) around the salt deposits has on the end product composition, a combination of techniques is necessary. This allows for the simultaneous collection of information about light-mass metallic elements at several hundred ppm to low % levels, non-metals, and trace heavy metals at sub-ppm levels. In this study, LIBS was optimized to detect some light elements, including K, Mg, and Ca. LA-ICP-MS analysis focused on elements with m/z $\geq$ 90, including non-metal, and trace heavy metal elements.

14.3.2.1 LIBS and LA-ICP-MS Measurements of the Salt Samples

The simultaneous LIBS and LA-ICP-MS measurements were performed using a commercial instrument (J200 tandem LA-LIBS instrument, Applied Spectra, Inc.) This instrument is equipped with a 266nm Nd:YAG laser. The conditions were optimized to obtain the best signal-to-noise ratio for both LIBS and LA-ICP-MS. For the LIBS measurements, the optical emission was collected by a fiber optic, and the light was delivered to a six-channel spectrometer covering a wavelength range from 190 to 1040 nm, with each channel equipped with Charged-Couple Device (CCD) detection. For the LA-ICP-MS measurements, a quadrupole based mass spectrometer (Plasma Quant MS Elite, Analytik Jena) was used. A total of 123 laser shots were used for the analysis and the LA-ICP-MS and LIBS acquisition were synchronized. For each salt sample, 18–23 pairs of accumulated LIBS and LA-ICP-MS spectra were collected. More details about the experimental conditions can be found in the original publication [15].

14.3.2.2 Mid-Level Data Fusion of LIBS and LA-ICP-MS of Salt Samples

As noted earlier, we evaluated the use of the mid-level data fusion approach. The steps of this process are outlined in Figure 14.3. The first step of a mid-level data fusion approach is selecting the features to build the model. In this case, this first step involves identifying the species detected by each technique and selecting the ranges to include. Figure 14.4 shows

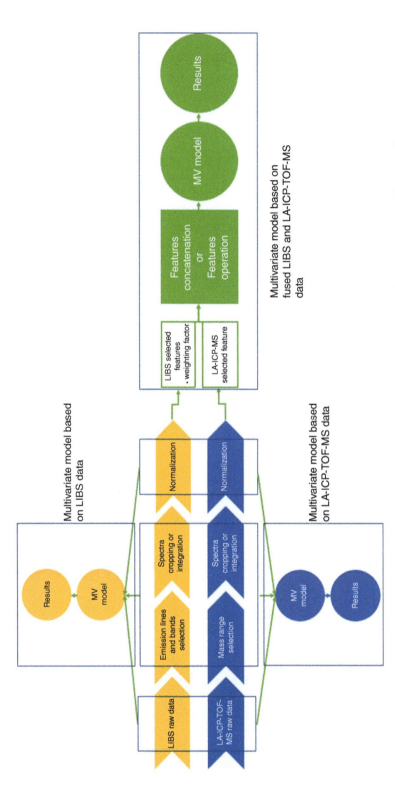

Figure 14.3 Schematic view of the steps to build an MV model based on the different data sources. Directly from the LIBS or LA-ICP-MS raw data or extracted features, as well as an MV model based on the mid-level data fusion approach applied to laser ablation-based techniques.

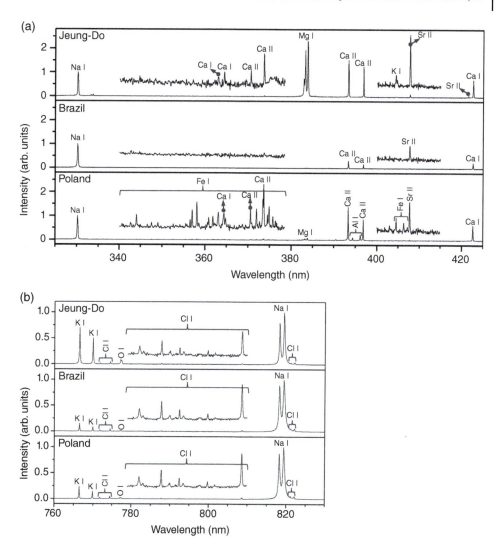

Figure 14.4 Averaged LIBS spectra of the salt samples from Jeung-Do, Brazil, and Poland in two different wavelength regions: 325–425 nm (a) and 760–830 nm (b) with assignments of the observed emission lines.

the averaged LIBS spectra of the salt samples from Jeung-Do, Brazil, and Poland, in wavelength regions 325–425 nm (a) and 760–830 nm (b), with assignments of the observed emission lines based on the NIST Atomic Spectra Database. Figure 14.5 shows LA-ICP-MS data for the salts from Haenam, Hokkaido, Chile, Brazil, India, and Poland. The mass peaks corresponding to Zr, Mo, Nb, Sn, I, Cs, Ba, La, Ce (and other lanthanides), W, and Pb isotopes were assigned.

The same approach was taken for every analyzed sample to account for the difference between sea and rock salts, where identification of detected species was performed for both techniques resulting in Table 14.1, where most of the identified species from both techniques are listed (excluding sodium, chlorine, and helium).

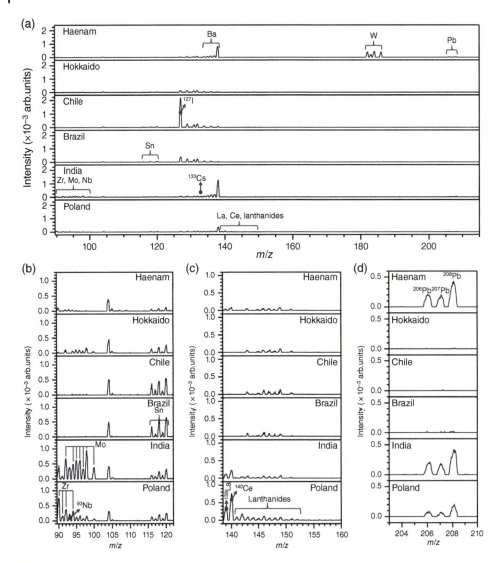

Figure 14.5 MS spectra of the salts from Haenam, Hokkaido, Chile, Brazil, India, and Poland recorded by the LA-ICP-MS in the range of m/z = 90–215 (a) and the expanded LA-ICP-MS spectra in the ranges of m/z = 89.5–122 (b), 138.5–160 (c), and 203–210 (d).

For the MVA of LIBS, it was decided to used two spectra regions. The first region included wavelengths from 362 to 410 nm where Na, Ca, Mg, Sr, Fe, and Al lines are observed, and the second region selected was 765–771 nm where K, Cl, O, and Na lines are observed. This step in the mid-level data fusion process is of extreme importance and has several consequences. First, it is meant to ensure that the most critical information is included in the model instead of noise or redundant data. Second, reduce the amount of information (reduction of data points), which lowers the impact on computational resources and, more importantly, reduces analysis processing time.

Table 14.1 Geographical origin and types of the sample salts and the elements identified in LIBS and LA-ICP-MS spectra. Na, Cl, and He, identified in all samples, were omitted.

Sample no.	Geographical origin	Type	Elements identified in LIBS spectra	Elements identified in LA-ICP-MS spectra
1	Laizhou, China	Sea salt	Si, Al, Mg, Ca, K, Sr, Li, O, H	Zr, Nb, Mo, Ag, Cd, Sn, I, Cs, Ba, La, Ce, W, Pb
2	Okinawa, Japan	Sea salt	Mg, Ca, K, Sr, Li, S, O, H	Mo, Sn, I, Cs, Ba
3	Hokkaido, Japan	Sea salt	Mg, Ca, K, Li, S, O, H	Mo, Sn, I, Cs, Ba
4	Kochi, Japan	Sea salt	Mg, Ca, K, Sr, Li, S, O, H	Mo, Ag, Cd, Sn, I, Cs, Ba, La, Ce, W, Pb
5	Jeung-Do, South Korea	Sea salt	Mg, Ca, K, Sr, Li, S, O, H	Zr, Nb, Mo, Ag, Cd, Sn, I, Cs, Ba, La, Ce, Pb
6	Haenam, South Korea	Sea salt	Si, Al, Mg, Ca, K, Sr, Li, S, O, H	Zr, Nb, Mo, Ag, Sn, I, Cs, Ba, La, Ce, W, Pb
7	Younggwang, South Korea	Sea salt	Si, Al, Mg, Ca, K, Sr, Li, S, O, H	Zr, Nb, Mo, Ag, Sn, I, Cs, Ba, La, Ce, W, Pb
8	Guerande, France	Sea salt	Si, Al, Mg, Ca, K, Sr, Li, S, O, H	Zr, Nb, Mo, Cd, Sn, I, Cs, Ba, La, Ce, Pb
9	Brazil	Sea salt	Mg, Ca, K, Sr, O, H	Sn, I, Cs, Ba
10	Chile	Sea salt	Mg, Ca, K, Sr, O, H	Sn, I, Cs, Ba
11	Mongolia	Rock salt	Ca, K, Li, O, H	Zr, Mo, Sn I, Ba, Pb
12	Poland	Rock salt	Fe, Si, Al, Mg, Ti, Ca, K, Sr, Li, S, O, H	Zr, Nb, Mo, Sn, I, Cs, Ba, La, Ce, Pb
13	Himalaya, India	Rock salt	Fe, Si, Al, Ti, Ca, K, Sr, Li, S, O, H	Zr, Nb, Mo, Sn, I, Cs, Ba, La, Ce, W, Pb
14	Himalaya, Pakistan	Rock salt	Si, Al, Mg, Ca, K, Sr, Li, S, O, H	Zr, Nb, Sn, I. Cs, Ba, La, Ce, Pb

The second step for the mid-level data fusion approach consisted of normalizing each dataset (LIBS and LA-ICP-MS) to reduce shot-to-shot fluctuations and improve precision within each sample measurement. The selected LIBS spectra regions were normalized to the strongest Na signal in each region. Since sodium and chlorine were not available for normalization of the ICP-MS data, the total intensity of each LA-ICP-MS spectrum was used as a normalization signal. Both of these approaches improved the precision of individual datasets, but there are still significant differences in the signal intensity, which stem from the distinctive natures of these techniques. Therefore, a scaling factor is necessary to optimize the concatenated dataset.

In this study, a simple mid-level data fusion approach was used to combine the selected data from LIBS and LA-ICP-MS. Each LA-ICP-MS spectrum was stitched at the end of the

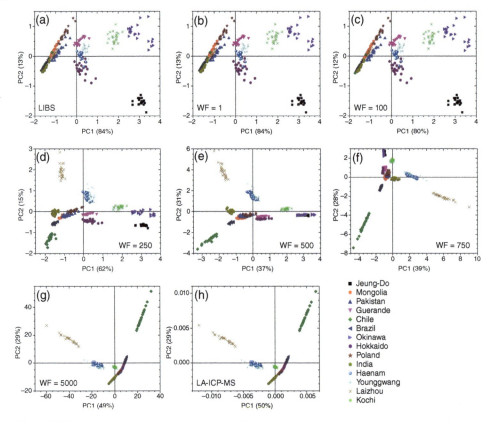

Figure 14.6 PC score plots obtained from the LIBS spectra (a), LA-ICP-MS spectra (h), and their fused data with the different WF values for the LA-ICP-MS spectra, 1 (b), 100 (c), 250 (d), 500 (e), 750 (f), and 5000 (g).

simultaneously acquired LIBS spectrum after being multiplied by a weighting factor (WF), that was varied for optimization to a factor between 1 and 5000. The WF is meant to increase or reduce the contribution to the Multivariate Model (MV) model of one of the fused datasets. This scaling or WF (in this case applied to the LA-ICP-MS dataset) is essential when the amount of information contained in one dataset is significantly less when compared to the other dataset. In this example, comparing the number of data points on the LA-ICP-MS dataset to the LIBS dataset shows more data points are available on the LIBS side, in part, because of the multiple emission lines per element being used in the model.

After the concatenation process, the normalized fused spectra were then mean-centered and analyzed using principal component analysis. Mean-centering is the act of subtracting a variable's mean from all the observations on that variable on the dataset, such that the variable's new mean is zero. This is usually recommended to reduce multicollinearity in the regression model. Figure 14.6 shows the Principal Component (PC) score plots obtained from LIBS spectra (a), LA-ICP-MS spectra (h), and their fused data (b–g). The PC score plots obtained from the fused data with the different WF values for the LA-ICP-MS spectra, 1, 100, 250, 500, 750, and 5000, are shown in Figure 14.4b–g, respectively.

Truth														
Brazil	100	0	0	0	0	0	0	0	0	0	0	0	0	0
Chile	0	100	0	0	0	0	0	0	0	0	0	0	0	0
Guerande	0	0	100	0	0	0	0	0	0	0	0	0	0	0
Haenam	0	0	0	67	0	0	0	0	0	0	0	0	0	33
Hokkaido	0	0	0	0	100	0	0	0	0	0	0	0	0	0
India	0	0	0	0	0	100	0	0	0	0	0	0	0	0
Jeung-Do	0	0	0	0	0	0	100	0	0	0	0	0	0	0
Kochi	0	0	0	0	0	0	0	100	0	0	0	0	0	0
Laizhou	0	0	0	0	0	0	0	0	100	0	0	0	0	0
Mongolia	0	0	0	0	0	0	0	0	0	100	0	0	0	0
Okinawa	0	0	0	0	0	0	0	0	0	0	100	0	0	0
Pakistan	0	0	0	0	0	0	0	0	0	0	0	100	0	0
Poland	0	0	0	0	0	0	0	0	0	0	5	0	95	0
Younggwang	0	0	0	0	0	0	0	0	0	0	0	0	0	100
	Brazil	Chile	Guerande	Haenam	Hokkaido	India	Jeung-Do	Kochi	Laizhou	Mongolia	Okinawa	Pakistan	Poland	Younggwang

Response

Figure 14.7 Confusion matrix from fused data (97.4%).

When the WFs are relatively small (1 and 100), the fused data PC scores show similar clustering patterns to the LIBS spectra. On the other hand, when large WF values are used (750 and 5000), the corresponding PC score plots are similar to those from the LA-ICP-MS spectra. Effective data fusion would utilize the intermediate WFs, in this case, 250 and 500, where 250 shows the best separation among the individual clusters.

14.3.2.3 PLS-DA Classification Model for Salt Samples

The following steps, which involve building an MV model based on partial least square-discrimination analysis (PLS-DA) for the weighted and fused data, are not different than implementing these MV analyses on individual technique (LIBS or LA-ICP-MS) data. A classification model built based on the fused data with a WF of 250 for all the samples was tested. The PLS-DA model performance was validated using a cross-validation (CV) method with a leave-one-out (LOO) algorithm. LOO CV is a particular case of CV where the number of folds equals the number of instances in the dataset. Thus, the learning algorithm is applied once for each instance, using all other instances as a training set and using the selected instance as a single-item test set. Figure 14.7 shows the confusion matrix generated from the CV using the LOO algorithm for the PLS-DA model using the fused data. This model's overall LOO CV correctness scores were 97.4%, which is significantly better than the results for either of the individual techniques, which were 87.2% for LIBS and 68.8% for LA-ICP-MS.

The presented data demonstrated the power of mid-level data fusion from laser ablation-based data to classify edible salts samples. It is clear that the mid-level data fusion of LIBS and LA-ICP-MS data provided improved performance when compared to LIBS and LA-ICP-MS models separately. The two datasets were fused to have a similar contribution to the main features with an optimized WF. The LIBS and LA-ICP-MS techniques have strengths in the analysis of different kinds of elements at different concentration levels. This enables the combination of LIBS and LA-ICP-MS to simultaneously provide more accurate chemical fingerprints for the classification of edible salts.

In this chapter, we are focusing on the process of building and using mid-level data fusion from laser ablation-based techniques; however, we encourage the reader to refer to the original publication for a more comprehensive discussion of the chemical-related explanations that help clarify the differences found between the individual technique MV models and the data fused-base MV model.

14.3.3 Coal Discrimination Analysis

Coal remains one of the world's most valuable resources despite the intensive development of alternative energy sources. Coal is a sedimentary rock with a complex composition containing several major elements including C, H, O, N, and S, plus a wide range of minor and trace elements. Coal is a heterogeneous material, and the chemical composition can be categorized into two major components: organic and inorganic.

Ideal operation conditions in a power plant vary based on the type of coal being used at any given time. Therefore, differentiating coal types for their appropriate use based on reliable chemical analysis is necessary. Coal classification plays a vital role in identifying washing processes and blending selections at power plants. This, in turn, affects the production efficiency, economic benefits, and the degree of combustion pollution released to the environment. The coal industry has experimented with many analytical techniques for the characterization of coal samples [21, 22] including X-ray fluorescence (XRF) [23, 24], inductively coupled plasma atomic emission spectroscopy (ICP-AES) [25, 26], atomic absorption spectroscopy (AAS) [27], and FTIR spectroscopy [28, 29], among others. There are advantages and disadvantages for each of the aforementioned techniques, ranging from not offering real-time measurements (ICP-AES, AAS, and FTIR) to insufficient sensitivity and selectivity (XRF). Other considerations such as price, safety, and strict regulatory requirements led the coal industry to consider other techniques that could be implemented in industrial environments.

Analysis of major elements of coal samples (organic content) provides information about their calorific heat values, volatile content, and other essential characteristics, which all directly impact the operating conditions in power plants. Chemical analysis pointing to the inorganic content is usually focused on providing information about the environmental impact of coal burning. These analyses, however, are traditionally done independently, increasing total analysis time and effort.

Recent advances in rapid solid sampling using LA-ICP-MS have shown it to be an attractive approach for direct trace element and isotopic analysis of solids in a wide variety of fields, including coal analysis [30]. LIBS has also been widely applied for coal property analysis [21]. By using LIBS for analysis of major elements, there is no longer a need to use the ICP-MS in a high count mode to detect elements like F, O, and N, which are difficult (or in some cases impossible) to analyze by ICP-MS. The combination of LIBS and LA-ICP-MS makes it possible to detect most of the elements in the periodic table and increase the detection reliability.

The present study focused on developing a comprehensive tandem LIBS/LA-ICP-TOF-MS method that is well suited for acquiring data for both the organic and inorganic portions simultaneously for coal characterization and discrimination. Different multivariate classification strategies were explored, ranging from using a simple linear classification method like principal component analysis (PCA) combined with K-means clustering to the complex linear and nonlinear classification methods like PLS-DA and SVM. These methods were compared to explore the efficacy for coal property discrimination analysis and the benefits of combining tandem LIBS/LA-ICP-MS measurements. As was the case in the previous example, the purpose of this exercise is to review the steps that were followed to implement a mid-level data fusion for coal samples analyzed by a tandem LIBS and LA-ICP-TOF-MS. This is not intended to be an exhaustive interpretation of the chemometric tools used after data has been fused; for that we refer the reader to the original publication [9].

14.3.3.1 LIBS and LA-ICP-TOF-MS Measurements of the Coal Samples

The tandem LIBS and LA-ICP-TOF-MS measurements were performed using a commercial instrument (J200 tandem LA-LIBS instrument, Applied Spectra, Inc.). This instrument was equipped with a 213nm Nd:YAG laser. For the LIBS measurements, the optical emission was collected by a fiber optic, and the light was delivered to a Czerny–Turner spectrometer with a spectral resolution between 0.8 and 1.25 nm, and captured via an Intensified Charged-Couple Device (ICCD) detector. For the LA-ICP-TOF-MS measurements, a time-of-flight-based mass spectrometer (GBC Scientific) was used. The mass spectrometer detector was operated in the temporal mode so that all the masses could be detected during sample ablation, and the integration time was the same as the ablation event. A total of 20 laser shots were used for the analysis, and the LA-ICP-TOF-MS and LIBS acquisition was synchronized. For each coal sample, data from nine locations were collected. More details about the experimental conditions can be found in the original publication [9]

14.3.3.2 Mid-Level Data Fusion of LIBS and LA-ICP-TOF-MS of Coal Samples

As described earlier, the first step of a mid-level data fusion approach is selecting the features with which to build the model. This first step involves identifying the species detected by each technique and selecting the corresponding ranges to include. Figure 14.8 shows representative mass and emission spectra measured simultaneously with the LIBS/LA-ICP-TOF-MS tandem instrument. The LA-ICP-TOF mass spectrum showed trace elements Li, Ti, Cu, Zn, Kr, Rb, Y, Zr, Nb, Mo, Cd, In, Sn, Sb, Lu, Hf, Po, Th, and U. The mass range between 12 and 43 (Figure 14.8a) was not recorded, or "blanked out," to avoid the effects of abundant species with high intensity from the major elements Al, Si, Mg, and Ca, as well as the interference of molecular argon species on some of the isotopes of these elements that also can reduce the detector lifetime. The 230–460 nm and 470–710 nm wavelength ranges of the emission spectrum show the major and minor elements C, Al, Ca, Mg, Si, Fe, Na, and H as measured by LIBS. The carbon-related molecular CN and C2 were also identified. According to the spectrograph characteristics of the plane grating, some second-order peaks were identified, as shown in Figure 14.8b.

Since the original LA-ICP-TOF-MS and LIBS data were extensive, overfitting would likely occur if they were used for multivariate models without filtering. Therefore, it is crucial to

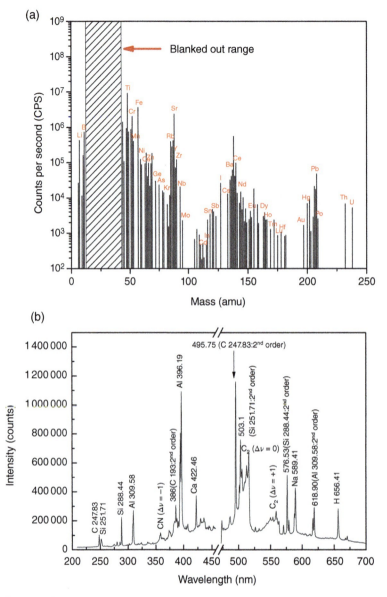

Figure 14.8 Tandem LIBS-ICP-LA-TOF-MS spectra. (a) LA-ICP-TOF-MS spectra and (b) LIBS emission (230–460 nm and 470–710 nm) measured from standard coal sample NIST 1632d.

identify the appropriate feature variables that should be used for the classification model. Figure 14.9 shows the correlation analysis performed between the TOF mass, LIBS emissions, and volatile content. Figure 14.9a shows that most trace element information had a negative correlation with the volatile content.

To determine of the appropriate number of variables for the model, the absolute value of correlation coefficient (r) was examined. Data with r values higher than 0.9 were used for the TOF data input variable, which resulted in a total of 24, shown in Table 14.2.

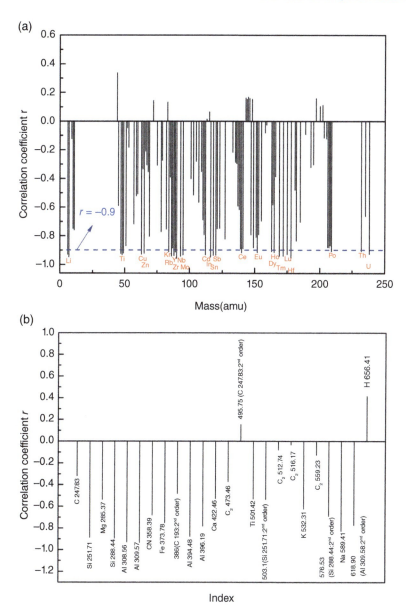

Figure 14.9 Correlation analysis between the (a) TOF mass and (b) LIBS emissions and volatile content.

This criterion was adopted as the method to determine the number of variables, based on the sample size and its influence on the model's robustness.

LIBS emission spectra, which provided a unique chemical signature for coal, were also analyzed for correlation with volatile content, as shown in Figure 14.9b. The inorganic element emissions had a higher correlation to the volatile content than the organic elements emissions. This is seen with C, H, CN, C_2, whose intensity saw a greater matrix effect, by comparison. The correlation between CN and C_2 was larger than that of C, which indicated

Table 14.2 Typical selected information from individual LIBS and TOF for the MV model.

Data types	Included elements/emission lines
LA-ICP-TOF-MS	Li, Ti, Cu, Zn, Kr, Rb, Y, Zr, Nb, Mo, Cd, In, Sn, Sb, Ce, Eu, Dy, Ho, Tm, Lu, Hf, Po, Th, U
LIBS	C 247.83nm, CN 358.39 nm, C_2 473.46 nm, C_2 512.74 nm, C_2 516.17 nm, C_2 559.23 nm, H 656.41 nm, Al 308.56 nm, Al 309.58 nm, Al 394.48, Al 396.19, Si 251.71 nm, Si 288.44 nm, Fe 373.78 nm, Na 589.41 nm

that the carbon molecular spectra better reflected the matrix properties. Common emission lines from C, C_2, CN, Al, Si, Ca, Fe, Na, and H are presented in Table 14.2.

The peak intensities were used for the MVA. A similar approach to the salt analysis was implemented, starting with spectrum normalization to reduce measurement uncertainty. In this case, due to the complexity of the chemical and physical composition of coal, the intensity range of each sample's spectral data at different pixel points can have massive differences, ranging from 10^{-1} to 10^{+7}. Therefore, instead of using a maximum signal or the total intensity for normalization, as was the case for the salts samples, the normalization scheme for the coal samples was defined as follows:

$$x^{norm} = \frac{2 \times (x - x_{min})}{(x_{max} - x_{min})} - 1 \qquad (14.1)$$

Equation 14.1: Normalization equation used for the coal analysis data.

Where x is the raw spectral intensity, the corresponding normalized value is x^{norm}, and x_{max} and x_{min} are the maximum and minimum LIBS and TOF mass spectral intensities of each sample, respectively.

In this study, two mid-level data fusion approaches were used to combine the selected data from LIBS and LA-ICP-TOF-MS. One approach involved each LA-ICP-MS spectrum being added to the end of the simultaneously acquired LIBS spectrum (LA-ICP-TOF-MS + LIBS). A second approach involved the summing of LIBS data to normalize the LA-ICP-TOF-MS data (LA-ICP-TOF-MS/LIBS). These results of these two approaches were compared with the results obtained by each LIBS and TOF dataset independently.

14.3.3.3 PCA Combined with K-means Cluster Analysis for Coal Samples

PCA and K-means clustering methods are effective techniques to reduce the input data dimensions and better describe the chemical structure with considerably fewer variables than the raw spectral data. When PCA was combined with K-means cluster analysis, PCA was able to reduce the spectral data of the original high-dimensional variable to just two dimensions, which expressed more than 90% of the total variance of the original input data. As a result, only principal component factors 1 and 2 (PC1 and PC2) were selected as variables. Then K-means cluster analysis was performed on the spectral data after dimensionality reduction. Figure 14.10 shows the two-dimensional K-means clustering analysis after PCA dimensionality reduction with different input data. The red circle is the clustering center for different categories. The accuracy of K-means cluster analysis of the TOF, LIBS,

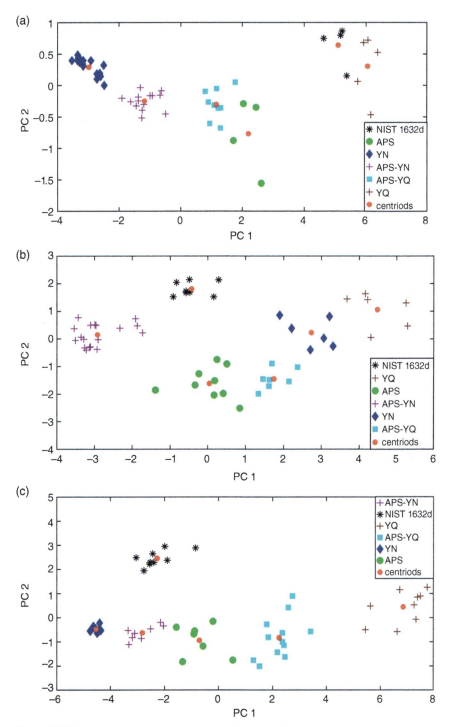

Figure 14.10 K-means cluster analysis after PCA reduction based on (a) TOF data, (b) LIBS data, (c) TOF with LIBS data, and (d) TOF normalized LIBS data.

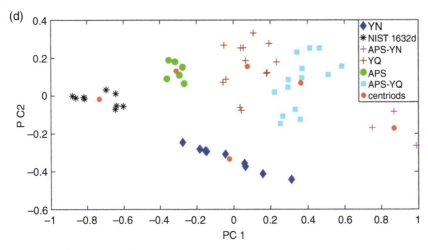

Figure 14.10 (Continued)

TOF+LIBS, and TOF/LIBS data were 59.26, 64.81, 92.59, and 62.96%, respectively. The analysis showed that the combination of TOF and LIBS data (TOF+LIBS) achieved significantly better discrimination among samples than the individual TOF or LIBS data. The K-means method classifies data samples based on the distance between the data sample cluster centers in the feature space and has a high requirement for the degree of separation between those sample clusters. At the same time, K-means is an unsupervised learning feature, making it challenging to establish complex discriminant boundaries between data point categories, especially for coal samples that contain complex spectral information.

14.3.3.4 PLS-DA and SVM for Coal Samples Analysis

PLS-DA is a classification model based on multiple partial least squares regression (PLSR) and simple discriminant analysis (DA). Support vector machine (SVM) is an approximate implementation of structural risk minimization that balances learning and generalization capabilities based on limited sample information. The SVM used in this work is based on Boser et al. [31] who proposed the C-support vector classification (C-SVC).

Table 14.3 summarizes the PLS-DA and SVM classification results for each model with the different datasets. From the calibration set, the models with different types of data achieved high classification accuracy (>97%), mainly because the number of samples often limits these MVA methods.

For the PLS-DA models (as shown in Table 14.3 and Figure 14.11a), the classification accuracy based on the combined data is greater than that of using individual TOF or LIBS data. A 7% improvement was demonstrated by using TOF/LIBS, and 15% improvement was shown by using TOF+LIBS data, when compared to the individual LIBS data, proving the significant value of the data fusion approach.

The validation set had 18 specimens, and the three PLS-DA models based on TOF+LIBS data accurately identified the number as 18, 17, and 16, respectively. The average number is 17 (as shown in Table 14.3), with an average accuracy of up to 94.44%, as shown in

Table 14.3 Comparison of classification results for each model with a different type of data.

Model type	Data type	Calibration set (36)		Validation set (18)	
PLS-DA	LIBS	35	97.22	14.3	79.63
	TOF	35	97.22	14.7	81.48
	TOF+LIBS	35.3	98.15	17	94.44
	TOF/LIBS	35	97.22	15.7	87.04
SVM	LIBS	35.7	99.07	16	88.89
	TOF	36	100	16.3	90.74
	TOF+LIBS	35.3	98.15	17.3	96.3
	TOF/LIBS	36	100	17.7	98.15

Figure 14.11a. For the SVM classification model (as shown in Table 14.3 and Figure 14.11b), the model's average accuracy based on TOF/LIBS data was increased to 98.15%, which is a significant improvement in coal classification when compared to recently published work [32, 33]. The SVM model based on TOF+LIBS data was second, and the classification model using individual TOF or LIBS data was the least effective. Figure 14.10c shows that data points from different coal samples had better aggregation in two-dimensional space after PCA dimensionality reduction, especially for the combined data (TOF+LIBS data). This is consistent with the linear classification model (PLS-DA) combining the TOF+LIBS data. When modeling with PLS-DA, the information from these two techniques was combined to carry out PCA to obtain the appropriate principal component number. This means each principal component contains more sample information, which helps to improve the accuracy of the model.

For the nonlinear classification model (SVM), the amount of information and the number of variables input affect the discrimination accuracy. The classification accuracy using individual TOF data was better than that of LIBS data alone, indicating a good correlation between trace elements and volatile content. The combination of TOF+LIBS data increases the amount of input information and increases the number of input variables in the model. When modeling with SVM, the original data needs to be mapped to high-dimensional feature space for classification. When the amount of input data increases, the required spatial dimensions will also become more extensive, affecting the model's accuracy. With TOF/LIBS, the amount of information compared to TOF is increased, but the number of input variables is the same, so the resulting model has better accuracy. These results help to further explain the requirements for feature selection while using the nonlinear SVM modeling method.

14.4 Comments and Future Developments

In general, the ability to perform data fusion with analytical data offers a significant step forward in data analysis methodology. Performing data fusion with considerably different analytical techniques requires assumptions related to the spatial resolution and the amount

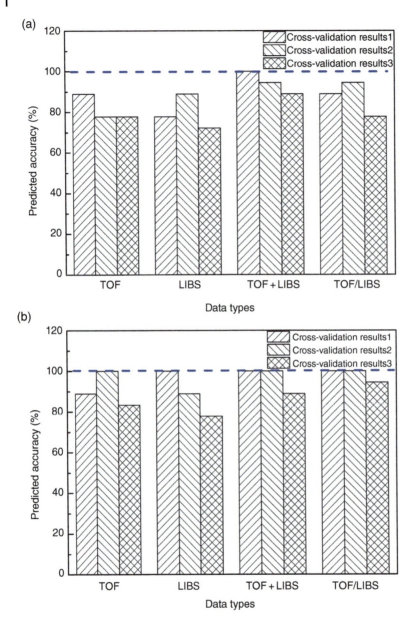

Figure 14.11 Cross-validation results based on (a) PLS-DA and (b) SVM model under different data types.

of material involved in the analysis that can impact the selection of scaling factors. However, one of the main advantages of performing data fusion from laser ablation-based techniques is that the data originate from the same event. This allows for more aggressive assumptions when planning data fusion, particularly with regard to the measurement of spatial resolution and amount of material analyzed. Despite the advantages offered by fusing data from laser ablation-based techniques, the individual datasets (LIBS and LA-ICP-MS) still need to be normalized (to improved analysis precision by compensating for shot-to-shot variation) and scaled (to compensate for differences in signals intensities) before fusion.

Laser ablation-based techniques in tandem mode LIBS/LA-ICP-MS offers an avenue to improve sample classification based on combined spectral information. This is an ideal tool for the chemical characterization of salt and coal samples. Data fusion showed significant improvement when compared to the classification using the individual techniques. Data fusion can combine spectral signals (raw, ratios, integrated values, etc.) from the different source techniques to produce a single new set or model. In this chapter, *mid-level fusion* was used to successfully classify edible salt and coal samples by extracting features from each technique's signal, combining them, and processing them together. Mid-level fusion has a positive impact when few features are sufficient to provide the necessary information, even though the same algorithms are used for processing after the features are extracted.

The expectation is that *mid-level and high-level data fusion* of LIBS and LA-ICP-MS, as well as the fusion of multiple LIBS sensors (LIBS^n, with different wavelength coverage, resolution, and sensitivities) will become an increasingly important methodology and continue to evolve and improve as the technology develops further and becomes more readily available.

Acknowledgments

The work was supported by the US Department of Energy (DOE) Small Business Innovative Research (SBIR) Office under contract number DE-SC0015787.

The author would also express gratitude to Dr. Meirong Dong and Dr. Yonghoon Lee for their substantial contribution to this review and Mr. Robb Hunt for his tremendous help editing this manuscript.

References

1 A. Lluveras-Tenorio, A. Spepi, M. Pieraccioni, S. Legnaioli, G. Lorenzetti, V. Palleschi, M. Vendrell, M.P. Colombini, M.R. Tinè, C. Duce, I. Bonaduce, A multi-analytical characterization of artists' carbon-based black pigments, *J Therm Anal Calorim*. **138** (2019), 3287–3299. https://doi.org/10.1007/s10973-019-08910-5.

2 D. Bazin, C. Jouanneau, S. Bertazzo, C. Sandt, A. Dessombz, M. Réfrégiers, P. Dumas, J. Frederick, J.-P. Haymann, E. Letavernier, P. Ronco, M. Daudon, Combining field effect scanning electron microscopy, deep UV fluorescence, Raman, classical and synchrotron radiation Fourier transform Infra-Red Spectroscopy in the study of crystal-containing kidney biopsies, *Comptes Rendus Chimie*. **19** (2015), 1439–1450. https://doi.org/10.1016/j.crci.2015.03.001.

3 K. Rammelkamp, S. Schröder, S. Kubitza, D.S. Vogt, S. Frohmann, P.B. Hansen, U. Böttger, F. Hanke, H. Hübers, Low-level LIBS and Raman data fusion in the context of in situ Mars exploration, *J Raman Spectrosc*. **51** (2020), 1682–1701. https://doi.org/10.1002/jrs.5615.

4 J. Forshed, H. Idborg, S.P. Jacobsson, Evaluation of different techniques for data fusion of LC/MS and 1H-NMR, *Chemom Intell Lab Syst.* **85** (2007), 102–109. https://doi.org/10.1016/j.chemolab.2006.05.002.

5 S. Roussel, V. Bellon-Maurel, J.-M. Roger, P. Grenier, Fusion of aroma, FT-IR, and UV sensor data based on the Bayesian inference. Application to the discrimination of white grape varieties, *Chemometr Intell Lab.* **65** (2003), 209–219. https://doi.org/10.1016/s0169-7439(02)00111-9.

6 A. Fernandez, X.L. Mao, W. Chan, M. Shannon, R.E. Russo, Correlation of spectral emission intensity in the inductively coupled plasma and laser-induced plasma during laser ablation of solid samples, *At Spectrosc.* **67** (1995), 2444–2450.

7 C. Latkoczy, T. Ghislain, Simultaneous LIBS and LA-ICP-MS analysis of industrial samples, *J Anal At Spectrom.* **21** (2006), 1152–1160. https://doi.org/10.1039/b607697c.

8 M. Bonta, J.J. Gonzalez, C.D. Quarles, R.E. Russo, B. Hegedus, A. Limbeck, Elemental mapping of biological samples by the combined use of LIBS and LA-ICP-MS, *J Anal At Spectrom.* **31** (2016), 252–258. https://doi.org/10.1039/c5ja00287g.

9 M. Dong, L. Wei, J.J. González, D. Oropeza, J. Chirinos, X. Mao, J. Lu, R.E. Russo, Coal discrimination analysis using tandem laser-induced breakdown spectroscopy and laser ablation inductively coupled plasma time-of-flight mass spectrometry, *Anal Chem.* **92** (2020), https://doi.org/10.1021/acs.analchem.0c00188.

10 M. Bonta, A. Limbeck, Metal analysis in polymers using tandem LA-ICP-MS/LIBS: eliminating matrix effects using multivariate calibration, *J Anal At Spectrom.* **33** (2018), 1631–1637. https://doi.org/10.1039/c8ja00161h.

11 O. Syta, B. Wagner, E. Bulska, D. Zielińska, G.Z. Żukowska, J. Gonzalez, R.E. Russo, Elemental imaging of heterogeneous inorganic archaeological samples by means of simultaneous laser induced breakdown spectroscopy and laser ablation inductively coupled plasma mass spectrometry measurements, *Talanta* **179** (2018), 784–791. https://www.sciencedirect.com/science/article/pii/S0039914017312134.

12 D. Oropeza, J. González, J. Chirinos, V. Zorba, E. Rogel, C. Ovalles, F. López-Linares, Elemental analysis of asphaltenes using simultaneous laser-induced breakdown spectroscopy (LIBS)–laser ablation inductively coupled plasma optical emission spectrometry (LA-ICP-OES), *Appl Spectrosc.* **73** (2019), 000370281881949. https://doi.org/10.1177/0003702818819497.

13 M. Bonta, S. Toeroek, B.D.S. Santa, 2017, Tandem LA-LIBS coupled to ICP-MS for comprehensive analysis of tumor samples, *Hungary Pure Elsevier.Com* (2017), https://doi.org/10.1063/1.4943278with.

14 B.T. Manard, C.D. Quarles, E.M. Wylie, N. Xu, Laser ablation – inductively coupled plasma – mass spectrometry/laser induced break down spectroscopy: a tandem technique for uranium particle characterization, *J Anal At Spectrom.* **32** (2017), 1680–1687. https://doi.org/10.1039/c7ja00102a.

15 Y. Lee, S.-H. Nam, K.-S. Ham, J. Gonzalez, D. Oropeza, D. Quarles Jr., J. Yoo, R.E. Russo, Multivariate classification of edible salts: Simultaneous laser-induced breakdown spectroscopy and laser-ablation inductively coupled plasma mass spectrometry analysis, *Spectrochim Acta B At Spectrosc.* **118** (2016), 102–111. https://doi.org/10.1016/j.sab.2016.02.019.

16 Y. Lee, J. Chirinos, J. Gonzalez, D.D. Oropeza, V. Zorba, X. Mao, J. Yoo, R.E. Russo, Laser-ablation sampling for accurate analysis of sulfur in edible salts, *Appl Spectrosc.* **71** (2016), 651–658. https://doi.org/10.1177/0003702817691288.

17 J. Chirinos, D. Oropeza, J.J. Gonzalez, Analysis of plant leaves using laser ablation inductively coupled plasma optical emission spectrometry: Use of carbon to compensate for matrix effects, *Appl Spectrosc.* **71** (2016), https://doi.org/10.1177/0003702816683686.

18 K. Subedi, T. Trejos, J. Almirall, Forensic analysis of printing inks using tandem laser-induced breakdown spectroscopy and laser ablation inductively coupled plasma mass spectrometry, *Spectrochim Acta B At Spectrosc.* **103–104** (2015), 1–8. https://doi.org/10.1016/j.sab.2014.11.011.

19 M. Dong, D.D. Oropeza, J. Chirinos, J.J. Gonzalez, J. Lu, X. Mao, R.E. Russo, Elemental analysis of coal by tandem LIBS and LA-ICP-TOF-MS, *Spectrochim Acta B At Spectrosc.* **109** (2015), 44–50. https://doi.org/10.1016/j.sab.2015.04.008.

20 J.R. Chirinos, D.D. Oropeza, J.J. Gonzalez, H. Hou, M. Morey, V. Zorba, R.E. Russo, Simultaneous 3-dimensional elemental imaging with LIBS and LA-ICP-MS, *J Anal At Spectrom.* **29** (2014), 1292–1298. https://doi.org/10.1039/c4ja00066h.

21 S. Sheta, MS Afgan, Z. Hou, S.-C. Yao, L. Zhang, Z. Li, Z. Wang, Coal analysis by laser-induced breakdown spectroscopy: a tutorial review, *J Anal Atom Spectrom.* **34** (2019), 1047–1082. https://doi.org/10.1039/c9ja00016j.

22 J.G. Speight, *Handbook of Coal Analysis*, (2015), 275–297. Wiley. https://doi.org/10.1002/9781119037699.ch11.

23 B.C. Pearce, J.W.F. Hill, I. Kerry, Use of X-ray fluorescence spectrometry for the direct multi-element analysis of coal powders, *Analyst* **115** (1990), 1397–1403. https://doi.org/10.1039/an9901501397.

24 X. Li, L. Zhang, Z. Tian, Y. Bai, S. Wang, J. Han, G. Xia, W. Ma, L. Dong, W. Yin, L. Xiao, S. Jia, Ultra-repeatability measurement of the coal calorific value by XRF assisted LIBS, *J Anal Atom Spectrom.* (2020), https://doi.org/10.1039/d0ja00362j.

25 D. Liu, Y. Zhang, Acid extraction of coal to determine major ash forming elements by ICP-AES, *Guang Pu Xue Yu Guang Pu Fen Xi Guang Pu* **20** (2000), 514–517.

26 H.G.M. Parry, L. Ebdon, J.R. Clinch, P.J. Worsfold, H. Casey, S.M. Smith, T.M. Khong, C.F. Simpson, O.O. Ajayi, D. Littlejohn, C.B. Boss, CJ Dowle, B.G. Cooksey, W.C. Campbell, S. Greenfield, M.S. Salman, M. Thomsen, J.F. Tyson, Coal analysis by analytical atomic spectrometry (ICP-AES and ICP-MS) without sample dissolution, *Anal Proc.* **25** (1988), 69–85. https://doi.org/10.1039/ap9882500069.

27 N. Mohamed, D.L. McCurdy, M.D. Wichman, R.C. Fry, J.E. O'Reilly, Rapid coal analysis. Part I: Particle size effects in slurry methods based on flame AA and swing-mill grinding, *Appl Spectrosc.* **39** (1985), 979–983. https://doi.org/10.1366/0003702854249457.

28 W. Geng, T. Nakajima, H. Takanashi, A. Ohki, Analysis of carboxyl group in coal and coal aromaticity by Fourier transform infrared (FT-IR) spectrometry, *Fuel* **88** (2009), 139–144. https://doi.org/10.1016/j.fuel.2008.07.027.

29 P.R. Solomon, M.A. Serio, R.M. Carangelo, R. Bassilakis, Z.Z. Yu, S. Charpenay, J. Whelan, Analysis of coal by thermogravimetry—Fourier transform infrared spectroscopy and pyrolysis modeling, *J Anal Appl Pyrolysis* **19** (1991), 1–14. https://doi.org/10.1016/0165-2370(91)80031-3.

30 A. Stankova, N. Gilon, L. Dutruch, V. Kanicky, Comparison of LA-ICP-MS and LA-ICP-OES to analyze some elements in fly ashes, *J Anal At Spectrom.* **26** (2011), 443–449. https://doi.org/10.1039/c0ja00020e.
31 B.E. Boser, I.M. Guyon, V.N. Vapnik, A training algorithm for optimal margin classifiers, Proceedings of the fifth annual workshop on Computational learning theory (1992), 144–152. https://doi.org/10.1145/130385.130401.
32 A. Metzinger, D.J. Palásti, É. Kovács-Széles, T. Ajtai, Z. Bozóki, Z. Kónya, G. Galbács, Qualitative discrimination analysis of coals based on their laser-induced breakdown spectra, *Energy Fuel* **30** (2016), 10306–10313. https://doi.org/10.1021/acs.energyfuels.6b02279.
33 D. Dou, W. Wu, J. Yang, Y. Zhang, Classification of coal and gangue under multiple surface conditions via machine vision and relief-SVM, *Powder Technol.* **356** (2019), 1024–1028. https://doi.org/10.1016/j.powtec.2019.09.007.

Part V

Conclusions

15

Conclusion

Vincenzo Palleschi

Applied and Laser Spectroscopy Laboratory, Institute of Chemistry of Organometallic Compounds, Research Area of the National Research Council, Pisa, Italy

The laser-induced breakdown spectroscopy (LIBS) technique is known to provide an incredible quantity of spectral information in short times, so that statistical methods for simplification and analysis of these large amounts of data are, nowadays, indispensable.

The practical approach of this book should make easy for the reader (typically a LIBS operator working in public or private organizations, or a PhD student) to recognize the LIBS applications of interest and find the examples on the chemometric techniques that could be applied to their specific case.

I am expecting that this book not only would be a useful support to consult according to the different problems that new users would face in their everyday LIBS activity but also would help to overcome the diffidence of "old school" LIBS spectroscopists against these methods, probably resulting from a scientific narrative that presents the chemometric tools as a kind of autonomous intelligent entities. In fact, the chemometric methods used in LIBS analysis are nothing else than mathematical algorithms and as such must be considered.

In the future, we should expect the introduction of more and more new techniques for data analysis, which would probably be transferred to LIBS, as it happened with principal component analysis, partial least squares, artificial neural network, and all the other information science methods that are now part of the fundamental chemometric tools used in LIBS spectral analysis. Likewise, specific techniques for the analysis of LIBS spectra might be developed. The modular structure of this book is planned to be able to accommodate, in future, the description of whatever new method will appear in the field of chemometrics for LIBS.

However, more important than the expandability of the specific topics discussed in this book is, in my opinion, the "take home" message underlying the discussion of all the chemometric methods presented, i.e. the good practice of using them only when it is really needed, in the first place, and then carefully checking and validating the results obtained, always considering the specific chemical and physical properties of the system under study.

I am sure that this way of working will always guarantee the best results, whatever new algorithm would enter the field of chemometrics for LIBS.

Chemometrics and Numerical Methods in LIBS, First Edition. Edited by Vincenzo Palleschi.
© 2023 John Wiley & Sons Ltd. Published 2023 by John Wiley & Sons Ltd.

Index

Page numbers in *italics* refer to Figures
Page numbers in **bold** refer to Tables

a

Abel inversion   14
ablation   1, 7–9, 20–21, 321–343
  factors affecting   11–14
  see also laser ablation
accuracy of ANN   225
activation function   214–215, *215*, 220, 224
adaptive gradient algorithm (AdaGrad)   219
adaptive moment estimation (Adam)   219, 227, 229
aluminum (Al)   33, 55, 58, 99, **287**
ambient gas   2, 8–10, 13–14
analog-to-digital converter (ADC)   26–28
analyte signals   26, 189, 192, 262
  data fusion   242
angle cosine distance   132
animal fats   168, *170–171*
archaeology   37
area under the curve (AUC)
artificial neural networks (ANN)   130, 213, 214–216, 224–233
  accuracy of ANN   225
  backpropagation   219–221, 224, 225
  confusion matrix   *146*
  cost functions and training   216–221, *217*, *218*, 228–229, 233
  evaluation and tuning   224–226
  feed-forward ANN   215
  fully connected feed-forward ANN   214
  history   213, *214*

ICA   141, **142**
KNN   145
LDA   152, **152**
LIBS classification summary   **230–232**
noise filtering   60
nonlinear algorithms   304–315
PCA   166, 174, 216, 251
regularization   219, 223, 227–229
state of the art   229, **230–232**, 233
validation   *147*
ash content of coal   34, 132, 136–137, *137*, 140–141, **142**, 283, **284**
asymmetric least squares correction (AsLs)   249
atomic absorption spectroscopy (AAS)   334
atomic emission spectroscopy (AES)   1, 22–23, 54, 334
  ICP-AES   19, 54, 334
atomic spectroscopy   132
à trous algorithm   53, 54
automatic spectra peak identification (ASPI)   151
autoscaling (AS)   246, *249*
average relative error (ARE)   54, 283

b

background noise   54, 55, 191, **284**
background subtraction   47, 55
backpropagation   219–221, 224, 225
Bacillus globigii   88–90, *90*

bacteria   88–90, *90*, 91, *91*, 249
Balmer alpha line   266, 315
basalt   84, *84*
baseline correction   47–50, *51*, 52–55, 59–60
Bayes formula   66, 324
Beer's law   260
biomass   132, 136–137, *137*
biomedical applications   34–35
black box models   180, 213, 229, 253, 312
blind source separation (BSS)   138, 193–206
  numerical examples   197–198, *199–203*, 204–205
block scaling   324
Boltzmann constant   263
Boltzmann equation   315
Boltzmann plot   15, 102–112, *113*, 114, *115*, 121–123, 266, 315
  3D   102–114, *115*, 119–123
  LTE condition assessment   114, *115*
Boltzmann surface   102–103, *104*, 108
brass   37, 260, *261*, 262, *262*, **287**
Bremsstrahlung radiation   15, 47, 189
building industry   37–38
butter and margarine   85–86

C

cadmium (Cd)   157
calcium (Ca)   63, 86, 87, 91, 168, 249
  teeth
calcium carbonate   157, **157**
calcium salt   178, *178*, **178**
calibration curves   259–260, 274
  ANN   227
  figures of merit   267–273
  inverse calibration   273–274
  linear vs nonlinear   264–267
  matrix effect   260–261
  nonlinear quantitative analysis   313–314
  normalization   *262*, 262–263
  polynomial algorithm   48
  zinc   260, *261*, *262*, 262
calibration-free LIBS (CF-LIBS)   33, 36, 263, 314–315
  3D Boltzmann plot   121–123
cancer   35, 166, *167*, 167–168
Caputo definition   61

carbon (C)   109–112, *113*, 114, *115*, *169*, 264, 283, 334–341
categorical cross-entropy   216–217
chain rule   219–220, 224
charge-coupled devices (CCD)   3, 24–28, 48, 327
ChemCam   57, 60, 84, 251–252, 285
chemical or biological threats   250, **292**
chemometrics defined   129–130
chirped pulse amplification (CPA)   22
classification   130–131, **131**, **148**, **230–232**, 242, *243*, 244–245, 247–251, 253, **341**
C-support vector   129–180, *138*, **131**, **148**, 153–161, 180, 229, 340, 333–334
cluster analysis (CA)   132–137, **134**, *138*
  applications   135–137, *138*
  theory   133–135
coal   34, 54, 59, 132, 136–137, *137*, 140–141, **142**, 283, **284**, 334–341, *339–340*, 343
coefficient of determination   270
coffee   58
coins   91, *93*, 93–94
collinearity   164, 245, 277, 291
colostrum and milk   86–87
concatenation   322–323, 331–332
concrete   **290**
confusion matrix   145, *146–147*, *333*, 333
continuous spine wavelet transform (CSWT)   61
continuous wavelet transform (CWT)   53–54
  features selection   70
  noise filtering   55–56
  overlapping peak resolution   *64*, 63–64
continuum background   48, *49*, 50, 54, 65, 189–192, 198
convolutional filter   222, *222*, 223–224
convolutional neural networks (CNN)   214, 221–224, 233, 310–312
convolution layer   221–222, *222*, 223–224
copper (Cu)   37, *57*, 58–59, 62, *83*, *105*, 105, 109, 145, *172*, **198**, 198, *199–203*, 204, **204**, **205**, 260–266, *261*, *262*, 262, *267*, **289**
cosine distance   **134**
cost   3, 21, 26, 28, 31–32, 48, 67, 81, 334
cost functions in ANN   216–217, *217*, 218–221, 228, 229, 233

Coulomb explosion   13
Covers theorem   279
cross-validation (CV)   152, 224, 227–229, 248, 250, 333, *342*
C-support vector classification (C-SVC)   340
cultural heritage   37
curve fitting method   61–64, 66
Czerny-Turner spectrometers   3, *23*, 23–24, 25, 335

d

dark noise   55
data fusion   180, 241–253, 321–343
   applications   248–251, *252*, 252–253
   background   242–243, *243*, 244
   data analysis and data fusion   247–248
   data clustering and data fusion   245, 248
   data treatment   244–245
   laser ablation and data fusion   321, 324
   multiple analytical techniques   322–324
   working with images   245–248, *249*
data model   193
data set splitting   224–226
Daubechies function   52, 54, 57, 59, 65–66
decision tree analysis   143, 145
decision tree-based classification methods   130, 131, **131**
decomposition level (DL)   53, 54–56, 65–66
deep learning   213, 215, 227
density gradients   10–11
detectors   22–23, 26–28
   detector noise   190, **190**
determination of transition probabilities   119, *120*
diode pumped solid state (DPSS) lasers   3, 20–21
discrete convolution operations   222–223
discrete wavelet transform (DWT)   *53*, 53–54, 57, 65
distance-based methods of classification   129–132, 180
   cluster analysis   132–137, *138*
   ICA   138–142
   KNN   143–145
   LDA   145–153
   PCA   161–174
   PLS-DA   153–161
   SIMCA   174–180
dolerite   *173*
dominant factor-based partial least squares   281–282, 285
Doppler broadening   14
dropout   228–229
dye lasers   21

e

early stopping   228
Earth   36, 60
echelle spectrometers   3, 25, 25–26, 27, 82
edible salts   326–334, 338, 343
eigenvalues   164–165, 176, 194–196, 281
   BSS   193–197, *202*, 205
eigenvectors   144, 161, 165, 176, 195, 280–281
   BSS   194, 195
electron density   14–15, 19, 64, 198, 263–266, 315
   time-dependent spectral analysis   97–99, 116, 121–122
EMSLIBS   226, 233
energy production   34
environmental applications   36–37
epochs   228, *228*
error compensation algorithm   62
Escherichia coli   89, *90*, *91*, 249
Euclidean norms   140, 142, 194, 280
explosives   133, 155, 291, **292**

f

fast clustering methods   132
fast Fourier transform (FFT)   55, 164
fast-independent component analysis (FastICA)   192, 196
   algorithm   138–140
features selection   66–68, **69**, **70**, 70–71, 180, 323
feed forward network   215
femtosecond (fs) lasers   12–13, 21–22
fiber lasers   3, **20**, 22
figures of merit (FOM)   225, 267–273
filter dimensions   222–223
Fisher linear discriminant (FLD)   145
flashlamp-pumped solid-state lasers   20–21, *21*

fly ash   140
food and feed applications   85–88, *89*, 153, **293–294**
forensic applications   91, *92*, *93*, 93–94, 248
Fourier transform (FT)   52
Fourier transform infrared (FTIR)   60, 249, 250, 334
frame-transfer charge coupled device   27, 28
Franzini-Leoni method   312–313
free-bound events   15
free model baseline correction   55
Friedrichs's algorithm   49–50, *51*, 52
full width half maximum (FWHM)   61, 63–64, 116

g

gabbro   84, *84*
garnet   **290**
Gaussian noise   197–198, *200–203*, **204**, 204–205, **205**
Gaussian variables   139, 190, 271, 279
GDS projection   175–176
generalizability   225–226
generalization   227, *228*
genetic algorithms (GA)   68, **70**, 250, 285, 309–310
geochemical applications *see* geological applications
GEOLIBS   82
geological applications   36–37, 83–85, 157, **159**, 159, **160**, 250–252, **290**, 312
glass and slags   **289, 290**
global positioning system (GPS)   313
Gornushkin's polynomial algorithm   55
gradient descent (GD)   *218*, 218–219
granite   *173*
granosa   152, 157
ground truth   216
Grunwald-Letnikov (GL) definition   61

h

handheld LIBS   3, 21, 33, 36
hard thresholding   56, 58
heavy metals   152, 157, 327
Heisenberg's uncertain principle   14
herbal medicine   174
hidden layers   214–215, *215*, 305, *306*, 307–309, 311

hierarchical clustering   132–135
  agglomerative   133, 134–135
  top-down splitting method   133, 134–135
high level data fusion   243–244, 322, *322*, 324, 343
high-power pulsed fiber lasers   **20**, 22
high repetition rate lasers   3, 19, 82
Hilbert feature space   279
H-J algorithm   138
homoscedasticity   268, 274
hyperbolic tangent (tanh) activation factor   214, *215*
hyperparameters   221, 222–225

i

Identity declaration   324
IFLAS-LDA   *151*, 151–152
igneous rocks   83, 85, 168–169, *173*
image intensifier   28
improved wavelet dual-threshold function (IWDTF)   58
independent component analysis (ICA)   130, 138–142, 196–198, *202–203*, 204–205, 253
  applications   140–141, **141, 142**, 142
  theory   138–140
independent component analysis artificial neural networks (ICA-ANN)   141, **142**
independent component analysis wavelet neuron networks (ICA-WNN)   140–141, 142
independent component scores (ICS)   140–141, 141
inductively coupled plasma atomic emission spectroscopy (ICP-AES)   19, 54, 334
inductively coupled plasma mass spectrometry (ICP-MS)   81, 259–261, 324–327, 331, 334
inductively coupled plasma optical emission spectroscopy (ICP-OES)   259, 261, 325, *325*
input layer   214, *215*
in-sample classification   226
intensified charge-couple devices (ICCD)   2–3, 48, 335
interference correction   65–66
interline charge-coupled devices (ICCD)   24, 26–28
interline photodiode array (IPDA)   27

International Union of Pure and Applied Chemistry (IUPAC)   267, 270–272
intrinsic patterns   221, 224
inverse Bremsstrahlung   10, 12
inverse calibration   273–274
iron (Fe)   33, 48, *51*, 52, 62, 66, *83*, *166*, *172*, 289
irradiation intensity   8, 10, 13
iterative discrete wavelet transform (IDWT)   66
iterative K-means clustering   132, 135, *136*

j
jewelry   285, 287

k
Kalman filter approach   313–314
Kernel partial least squares (KPS)   279–280, 291
K-fold support vector machine recursive feature elimination (K-SVM—RFE)   58–59
kidneys   35, 322
K-means clustering analysis   132, 133–137, *138*, 335, 338, *339–340*, 340
K-nearest neighbor (KNN)   143–145, *146*, 245
  application   143–145, **148**, 145, 177, **179**, 180
  theory   143–144, *144*
  validation   **147**

l
LACOMORE   32
Langmuir probes   14
laser ablation   7–9, 20–21, 321–343
  factors affecting   11–14
laser ablation inductively coupled plasma mass spectroscopy (LA-ICP-MS)   81, 323–327, *328–329*, 330, **331**, 331–335, 338, 343
laser ablation inductively coupled plasma time of flight mass spectrometry (LA-ICP-TOF-MS)   335, *336*, **338**, 338
laser ablation inductively coupled plasma optical emission spectroscopy (LA-ICP-OES), *325*, 326

laser energy   11–13, 20, **20**, 22
laser-induced breakdown spectroscopy (LIBS) 2D map   190–191, *191*
laser-induced breakdown spectroscopy (LIBS) classification summary   **230–232**
laser-induced breakdown spectroscopy (LIBS) and Raman sensor (LIRS)   252
laser induced fluorescence (LIF) spectroscopy   252
laser-induced plasma (LIP)   3–4, 7–9, 19, 26, 23, 104
  factors affecting formation   11–14
laser-matter interaction   7–12
laser parameters   2, 8, 11–12
laser photoacoustic spectroscopy (LPAS)   250
laser pulse duration   11, 12–13, 20, **20**, 21–22
laser-supported combustion (LSC) waves   9–10
laser-supported detonation (LSD) waves   9–10
laser-supported radiation (LSR) waves   9–10
laser wavelength   11, 12, **20**, 20–22
latent variables (LV)   277–281
lead (Pb)   157, 198, **198**, *199–203*, **204**, 204–205, **205**
learning rate (LR)   *218*, 218–219
least squares regression   253
leave-one-out (LOO) algorithm   309, 333
Leibniz's fractional differential theory   61
leucogtanite   *173*
Levenberg-Marquadt (L-M) algorithms   61, 219, 310
limestone   84, *84*
limit of detection (LOD)   242, 264, 270–273, *273*
limit of qualification (LOQ)   264, 272–273, *273*
linear calibration   264–267
linear discriminant analysis (LDA)   145, 148, **148**, 150, *150*, 152, **152**, *153*, 174, 245, 250
  applications   151–153
  theory   148–150
liquid chromatography/mass spectrometry (LC/MS)   323
lithium (Li)   63

locally weighted partial least square
(LW-PLS) 280–281
local thermodynamic equilibrium (LTE) 97,
103, 108, 116, 119, 121, 123, 263, 266, 314
assessment of condition 114, *115*
Lorentzian function 61–62, 64, 116, 265
Lorentzian profile 116
loss landscape *217*, 217–218
low-level data fusion 243–244, 248, 251, *322*, 322–323

m

machine learning (ML) 213–214, 221–222, 228, 231, 233
Mahalanobis distance 141, 142
Malaga group 31–32
Mallat's algorithm 53, 54
manganese (Mn) 161, 168, *169*
Manhattan distance 133, **134**, 141, 142
margarine and butter 85–86
Mars exploration 36, 60, 63, 84, 250–251, 285, **288**, 323
Mars Science Laboratory (MSL) 84
matrix effects 63, 67, 81, 89, 91, 260–261, 283, 291
maximum negative entropy 139
maximum noise fraction (MNF) 194–198, *200–201*, 205
max-pooling 223, *223*
Maxwell-Boltzmann distributions 97
Maxwell equation 314
mean-pooling 223, *223*
mean square error (MSE), *140*, 140–141, *141*, **141**, 217
meat 86–88, *88*, *89*, **294**
metal analysis 285, **286**, **287**
metal industry 31–33
metallurgy applications 31–33, 36–37, 140, 157, 164, *166*, 233
metamorphic rocks 85, 151–152
microbiological applications 88–89, *90*, *91*, 91
microchannel plate (MCP) 28
mid-level data fusion 243–244, 250, *322*, 322–323, 327–334, 343, 335

milk and colostrum 86–87
minerals 240, 250–251, 285, **288**, **289**, **290**, 312, 326–327, 330–331, 331
mini batch 227, 229
minimum noise fraction (MNF) 192
Minkowski distance 133
MLS-LDA *151*, 151–152
mobile phone waste 177, **179**, 180
model-free algorithm 49–50, *51*, 52
mode-locking 21–22
molten steel 285, **286**, **287**
momentum 219
motor oil **294**
multicollinearity 155, 277, 332
multilayer perceptron (MLP) 214, 221, 222, 224
multiple analytical techniques 322–324
multivariate analysis (MVA) 177, 324, 330, 338, 340
multivariate model *328*, 332, 333–334, 335–336, **338**
multivariate nonlinear algorithms 304–315
municipal sludge 132, 136–137, *137*
mutual subspace method (MSM) 175

n

nanoparticle-enhanced LIBS (NELIBS) 84, 87–88, 168–169, *172*, *173*
nanosecond (ns) lasers 12–13, 20, 22
Nd:YAG lasers **20**, 20–21, *21*, 327, 335
Nd:YAG wavelength 260
near-infrared (NIR) region 22–23
near-infrared (NIR) spectra 54
neurons 214, *215*
noise 49–50, 55–60, 189–198, *200–203*, 204–206
see also signal-to-noise ratio (SNR)
noise adjusted principal component (NAPC) transform 195–196, 197
nonlinear calibration 264–267
nonlinear iterative partial least squares (NIPALS) 249, 278, 279
normalization 246, 248–249, *249*, 251, 261–264, 331–332, 338, 343
nuclear industry 38

nuclear magnetic resonance (NMR)   49–50
number of filters   222–223

o

obsidian   84, *84*, 155
one-hot encoding   216
optical emission spectroscopy (OES)   19, 22
organic content of coal   334, 337
organics   291, **292**, **293**, **294**
out-of-sample classification   226
output layer   214, *215*
overfitting   227–228, 242, 304, 308–309, 312, 315–316, 335
overlapping peak resolution   60–66, *63*, 192
overtraining   227, 228

p

padding   222, *222*, 224
parallel factor analysis (PARAFAC)   99
partial least squares (PLS)   33, 130, 153–156, 277–294, 307, 313
  results in LIBS   283, **284**, 285, **286–290**, 291, **292–294**
partial least squares discriminant analysis (PLS-DA)   37, 152–163, *158,* 177, **179**, 180, 242, 245, 248, 250–251, 278, 282, 285, 291, 333–335, 340–341, **341**, *342*
  applications   157, **158**, *158*, 159–163
  classification results   **157**, **158**, **159**, 159–160, **160**, 161–163
  theory   155–157
  validation   145, **147**
partial least squares regression (PLSR)   59–60, 63, 249, 253, 277, 278–282, 285
partition function   104, 121–122, 263
Paschen-Runge design   *24*, 24–25
performance metrics   224–225
pharmaceuticals   177, *178*, **179**, 292
phosphor screen   28
photocathode   26, 28
photodiode array (PDA)   27
photoionization   10, 12
photomultiplier detectors   2, 26–27
photomultiplier tubes (PMT)   2, 24, 26

Pisa group   31–33, 37, 116, 121, 266, 274
plasma   1–4, 7–14, 81
  physics   2, 7, 8, 191, 206
  processes   10–11, 116
  emission spectra   14–15
  shielding   11–12, 13
  temperature   9–11, 14–15, 97–98, 102, 104, 109, 111–123, 263, 265, 266
plastic   133, 137, 229, 291, **292**
Pockels cell   20, *21*
Poisson distribution   268
pollen   88–89, *90*
polymers   132–133, 135–137, 177, 179, 180, 293
polynomial algorithm   48, *49*, 55
pooling   223, *223*
portable LIBS   22, 33, 36, 81, 86–87
potassium salt   177, *178*, *178*
precision in ANN   225
principal component analysis (PCA)   68, 69, 81–82, 98, 130, 132, 136, 138, 140–141, 145, 150, *150*, 152, *153*, 161, 164–174, 192, 194–196, 198, 216, 242, 245, 248–251, 335, 388, 338, *339–340*, 340
  applications   83–94, *83*, *84*, *85*, 85, *86*, *88*, 88–89, *89*, *90*, *91*, 91, 166–174, 177
  theory   164–165
principal component analysis linear discriminant analysis (PCA-LDA)   153–155, 157, 245
principal component analysis support vector machine (PCA-SVM)   166, 168, 233
principal component fractions (PCS)   166–168, *172–174*
principal component regression (PCR)   253
principal component (PC) scores   *332*, 332–333
printed circuit boards (PCB)   155
proton nuclear magnetic resonance (^{1}H-NMR)   323
pulse duration   11, 12–13, 20, **20**, 21–22

q

quadrature mirror filters (QMF)   57
quantum efficiency   23, 26, 28

quasi-whitening   197, 205
Q-switching   20, **20**, *21*, 22

r

radial basis function artificial neural network (RBFNN)   215
Raman spectroscopy   60, 89, *91*, 321, 323
  data fusion   241–242, 244–246, 248–253
random forest (RF)   130, **148**, 157, *158*
rare-earth elements   20, 91, *92*
receiver operating characteristic (ROC)   225
recombination   15, 47
rectified linear unit (ReLU)   214–215, *215*
recycling metal   32–33
regression coefficient of determination on cross-validation (R^2CV)   250
regression coefficient of determination on prediction (R^2P)   250
regression coefficients   156, 250, 277, 278, 281
regularization   219, 223, 227–229
relative abundances   **198** 198, 206
relative standard deviation, (RSD)   54–55
resonant lines   114, 118, 264
rice   145
Richardson-Lucy deconvolution (RLD)   66
Riemann-Liouville definition   61
rocks   83–85, 148, 151–152, 168, 251, *252*, 285, **288**, **289**, **290**, 312
root mean square error (RMSE)   63, 283, 285
root mean square error of calibration (RMSEC)   59, 65–66, 270, *271*, 304–305, 307–310
root mean square error of cross-validation (RMSECV)   59, 63, 250, 283
root mean square error of prediction (RMSEP)   54, 57, 58, 250, 283, *284*, **283**, 283, 285, **287–290**, 291, **293–294**, 309–310
roots of *Angelica pubescens*   174
rule induction algorithm classification method   130, 131, **131**

s

Saha-Boltzmann equation   122
  see also Saha-Eggert equation, Saha's equation
Saha-Boltzmann plot   102, 266, 315
Saha-Eggert equation   97, 112, 263, 315
Saha's equation   15
salts   **331**, 323, 326–334, 338, 343
sample pretreatment   180
Savitzky-Golay algorithm   244
scaling   322–324, 331–332, 342–343
scaling factor   54, 58, 65–66
Schlieren technique   14
second-order BSS   193–194, 195, 196
sedimentary rocks   83, *83*, 85, 99, 106, 114–118, 168–169, *172*, 334
self-absorption   64, 67, 81, 114–118, 119, 192, 206, 264–266, 270, 291, 312
self-mode-locked Ti:sapphire lasers   **20**, 21–22
self-organizing maps (SOM)   137, 249, 315
self-reversal   99, *103*
semi-soft thresholding   56
sensitivity   9, 23, 26–28, 36, 87, 142, 145, **147**, **148**, 160, 177, **179**, 180, 225, 241–242, 321, 325–327, 334
shape of filter   222, *222*, 223–224
shellfish   152, 157
shock wave   9–10, 263
shot noise   190, **190**, 250
SHREDDERSORT   33
sigmoid activation function   214–215, *215*, 307, 307
signal-to-background ratio (SBR)   48, 54
signal-to-noise ratio (SNR)   23, 47, 52, 55, 57, 64, 87, 89, 140, 169, 189, 192, 194–198, *200–203*, 204, **204**, **205**, 205–206, 244
siltstone   83, *172*
silver (Ag)   99, *100–103*, 118
SIMPLS   278, 279
SIMCA see soft independent modelling of class analogy (SIMCA)
smoothing   55
sodium (Na)   86, 91
sodium salt   177, **178**, *178*
soft independent modelling of class analogy (SIMCA)   37, 145, 152, *154*, 174–180, 248
  applications   177, **178**, *178*, **179**, 180
  theory   174–177
  validation   **147**

soft thresholding 56, 57, 58–59
soil 36, 59, 285, **288**, **289**, **290**
solid-state detectors 27
source noise 190, **190**
soybean 148
specificity 228
spectral angle mapping (SAM) 206
spectral treatment 47–71, 180
  baseline correction 47–55, 59–60
  noise filtering 55–60
  overlapping peak resolution 60–66
spectrometers 23–26, 33
spline interpolation 48
Staphylococcus, 89, *90*, *91*, 249
Stark coefficient 265
Stark effect 14–15, 19, 99, 116, 265–266, 315
stationary discrete wavelet transform (SWT) 54, 55–56
steel 32, 59, 285, **286**, **287**
stochastic gradient descent (SGD) 219, 227
stride 222, *222*, 224
strontium (Sr) 91
sulfates 323
sulfur minerals **288**
support vector machine (SVM) 130, 132, 136–137, 145, **148**, 152, 157 *158*, 159, **159**, **160**, *160*, 166, 167–168, 174, 245, 248, 250, 285, 335, 340–341, **341**, *342*
surface-enhanced Raman spectroscopy (SERS) 250
Symlet function 59

t

Tag-LIBS 35
tandem laser ablation *325*, 325–326
tandem LIBS/LA-ICP-TOF-MS 335, *336*
tanh 307, *307*
teeth 160–161, *162*, *163*, 168, *169*, **293**
temperature *see* plasma temperature
testing 224–226
thermal drift 190, **190**
thin plasma 97, 103, 119
Thompson scattering 14
time-dependent spectral analysis 97–123
  3D Boltzmann plot 102–108
  applications 109–123
  ICA 98–99, *100–101*, *102*, *103*
time of flight mass spectrometry (TOF-MS) 326, 335–337, **338**, 338, *339–340*, 340, **341**, 341, *342*
tin (Sn), **198**, 198, 199, 201, 203, **204**, 204–205, **205**, 266
titanium (Ti) 91
Townsend effect plasma spectroscopy (TEPS) 250
  Raman explosive detection system (TREDS-2) 250
trace elements 326, 334, 335, **338**
trachy-andesite 84, *84*
trachyte 84, *84*
training 224–225, 227–229, 242, 245, 305, 307, 310–312
transient plasma 102, 118
transition probabilities 118–119, *120*, 120–121
Trust-Region algorithm 63–64
tungsten (W) 119

u

ultrasonic autofocusing 34
ultraviolet Raman spectroscopy (UVRS) 250
ultraviolet region (UV) 22–23, 28
ultraviolet-Vis absorption spectroscopy 60
unit vector normalization 244
univariate linear analysis 259–274
  figures of merit 267–273
  inverse calibration 273–274
  linear vs nonlinear calibration curves 264–267
  standards 259–260
upper calorific value (UCV) of coal 34
uranium (U) 91, *92*

v

validation data set 225–226
vectors co-addition 246
vectors concatenation 246
vectors outer product 247
vectors outer sum 246–247
Voigt function 61–62, 63, *64*

W

wavelet neural network (WNN)   140–141
wavelet threshold de-noising (WTD)   55–59, *60*, **71**
wavelet transform   52–55, 57, 64–66, *67*, 68, 70–71, **71**
weighting   194–195, 220, 222-225, 312
  weighting factors (WF), *332*, 332–334
whitening   194–195, 197, **204**
white noise   50, 58

X

X-ray fluorescence (XRF)   36, 81, 145, 259, 312, 334
  applications   32, 33, 36
X-ray powder diffraction   321
X-rays   11, 13

Z

zinc (Zn)   58, 109, 157, **198**, 198, *199–203*, 204, **204**, **205**, 260, *261*, *262*, 262